# Microorganisms and Minerals

# MICROBIOLOGY SERIES

*Series Editor*
ALLEN I. LASKIN
EXXON Research and Engineering Company
Linden, New Jersey

*Other Volumes in Preparation*

# Microorganisms and Minerals

edited by

## EUGENE D. WEINBERG

*Department of Microbiology
and Program in Medical Sciences
Indiana University
Bloomington, Indiana*

MARCEL DEKKER, INC.                              New York • Basel

**Library of Congress Cataloging in Publication Data**
Main entry under title:

Microorganisms and minerals.

   (Microbiology series ; v. 3)
   Includes bibliographical references and indexes.
   1.  Micro-organisms--Physiology.  2.  Mineral
metabolism.  I.  Weinberg, Eugene D.
QR92.M5M53          576'.11'9214          76-53191
ISBN 0-8247-6581-8

MARCEL DEKKER, INC.

270 Madison Avenue, New York, New York 10016

Current printing (last digit):
10 9 8 7 6 5 4 3 2 1

PRINTED IN THE UNITED STATES OF AMERICA

In memory of

HENRY A. SCHROEDER

Dr. Schroeder was for many years Professor of Physiology at Dartmouth Medical School and had built and operated a Trace Element Laboratory at Brattleboro, Vermont. His numerous and systematic publications on each of several dozen elements plus his leadership in trace element research and applications were major factors in the worldwide establishment of trace element biomedical research programs and awareness of health applications since 1955. Although not a microbiologist, Dr. Schroeder was quite cognizant of microbial roles in mineral cycling, and he was always enthusiastic about the use of microorganisms as models in cell biology.

## PREFACE

Textbooks of microbiology generally devote few paragraphs to the critical roles of minerals in the growth, differentiation, and death of microbial cells. Moreover, many microbiologists assume that either intact hosts or tissue extracts automatically provide microorganisms with correct qualitative and quantitative balances of minerals in available forms. We hope that the present book will persuade the reader that individual minerals not only have unique functions but must be acquired, stored, and cycled by microbial cells with care and precision. Indeed, all persons who wish successfully to detect, propagate, exploit, or destroy microorganisms should keep aware of the levels and balances of minerals and of organic compounds that bind or transport minerals in the systems with which they are working.

The reader is referred to the <u>Introduction</u> on pp. 1-4 for a general overview of the chapters in this volume.

Eugene D. Weinberg
Bloomington, Indiana

CONTRIBUTORS

Jean E. L. Arceneaux, Department of Microbiology, University of Mississippi Medical Center, Jackson, Mississippi

B. Rowe Byers, Department of Microbiology, University of Mississippi Medical Center, Jackson, Mississippi

Mark L. Failla,* Department of Microbiology, Indiana University, Bloomington, Indiana

William O. Foye, Department of Chemistry, Massachusetts College of Pharmacy, Boston, Massachusetts

Paula Jasper, †Department of Biology, Washington University, St. Louis, Missouri

Ivan Kochan, Department of Microbiology, Miami University, Oxford, Ohio, and Department of Microbiology and Immunology, Wright State University School of Medicine, Dayton, Ohio

Walter A. Konetzka, Department of Microbiology, Indiana University, Bloomington, Indiana

Simon Silver, Department of Biology, Washington University, St. Louis, Missouri

Michael R. Tansey, Department of Plant Sciences, Indiana University, Bloomington, Indiana

Eugene D. Weinberg, Department of Microbiology and Program in Medical Sciences, Indiana University, Bloomington, Indiana

---

Present Address
*Department of Nutrition, Cook College, Rutgers University, New Brunswick, New Jersey
†Department of Biochemistry, St. Louis University School of Medicine, St. Louis, Missouri

# CONTENTS

Chapter 3

MANGANESE TRANSPORT IN MICROORGANISMS    105
Simon Silver and Paula Jasper

Chapter 4

ZINC: FUNCTIONS AND TRANSPORT IN MICROORGANISMS 151
Mark L. Failla

Chapter 5

MICROBIAL TRANSPORT AND UTILIZATION OF IRON  215
B. Rowe Byers and Jean E. L. Arceneaux

## ABBREVIATIONS

| | |
|---|---|
| AAS | atomic absorption spectroscopy |
| 5′-AMP | adenosine 5′-monophosphate |
| ATP | adenosine triphosphate |
| CCCP | $\underline{m}$-chlorophenyl carbonylcyanide hydrazone |
| CD | circular dichroism (spectroscopy) |
| cyclic AMP | adenosine 3′:5′-cyclic monophosphate |
| cyt | cytochromes |
| DNP | 2,4-dinitrophenol |
| DTNB | 5,5′-dithiobis(2-nitrobenzoic acid) |
| DTT | dithiothreitol |
| EDTA | ethylenediamine tetraacetic acid |
| EGTA | ethyleneglycol-bis-($\beta$-aminoethyl ether)-N′ tetraacetic acid |
| FAD | flavin adenine dinucleotide |
| FCCP | p-trifluoromethoxy carbonylcyanide phenylhydrazone |
| HEPES | N-2-hydroxyethylpiperazine-N′-2-ethanesulfonic acid |
| $K_i$ | inhibition constant |
| $K_m$ | concentration for half-saturation of rate |
| mRNA | messenger ribonucleic acid |

| | |
|---|---|
| NAD | nicotinamide adenine dinucleotide |
| NADPH | reduced nicotinamide adenine dinucleotide phosphate |
| NEM | N-ethylmaleimide |
| ORD | optical rotary dispersion |
| pCMB | p-chloromercuribenzoate |
| PHA | phytohemagglutinin |
| $p_i$ | inorganic phosphate |
| PIPES | piperazine-N,N'-bis(2-ethanesulfonic acid) |
| $Q_{10}$ | temperature coefficient |
| SDS | sodium dodecylsulfate |
| $V_{max}$ | maximum velocity of uptake |

# Microorganisms
# and
# Minerals

# INTRODUCTION

Eugene D. Weinberg

> What we learn from Nature's smallest living things often emerge
> as principles of general biology. [From Ref. 1.]

> It is clear that the initial inorganic matrix in the earth's crust
> has left its stamp, because metals are still involved in every
> aspect of biosynthesis, degradation, and macromolecular
> assembly. [From Ref. 2.]

The scientific practice of microbiology is presently entering its second
century. Well before the beginning of this practice, however, mineral
elements were employed to influence microbiological processes. Since
early medieval times, for example, mercury and its salts have been important therapeutic agents for skin and intestinal infections. The remarkable selective toxicity of silver for microorganisms was utilized by physi-
cians as early as the 1500s, three centuries prior to an understanding of
the microbial etiology of infectious disease. A proper balance of inorganic
elements in soils and waters has long been recognized to be a critical factor in cultivation of plants from which desirable products of microbial
fermentation are to be obtained.

Nevertheless, identification of the precise roles of minerals in the
physiology of microbial cells occurred slowly and sporadically during the
first century of scientific microbiology. For example, at the start of the
century, Raulin, a student of Louis Pasteur, reported that zinc is required
for growth of fungi [3]; in the 1940s, industrial scientists became aware
of the importance of this metal in mycotic fermentations. Only within the
past decade, however, have microbiologists begun to examine intensively
the many metabolic aspects of zinc in microbial systems.

Although, as will be evident in the book, knowledge concerning molecular aspects of mineral roles and requirements has been gained relatively
recently, a few significant discoveries were made in the early 1930s. At
that time, for example, molybdenum was discovered to be needed for
microbial nitrogen fixation [4]. It also became apparent in that decade

1

that the quantity of a metal that is needed and/or tolerated for growth of microbial cells can be quite different from that required and/or permitted for the organisms to carry on a specific physiologic process [5-7]. This important principle of quantitation has only recently been extrapolated to cells of plants and animals [8]. Also recognized in the 1930s and 1940s was the attempt by vertebrate hosts to withhold growth-essential iron from invading microorganisms [9,10]. This phenomenon, now termed "nutritional immunity" [11,12], is finally achieving well-merited, widespread attention.

A number of books [e.g., 8,13-21] have appeared within the past decade on the roles of mineral elements in living matter. In each of these, plants and/or animals have been the focus of concern. This is the first book on biological aspects of a spectrum of mineral elements that is devoted primarily to microorganisms.

The subject matter of the book is divided into four topics. The first concerns the manner in which microbial cells acquire and store selected mineral elements. This topic is important for several reasons. Its study contributes knowledge at the cellular level on mechanisms, active compounds, sites, and homeostatic regulation of acquisition of growth-essential nutrilites. Moreover, specific metal-transport and possibly metal-storage compounds that are being and will be isolated have actual or potential utility in prophylaxis and therapy of metabolic and infectious diseases of animals and plants. Additionally, knowledge of the ability of some intestinal microorganisms to facilitate and others to retard assimilation of essential mineral elements by cells of the gut wall is of potential importance in human and animal nutrition.

Such mineral elements as magnesium, calcium, manganese, zinc, iron, silicon, potassium, cobalt, copper, and molybdenum are important in the life of microbial cells. Because of space limitations and because the first five have been most intensively studied, only these five will be discussed here (Chapters 1 to 5). A review of the momentous competition for iron between invading microorganisms and animal hosts is contained in Chapter 6.

The second topic is concerned with the functions of magnesium, calcium, manganese, zinc, and iron in microbial cells. As with cells of plants and animals, the catalytic function of minerals in microorganisms has become well recognized within the past three decades. The structural and stabilizing functions of minerals in microbial cells presently comprise exciting areas of research. The various roles of the five selected elements in microbial systems are discussed in Chapters 1 to 5 and, where pertinent, are compared with plant and animal systems. A possible novel function, thus far studied mainly in microorganisms, is that of the ability of a few mineral elements (e.g., manganese, zinc, iron) to induce and suppress the processes of secondary metabolism and differentiation (Chapter 7). Comparisons of mineral requirements for growth with those needed for specific functions in plant and animal cells are only now beginning to be made.

The third topic, microbial roles in cycling of mineral elements, is divided into two aspects: geochemical (Chapter 8) and phytochemical (Chapter 9). The first deals with microbial conversion of elements between inorganic and organic forms and among their various oxidation states. The second aspect concerns the role of microorganisms in facilitating assimilation of minerals in plants. Each aspect is indispensable to the economy of life on earth.

The fourth topic is that of the antimicrobial action of mineral elements (Chapter 10). Not only are silver, mercury, copper, and zinc still employed to suppress and kill unwanted microorganisms, but metals are now also important activators of various organic antimicrobial agents.

Within the past half century, the doctrine of the biochemical unity of living matter has become well established. Metabolic pathways as well as methods of storage, transfer, and use of energy and of genetic information are found to have strong similarity in the cells of microorganisms, plants, and animals. The qualitative and quantitative equivalencies in requirements for and roles of minerals in these three forms of life are examples of such unity. As we will see, in some cases principles of mineral aspects of biology were learned first with cells of higher organisms and then applied to microbial cells; in others, microorganisms served as model systems in inorganic cell biology.

But within the overall theme of bioinorganic unity lie variations of fascinating diversity. Sufficient examples of uniqueness in the metabolism and roles of minerals exists in specific groups of microorganisms to provide a wealth of productive areas of research. As importantly, the examples of uniqueness described in this book may suggest potentially useful and novel methods for (a) environmental enhancement of desirable microbial groups; (b) suppression of microbial pests; (c) microbial detoxification of poisoned local environments; and (d) microbial accumulation of elemental metals. Thus we hope that the material will be useful not only to bioinorganic scientists but also to environmental, medical, molecular, and physiological microbiologists.

## REFERENCES

1. J. W. Foster, First Henrici Memorial Lecture, Department of Microbiology, University of Minnesota, Minneapolis, 1963.
2. J. M. Wood, in Biochemical and Biophysical Perspectives in Marine Biology (D. C. Malins and J. R. Sargent, eds.) vol. 3, Academic Press, New York, 1976, p. 408.
3. J. Raulin, Ann. Sci. Nat. Botan. Biol. Vegetale, 11, 93 (1869).
4. H. Bortels, Arch. Mikrobiol., 1, 333 (1930).
5. A. Locke and E. R. Main, J. Infect. Dis., 48, 419 (1931).
6. E. D. Weinberg, Persp. Biol. Med., 5, 432 (1962).

7.  E. D. Weinberg, Advan. Microb. Physiol., 4, 1 (1970).
8.  H. A. Schroeder, The Trace Elements and Man, Devin-Adair, Old
    Greenwich, Connecticut, 1973.
9.  A. Locke, E. R. Main, and D. O. Rosbach, J. Clin. Invest., 11,
    527 (1932).
10. A. Schade and L. Caroline, Science, 104, 340 (1946).
11. I. Kochan, Curr. Top. Microbiol. Immunol., 60, 1 (1973).
12. E. D. Weinberg, Science, 184, 952 (1974).
13. H. J. M. Bowen, Trace Elements in Biochemistry, Academic Press,
    New York, 1966.
14. V. Sauchelli, Trace Elements in Agriculture, Van Nostrand Reinhold,
    New York, 1969.
15. E. J. Underwood, Trace Elements in Human and Animal Nutrition,
    3rd ed., Academic Press, New York, 1971.
16. D. R. Williams, The Metals of Life, Van Nostrand Reinhold, New
    York, 1971.
17. I. J. T. Davies, The Clinical Significance of the Essential
    Biological Metals, Charles C. Thomas, Springfield, Illinois, 1972.
18. H. G. Gauch, Inorganic Plant Nutrition, Dowden, Hutchinson, &
    Ross, Stroudsburg, Pennsylvania, 1972.
19. M. N. Hughes, The Inorganic Chemistry of Biological Processes,
    Wiley, New York, 1972.
20. S. K. Dhar (ed.), Metal Ions in Biological Systems, Plenum Press,
    New York, 1973.
21. W. G. Hoekstra, J. W. Suttie, H. E. Ganther, and W. Mertz (eds.),
    Trace Element Metabolism in Animals — 2, University Park Press,
    Baltimore, Maryland, 1974.

INTRODUCTORY NOTE TO CHAPTERS
1, 2, AND 3

Our laboratory has been concerned with bacterial divalent cation metabolism for the last 8 years or so. Whereas an enormous body of information
about microbial iron metabolism is available because of the importance of
iron compounds to microbial metabolism and also because of the clinical
use of microbial extracellular iron chelates, the other essential divalent
cations of all living cells have experienced an era of benign neglect. We
have attempted in the following three chapters to redress the balance with
regard to microbial magnesium, calcium, and manganese metabolism.
Each of the three chapters represents the first attempt that we are aware
of to organize and rationalize what is known about transport of the divalent
cation as well as about nutritional requirements and cellular roles. The
three cations (and transport systems) are very different. Magnesium is
the major intracellular divalent cation, and active transport systems for its
accumulation and maintenance have evolved and been retained by all living
cells. Calcium is the major extracellular divalent cation, and the different
locations and properties of $Mg^{2+}$ and $Ca^{2+}$ are frequently used to regulate
metabolic activities. All cells, we believe, actively extrude calcium;
Chapter 2 presents the first explicit exposition of this hypothesis. Already
much supportive data are available. Manganese differs from $Ca^{2+}$ and
$Mg^{2+}$ in that it usually is found as a "trace" element. For most, if not all,
living cells, $Mn^{2+}$ is an essential micronutrient; therefore highly specific
active transport systems evolved to provide cells with this needed material
in the face of a $Mn^{2+}$-famine environment rich in other divalent cations,
especially $Mg^{2+}$ and $Ca^{2+}$. It has proved easier to demonstrate the universal existence of $Mn^{2+}$ transport systems than it has been to prove
universal requirements for $Mn^{2+}$ for cell growth. This, of course, is a
tribute to the potency or high efficiency of these cellular systems.

Simon Silver

5

Chapter 1

MAGNESIUM TRANSPORT IN MICROORGANISMS

Paula Jasper[1] and Simon Silver

Biology Department
Washington University
St. Louis, Missouri

---

[1] Present address: Department of Biochemistry, St. Louis University, St. Louis, Missouri.

## I.  INTRODUCTION

Magnesium is essential to all living cells and is the most abundant intracellular divalent cation.  Because magnesium constitutes 2.2% of the
earth's crust [28], it is likely that microorganisms in many environments
will have sufficient magnesium for growth.  Studies of the means by which
organisms regulate magnesium are important for several reasons:  (a)
Magnesium is required at a considerable concentration for essential cellular processes.  (b) Cells are able to accumulate magnesium selectively;
their ability to discriminate between the various cations suggests complex
systems for transport which should be under both genetic and physiological
control.  (c) Magnesium transport is an active process and thus provides
yet another system in which the coupling to energy metabolism must be
understood.

In this chapter we shall first consider the functions of magnesium in
microbial cells and the effects of magnesium deficiency.  Next we will
describe the transport of magnesium by bacterial cells and what is known
about magnesium transport systems from experiments with whole cells,
mutants, and subcellular vesicles.  The accumulation of magnesium by
fungi and algae will be included, as will a brief discussion of magnesium
transport by nonmicrobial systems.

## II.  MAGNESIUM IN THE CELL

### A.  Functions

Many cellular processes require magnesium.  The bulk of bacterial magnesium is associated with ribosomes and is needed for the preservation of
the structure of ribosomes [94].  Most of the magnesium associated with
ribosomes is tightly bound; only about one-third of the magnesium bound to
ribosomes is exchangeable in less than 1 min at $0°$ C; the remaining two-
thirds exchanges slowly at 4-5% per hour during dialysis at $0°$ C [73].

Some magnesium is associated with the bacterial cell wall [24].  Magnesium stabilizes spheroplasts, suggesting its involvement in the integrity
of the cellular membrane [48,108].  Studies of the effects of osmotic shock
[64] and ethylenediaminetetraacetate (EDTA) [49] also suggest that magnesium is important in maintaining the bacterial permeability barrier.  The
presence of 10 mM $MgCl_2$ caused a small but significant decrease in the
porosity of intact Bacillus megaterium cell membranes, but not in that of
cell walls [80].  "Porosity" was measured in these studies by the ability to
exclude neutral sugars and glycols of graded sizes; 10 mM $Mg^{2+}$ caused a
reduction of the dextran-excluded space of packed cells.  Differences occurred in the uptake of small molecules across the cell membrane in the
presence of $MgCl_2$.  No such differences were found in uptake across the

cell wall. Ordal [67a] recently found a highly specific role for cell-surface $Mg^{2+}$ in bacterial chemotaxis.

Many enzymes are activated by magnesium, including phosphohydrolases and phosphotransferases. Although the amount of magnesium bound by enzymes is quantitatively small in comparison to that bound to ribosomes or to the cell envelope, variations in intracellular nonribosomal magnesium undoubtedly affect the activity of these enzymes.

There are additional magnesium functions which are specific to certain organisms but are not functions of magnesium in all microbes. For instance, magnesium was reported to be essential for the normal cell division of yeast [22] and of bacilli growing in complex media [98]. Photosynthetic microorganisms, including photosynthetic bacteria, blue-green algae, and other protists (such as Euglena) require magnesium as a com-component of their chlorophyll [31].

The growth requirement for magnesium is absolute — no other ion can be substituted for magnesium [56,90,101]. The magnesium requirements of B. megaterium and Bacillus subtilis were reportedly reduced, however, by the addition of 25 μM manganese; and, in low concentrations, manganese stimulated magnesium accumulation in these organisms [101]. On the other hand, $Cu^{2+}$, $Zn^{2+}$, and $MoO_4^{2-}$ increased the magnesium requirement of these two organisms [101]. Similarly, the amount of magnesium required by the marine bacterium B-16 was dependent upon the level of calcium or strontium present in the medium, although calcium was not required for growth [56]. On the other hand, a mutant of Escherichia coli which required 100 times more magnesium for growth than did wild-type cells was able to grow at the same low levels of magnesium as needed by wild-type cells when the medium was supplemented with calcium or strontium [55].

B. Concentrations

The intracellular magnesium concentrations of Gram-positive and Gram-negative bacteria are not very different [77,91,100], although Gram-positive bacilli require appreciably more magnesium for growth than do Gram-negative rods [99]. Estimates of free intracellular $Mg^{2+}$ have been made in each case from measurements of free $Mg^{2+}$ after cellular disruption by a French-Press chamber or by sonication. Since it is generally agreed that 90% or more of the cellular magnesium is bound — largely in ribosomes and with other polyanions — redistribution of the magnesium during and after disruption makes true measurements of intracellular concentrations impossible. Two laboratories estimated the upper limits of free intracellular $Mg^{2+}$ at 1-4 mM [32,55], much higher than the extracellular concentration in the growth medium and representing 3-10% of the total cellular $Mg^{2+}$. Günther and Dorn's measurements [32] of free $Mg^{2+}$ were probably inaccurate because they were obtained by use of a $Mg^{2+}$ electrode, which is unable to discriminate between $Mg^{2+}$ and polyamines in the extract [45]. Radioisotope experiments described in Section III. A established the existence of energy-dependent accumulation of $Mg^{2+}$ by transport systems that could maintain a concentration gradient. The total cellular $Mg^{2+}$ content of E. coli was relatively constant and increased by less than a factor of three when the growth medium concentration was raised by a factor of one-hundred thousand from $10^{-6}$ to $10^{-1}$ M $Mg^{2+}$ [55, 85]. Similarly, with magnesium-limited chemostat cultures of Aerobacter aerogenes, the level of cell growth was directly proportional to the culture magnesium level [92,93], indicating a constant cellular $Mg^{2+}$ content. However, still another laboratory has obtained results that are considered to be inconsistent with the aforementioned picture; and, in the interest of openmindedness, we cite these here. Hurwitz and Rosano [34,35] also measured free $Mg^{2+}$ after disruption of E. coli cells. After correction of total free $Mg^{2+}$ for an extracellular fraction entrained in the cellular pellet during centrifugation (not needed in the other measurements which were on washed cells), these investigators calculated that the free internal $Mg^{2+}$ exactly equaled the external concentration over the concentration range from $6 \times 10^{-6}$ M to $10^{-2}$ M $Mg^{2+}$. Hurwitz and Rosano [34,35] concluded from these results that there was no need to invoke an energy-dependent magnesium transport system since passive equilibration would suffice. Similar results with equal free internal and external $Mg^{2+}$ levels were obtained with a pseudomonad [74], but with cells of Bacillus cereus Schmidt et al. [81] found a stable internal free $Mg^{2+}$ level of about 6 mM independent of the lower external $Mg^{2+}$ level. There are two problems with these results: First, the same question of redistribution of $Mg^{2+}$ after cellular disruption that limits conclusions from the other studies is equally a problem here. Second is the question of the correction for extracellular entrained $Mg^{2+}$. This was approximately equal to the calculated free intracellular $Mg^{2+}$, leading to a large experimental uncertainty. What one can presently conclude with sureness is (a) that often more than 90% of intracellular $Mg^{2+}$ is bound and not osmotically free; (b) that total cellular $Mg^{2+}$ is generally in the narrow range of 15–35 mmol/kg wet cells, even when the growth medium range varies by a factor of 100,000; and (c), as will be described in Section III. A, that $Mg^{2+}$ uptake by E. coli as well as many other bacteria is an energy-requiring process.

C.  Magnesium-Deficient Conditions

Magnesium starvation results in morphological changes in many bacteria. During short periods of magnesium starvation, Escherichia coli underwent filamentation [15]. After extended periods of magnesium starvation in rich media, Gram-positive bacteria also formed filaments [98], although such forms did not appear when these cells were grown in magnesium-deficient, chemically defined minimal media [99]. After several hours of

$Mg^{2+}$ starvation, two classes of A. aerogenes cells were found: (a) a population of filaments constant in number and able to form colonies upon placement in $Mg^{2+}$-containing media; and (b) a population of very small DNA-containing inviable cells. The numbers of these small inviable cells increased with time during $Mg^{2+}$ starvation [43]. The production of these small cells accounts for the increase in DNA in the absence of an increase in number of colony-forming units during $Mg^{2+}$ starvation of A. aerogenes [43]. Normally, up to 26% of the total $Mg^{2+}$ of A. aerogenes was loosely bound to the bacterial surface; this adsorbed $Mg^{2+}$ was removable upon washing with 0.85% (v/v) NaCl [93]. Magnesium-limited bacteria lacked the adsorbed $Mg^{2+}$. Extracellular polysaccharide synthesis was stimulated by the presence of the adsorbed $Mg^{2+}$. Surface-adsorbed $Mg^{2+}$ also inhibited loss of viability due to starvation, heat treatment, or cold shock [93].

The effects of magnesium starvation on RNA metabolism vary depending on bacterial species, mode of growth (continuous or batch culture), and concentration of carbon source. In magnesium-limited continuous cultures of A. aerogenes, the ratio of RNA to magnesium content was growth-rate independent, whereas the protein-to-RNA ratio varied, dependent on the growth rate [92]. In batch cultures, extensive degradation of E. coli ribosomes was observed under conditions of magnesium deprivation [18,59]. Less extensive degradation of ribosomes was observed in A. aerogenes during equivalent magnesium starvation; and it was noted that in this organism the combined deprivation of magnesium and glucose resulted in a more rapid degradation of ribosomes than did deprivation of magnesium alone [41,44]. The breakdown of ribosomes in the Gram-positive B. megaterium in the absence of magnesium was even more rapid than in E. coli [102]. Ribosomal proteins of A. aerogenes synthesized during exponential growth preceding $Mg^{2+}$ starvation were slowly lost from ribosomes [57]. These proteins were ultimately degraded to acid-soluble products and accounted for all the protein lost by $Mg^{2+}$-starved cells [57].

Protein synthesis during magnesium starvation depends, of course, upon the presence of functional ribosomes. Thus E. coli, whose ribosomes are more sensitive to magnesium starvation than are those of A. aerogenes, stopped growing within a few hours after the beginning of magnesium starvation [59,61]; net protein synthesis in A. aerogenes continued for 20 hr or longer of magnesium starvation [44], depending on the absence or presence of additional carbon sources. DNA synthesis, on the other hand, was found to continue indefinitely during magnesium starvation of A. aerogenes [42]. During magnesium starvation, E. coli strain B synthesized DNA at only 3% of the exponential rate [59], much less than with A. aerogenes.

It is now thought that the loss of salt tolerance by Staphylococcus aureus after sublethal heat injury is due to the loss of cellular $Mg^{2+}$ [33]. Kinetic studies in phosphate buffer showed good correlation between loss of salt tolerance and cellular $Mg^{2+}$ loss, but additional evidence on the reaccumulation of $Mg^{2+}$ to normal levels and the recovery of salt tolerance are necessary so as to clearly establish the function of $Mg^{2+}$ in salt tolerance.

In summary, there are numerous examples of specific effects of $Mg^{2+}$ deficiency on bacterial cells, but sometimes there are different "sensitive sites" with different types of bacteria. Macromolecular synthesis, especially that of nucleic acids, seems most sensitive to $Mg^{2+}$ deprivation.

## III.   MAGNESIUM TRANSPORT IN INTACT BACTERIAL CELLS

### A.   Gram-Negative Bacteria

Lusk and Kennedy [52] and Silver [84] first demonstrated active transport of magnesium by E. coli. Since then, additional studies of magnesium transport have been carried out in E. coli, and magnesium active transport has been demonstrated in B. subtilis [25,82,88], Rhodopseudomonas capsulata [38], S. aureus [104], and A. aerogenes [102].

Before proceeding through the various studies of E. coli $Mg^{2+}$ transport in detail, it is useful to present our current understanding briefly so as to place the results in context. Available information on E. coli $Mg^{2+}$ transport is best considered in terms of two kinetically distinct $Mg^{2+}$ transport systems (Table 1-1), as was first proposed by Nelson and Kennedy [63]. Studies with mutants defective or altered in these transport systems have provided the basis for this picture (see Section III. C). System I has a high affinity for $Mg^{2+}$ (Table 1-1) and also affinities for $Co^{2+}$, $Mn^{2+}$, $Ni^{2+}$ (D. L. Nelson, personal communication) and perhaps $Zn^{2+}$; it is synthesized constitutively by the cells. System I is missing in corA mutants and is probably altered in the mng mutants discussed below. System II is thought to be specific for $Mg^{2+}$ as a substrate for the transport system, although $Co^{2+}$ inhibits System II $Mg^{2+}$ transport [68]. System II is repressible by growth in high $Mg^{2+}$. It is only studies with mutants lacking System I, System II, or both [68] that allow the compilation of the properties in Table 1-1. With wild-type E. coli, $Mg^{2+}$ transport is fully constitutive [68,85] and there is no sign of additivity of the two systems. It is as if an additional level of regulation of total $Mg^{2+}$ transport activity occurs.

Several different methods for the measurement of magnesium in E. coli have been used. Hurwitz and Rosano determined magnesium by fluorometric analysis of magnesium-8-hydroxyquinolate [34,35,74,81]. Günther and Dorn [32] used a magnesium-sensitive electrode, which has a (then unknown) complication of being sensitive also to polyamines [45]. Magnesium was also measured by atomic absorption spectrometry [55,85]. However, measurement of magnesium by all of these methods is difficult

TABLE 1-1   Characteristics of the Two $Mg^{2+}$ Transport Systems of
E. coli K-12 [a]

| | |
|---|---|
| System I | Constitutive |
| | Substrates $Mg^{2+}$, $Co^{2+}$, $Mn^{2+}$, and $Ni^{2+}$ |
| | $K_m = 30\ \mu M\ Mg^{2+}$; $V_{max} = 11$ nmol/min per milligram of protein |
| | Missing in corA mutants; altered in mng mutants (?) |
| System II | Repressible by growth in 10 mM $Mg^{2+}$ |
| | $Mg^{2+}$ substrate specific |
| | $K_m = 30\ \mu M\ Mg^{2+}$; $V_{max} = 8$ nmol/min per milligram of protein |
| | Missing in mgt mutants |

[a] Here $K_m$ represents the concentration for half-saturation of rate,
and $V_{max}$ the maximum velocity of uptake.


and limited to net uptake and intracellular content.  Measurements of total
magnesium uptake are most readily and accurately made with the radioactive isotope $^{28}Mg$.  Magnesium-28 has a half-life of 21 hr and is produced regularly, as far as we know, only by the Brookhaven National
Laboratory, Upton, Long Island, New York.

Experiments with $^{28}Mg$ have clearly demonstrated the existence of
active transport systems for magnesium.  Silver [84] showed that $^{28}Mg^{2+}$
was accumulated by E. coli by a temperature-dependent process which
was inhibited by the uncouplers dinitrophenol (DNP) and m-chlorophenyl
carbonylcyanide hydrazone (CCCP) [85] and by cyanide.  Lusk and Kennedy
[52] also showed that magnesium accumulation was energy dependent; they
reported that magnesium accumulation was inhibited at 0° C and in the
presence of sulfhydryl poisons such as N-ethylmaleimide.  Both laboratories found that the rate of magnesium accumulation was concentration dependent but reported very different values for the $K_m$: Silver [84] suggested
a $K_m$ of 500 μM, and Lusk and Kennedy suggested a $K_m$ of 4 μM.  This
rather large discrepancy was largely explained later by the dependence of
the kinetic parameters of magnesium transport on the medium in which
they are determined.  Lusk and Kennedy made their measurements in
Tris-casamino acids medium at 37° C, whereas Silver used tryptone broth

at $22°$ C.  Subsequent determinations of the $K_m$ for magnesium accumulation in E. coli yielded values of 18 μM in Tris-glucose and 31 μM in dilute
tryptone broth, and $V_{max}$ was found to be 0.56 μmol/min per $10^{12}$ cells in
Tris-glucose and 0.20 μmol/min per $10^{12}$ cells in tryptone broth at $22°$ C
[85].  Park et al. [68] provide the currently best available kinetic analysis of the magnesium transport systems of E. coli and report a $K_m$ of from
30 to 60 μM depending on conditions.

Magnesium accumulation by E. coli is specific in that neither calcium
nor strontium nor potassium inhibited magnesium uptake [52,84,85].
Manganese and cobalt, however, competitively inhibited magnesium accumulation [62,85].  The inhibition of magnesium uptake by manganese
followed classical competitive inhibition kinetics with a $K_i$ (inhibition
constant) of 0.5 mM manganese in Tris-glucose and of 2 mM in dilute
tryptone broth [85].  The $K_i$ for the inhibition of magnesium accumulation by cobalt was approximately 0.4 mM [62].  More recent measurements suggest that the $K_m$ for accumulation of any of the three
ions — $Mg^{2+}$, $Co^{2+}$, or $Mn^{2+}$ — is about the same (J. E. Lusk, personal
communication); the $K_m$ and $K_i$ values clearly vary with the conditions of
the experiments.  Evidence suggested that manganese and cobalt were
actually transported by the magnesium transport system.  First, the $K_m$
for $Mg^{2+}$ uptake was approximately equal to the $K_i$ for $Mg^{2+}$ as an inhibitor
of $Co^{2+}$ uptake; and, conversely, the $K_m$ for $Co^{2+}$ uptake was roughly the
same as the $K_i$ for $Co^{2+}$ in inhibiting $Mg^{2+}$ uptake [62].  Second, if (as
will be discussed shortly) the influx and the efflux of $Mg^{2+}$ are coupled,
then the data which show that $Mn^{2+}$ and $Co^{2+}$ as well as extracellular
$Mg^{2+}$ can promote $Mg^{2+}$ egress are evidence that $Mn^{2+}$ and $Co^{2+}$ act as
alternative substrates for the $Mg^{2+}$ accumulation system [62,85].  Finally,
mutants isolated as resistant to $Co^{2+}$ proved to be defective in both $Co^{2+}$
and $Mg^{2+}$ uptake [62].

Entry and exit of magnesium in E. coli are apparently both carrier
mediated, since efflux as well as influx of magnesium is energy dependent
[32a,52,62,85].  Cations which affected influx of $^{28}Mg^{2+}$ also affected its
efflux: $Co^{2+}$, $Mn^{2+}$, and nonradioactive magnesium (but not $Ca^{2+}$) promoted the release of accumulated $^{28}Mg^{2+}$ [62,85].  The half-saturation
concentration for added external $Mn^{2+}$ accelerating $^{28}Mg^{2+}$ efflux was in
good agreement with the $K_i$ for $Mn^{2+}$ as a competitor for $^{28}Mg^{2+}$ uptake,
i.e., 2 to 5 mM and 2 mM $Mn^{2+}$, respectively, in tryptone broth [85].
Furthermore, as expected if manganese enters the cells via the normal
magnesium accumulation system, added nonradioactive magnesium competitively inhibited manganese-stimulated loss of $^{28}Mg^{2+}$ [85].  Finally,
in a mutant defective in the cobalt-magnesium accumulation system,
cobalt did not induce magnesium efflux [62,63].

Nickel ($Ni^{2+}$) also appears to be transported by the system primarily
responsible for the accumulation of $Mg^{2+}$.  D. L. Nelson (personal communication) demonstrated that accumulation of $^{63}Ni^{2+}$ by E. coli is

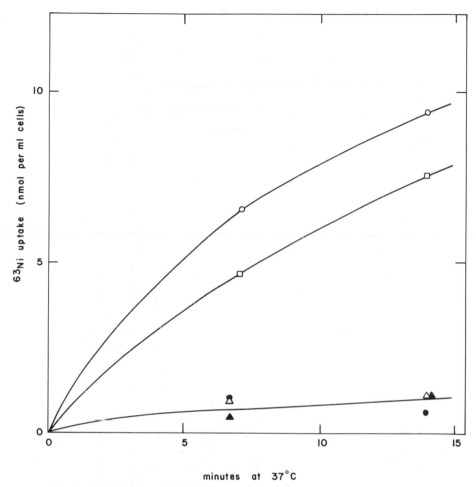

**minutes at 37°C**

FIG. 1-1   Nickel uptake by wild-type but not by corA mutant E. coli. Overnight cells were grown in broth for 1 hr, centrifuged and resuspended in assay buffer with glucose as energy source.   Uptake of 100 μM $^{63}Ni^{2+}$ by the cells was determined by millipore filtration and washing.   Wild-type K-12 strain A324 control (O) or in the presence of 1.0 mM $Mg^{2+}$ (●) or $Ca^{2+}$ (□); corA53 mutant control (△) or in the presence of 1.0 mM $Mg^{2+}$ (▲). (Data from D. L. Nelson, unpublished experiment.)

energy dependent since such accumulation was reduced at 0° C and inhibited by NaCN and NaN$_3$.   This investigator also found that both $Mg^{2+}$ and $Co^{2+}$ competitively inhibited $^{63}Ni^{2+}$ accumulation.   A mutant defective in $Co^{2+}$ transport (corA) was defective in $Ni^{2+}$ uptake as well (Fig. 1-1).

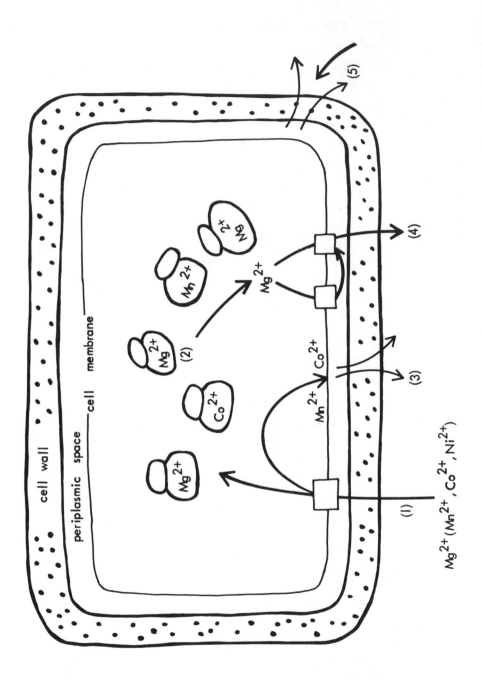

The displacement of $^{28}Mg^{2+}$ from the cell by manganese and cobalt and the effects of these ions on growth of the cells (i.e., manganese is growth inhibitory, cobalt is lethal) must be explained.  Silver and Clark [85,86] suggested that accumulated manganese readily exchanges with ribosome-bound magnesium, thus increasing the intracellular "free" magnesium level and allowing this magnesium to leave the cells by the magnesium transport system.  Nelson and Kennedy [62] have suggested, alternatively, that if the cell wall hinders the diffusion of $Mg^{2+}$ away from the periplasmic space, $^{24}Mg^{2+}$ or $Co^{2+}$ (or for that matter, $Mn^{2+}$) which is present in the medium could diffuse into the periplasmic space, where it would immediately dilute any $^{28}Mg^{2+}$ shuttled out across the membrane and would thus reduce the probability of $^{28}Mg^{2+}$ being recaptured by the uptake system.  At high concentrations of manganese or cobalt, magnesium involved in membrane stability might be displaced also.

Our current picture of the magnesium transport system of E. coli and the effects of other divalent cations on cellular magnesium is summarized in Figure 1-2.  Magnesium is taken up by E. coli cells by a high-affinity transport system (step 1).  Alternative but lower affinity substrates for either or both of these systems include $Mn^{2+}$, $Co^{2+}$, and $Ni^{2+}$.  At micro-nutrient levels, however, these cations are probably accumulated by separate micronutrient systems (see Chapter 3).  Once within the cells perhaps 90% of the magnesium is associated with nucleic acids, largely in ribosomes.  In equilibrium exchanges, newly accumulated $Mg^{2+}$ can displace $^{28}Mg^{2+}$ from the ribosomes (step 2), as can newly accumulated $Mn^{2+}$ or $Co^{2+}$.  This causes an accelerated egress of $^{28}Mg^{2+}$ via the normal magnesium transport system (step 4).  Under low $Mg^{2+}$ growth conditions, $^{28}Mg^{2+}$ lost from the cell interior into the periplasmic space may be recaptured by the high-affinity membrane-bound transport system without coming into equilibrium with extracellular magnesium.  However, periplasmic $Mn^{2+}$ or $Co^{2+}$ would inhibit $Mg^{2+}$ uptake and stimulate apparent $Mg^{2+}$ loss at this level (step 4).  In addition, accumulated $Mn^{2+}$ or $Co^{2+}$ might induce nonspecific cellular leakage by displacing $Mg^{2+}$ involved in stabilizing the cell membrane (step 3).  The fact that $Mn^{2+}$ and $Co^{2+}$ are concentrated to high intracellular levels by the $Mg^{2+}$ transport system

---

FIG.  1-2   Model for transport and inhibitory effects of divalent cations. Steps:  (1) uptake of various cations by the magnesium transport system(s); (2) displacement of $Mg^{2+}$ from intracellular polyanions, primarily on ribosomes; (3) nonspecific membrane leakage induced by $Mn^{2+}$ or $Co^{2+}$ from within; (4) egress of $Mg^{2+}$ (or $Mn^{2+}$) by way of the magnesium transport carrier with "recapture loop" within the periplasmic space; and (5) non-specific leakage induced by $Mn^{2+}$ or $Co^{2+}$ from without (this step is not thought to occur).

makes intracellular destabilization more likely than a similar mode of attack from without (step 5). The $Mn^{2+}$ and $Co^{2+}$ accumulated by the cells via the $Mg^{2+}$ transport system would also be found predominantly associated with the RNA of ribosomes. The $Mn^{2+}$ could be displaced by later accumulated $Mg^{2+}$, accounting for the nonlethal inhibitory effects of high $Mn^{2+}$ [86]. The cobalt is essentially irreversibly bound to the ribosomes, accounting for the lethal effects of high $Co^{2+}$ [62]. Only in the presence of $Mg^{2+}$ and EDTA can some of the $Co^{2+}$ be displaced (J. E. Lusk, personal communication).

Magnesium transport of wild-type E. coli is constitutive since the $^{28}Mg^{2+}$ uptake rate is independent of the concentration of magnesium in the growth medium [85]. However, studies with mutants indicate that two systems function in the transport of $Mg^{2+}$ by E. coli, one of which (System II) is almost completely repressed during growth in media with high concentrations of magnesium [63]. System I is nonrepressible and is present in wild-type but not in cobalt-resistant mutants (Table 1-1), whereas System II is present in wild-type and corA mutant strains and is repressible. System I, unlike System II, catalyzes the uptake of cobalt as well as that of magnesium. Double mutants lacking both Systems I and II exhibit no saturable, energy-dependent transport of $Mg^{2+}$ and require 10 mM $Mg^{2+}$ for optimal growth [68].

The effects of membrane-perturbing agents on cellular uptake and retention of $^{28}Mg^{2+}$ has been studied in some detail. However, in each case the effects have been considered indirect and relatively nonspecific. Silver et al. [87] measured leakage of $^{28}Mg^{2+}$ from E. coli during early stages of infection with bacteriophage T2. Potassium was also lost from the cells. As the amount and rate of loss was proportional to the number of infecting virus particles per cell, these results were interpreted as being due to a transient nonspecific leakage around the site of virus penetration, followed by a "sealing" process. Related and nonspecific loss of $^{28}Mg^{2+}$ was found when E. coli cells were killed by the class of protein colicins that are known to interfere specifically with membrane function and energy metabolism [54]. Colicins E1 and K cause net loss of cellular $Mg^{2+}$ at rates proportional to the number of killing particles added per cell. In addition, these colicins increase passive permeability to $Mg^{2+}$ (and to $Co^{2+}$) by routes other than the normal transport systems [54]. Since colicin K causes extensive loss of cellular $Mg^{2+}$ and $K^+$, Kopecky et al. [46] asked whether cells could be protected from the killing effects of the colicin by maintenance in very high concentrations of cations. Plating on media containing 100 mM $K^+$ and 1-2 mM $Mg^{2+}$ afforded appreciable protection against the lethal effects of the energy-metabolism colicins (E1 and K) but had no effect on survival after treatment with other colicins (E2 and E3) that have completely different modes of action [46].

The third type of membrane-perturbing agent that has been studied in some detail consists of lipid-soluble compounds. Devynck et al. [20]

TABLE 1-2   Kinetic Parameters of Cation Transport in Phototrophically
Grown R. capsulata

| Cation | $K_m$ ($\mu$M) | $V_{max}$ (nmol/min per mg) | Specific inhibitors |
|--------|------|------------------|---------------------|
| Potassium | 200. | 8.0 | Competitive<br>$K_i(Rb^+) = 560\ \mu M$<br>$K_i(Cs^+) = 2,700\ \mu M$ |
| Magnesium | 55. | 1.8 | Competitive<br>$K_i(Co^{2+}) = 200\ \mu M$<br>$K_i(Mn^{2+}) = 300\ \mu M$<br>$K_i(Fe^{2+}) = 390\ \mu M$ |
| Manganese | 0.48 | 0.02 | $Co^{2+}$, $Fe^{2+}$ |

found that levallorphan, a morphine analog, inhibited cellular uptake of
$^{28}Mg^{2+}$ and stimulated $^{28}Mg^{2+}$ loss from E. coli cells.  Magnesium protected the cell surface by displacing levallorphan in a process wherein one
magnesium ion replaced two levallorphan molecules.  However, calcium
also afforded protection to the cells, and the action of levallorphan was
considered a relatively nonspecific membrane perturbation [20].  The
effects of levallorphan appeared similar to those we described a little
earlier with a variety of membrane-perturbing agents including phenethyl
alcohol, steroidal diamines, and fatty acid-containing N-acyl triamines
[89].

Recently the magnesium transport system of another Gram-negative
bacterium, Rhodopseudomonas capsulata, was studied [38].  R. capsulata
is a purple nonsulfur photosynthetic bacterium capable of heterotrophic
growth in the dark as well as of phototrophic growth.  Magnesium accumulation has been studied in organisms grown both ways.  As with the mag-
nesium-accumulation system of E. coli, the accumulation of magnesium
by R. capsulata followed Michaelis-Menten saturation kinetics (Table 1-2).
The $K_m$ for $Mg^{2+}$ uptake by phototrophically grown cells (i.e., those grown
anaerobically with light) was the same as that found with cells grown heterotrophically (aerobically in darkness): 55 and 52 $\mu$M $Mg^{2+}$, respectively.
Although the $V_{max}$ for magnesium accumulation varied from experiment to
experiment with dark-grown cells (0.6, 0.9, 2.0 nmol/min per milligram
dry weight in three separate experiments), the $V_{max}$ found for phototrophic
cells was relatively constant (1.8 and 2.0 nmol/min per gram dry weight
in two experiments).

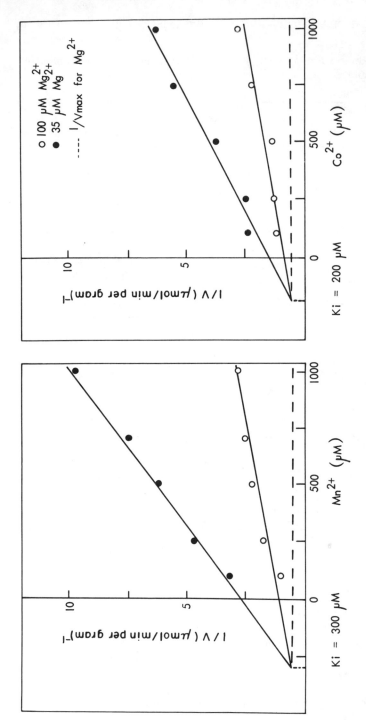

FIG. 1-3 Competitive inhibition of magnesium transport by manganese and by cobalt in R. capsulata. The rate of $^{28}Mg^{2+}$ accumulation was determined by millipore filtration. The reciprocals of the rates of $^{28}Mg^{2+}$ uptake at two different substrate concentrations (35 μM and 100 μM $Mg^{2+}$) are plotted against the concentration of inhibitor ($Mn^{2+}$ or $Co^{2+}$) present during the assay. With this method of plotting, competitive inhibition is indicated by the intersection of the lines at a point above the x axis with a y coordinate equal to $1/V_{max}$ in the absence of inhibitor and an x coordinate equal to $K_i$, the inhibition constant. (From Ref. 38.)

As expected, magnesium accumulation by R. capsulata was energy dependent. Inhibitors such as the uncouplers DNP and CCCP, and sodium cyanide (in aerobically grown cells only), and low temperature ($0°$ C) blocked accumulation of magnesium by the cells [38].

Both cobalt and manganese, divalent cations which were found to inhibit the accumulation of magnesium by E. coli, also competitively inhibited magnesium accumulation by R. capsulata (Fig. 1-3). In addition, iron and calcium (without effect on E. coli [84]) inhibited magnesium uptake in R. capsulata. The inhibition by iron was competitive, but the nature of the inhibition by calcium was unclear [38]. Constants for the competitive inhibitors were as follows: $K_i(Co^{2+}) = 200 \mu M$; $K_i(Mn^{2+}) = 300 \mu M$; and $K_i(Fe^{2+}) = 390 \mu M$ (cf. Table 1-2 and Fig. 1-3). These values demonstrate that the transport system has a higher affinity (lower $K_m$) for $Mg^{2+}$ than for the other divalent cations. Much of the magnesium accumulated by phototrophically grown R. capsulata was not firmly bound within cells. Nonradioactive $Mg^{2+}$ and $Mn^{2+}$ (but neither $Co^{2+}$ nor $Fe^{2+}$) induced the loss of accumulated $^{28}Mg^{2+}$ in the absence of added nonradioactive magnesium (P. Jasper, unpublished data).

The kinetic parameters of various ion transport systems frequently reflect the intracellular concentrations of each ion required by the cells. For comparison, Table 1-2 gives the $K_m$ and $V_{max}$ values for three cation transport systems of R. capsulata. It can be seen that the transport system for $K^+$, the most abundant cation, has a high $V_{max}$ and a relatively low affinity (high $K_m$), whereas the transport system for $Mg^{2+}$, the second most abundant intracellular cation, has a higher affinity (lower $K_m$) but also lower $V_{max}$. Finally, the micronutrient transport system for manganese (see Chapter 3) has a very high affinity ($K_m = 0.48 \mu M$), inasmuch as manganese is found only in trace quantities in nature.

The reaccumulation of magnesium liberated from a wide variety of Gram-positive and Gram-negative microorganisms has been studied by Webb, who showed that resting and exponentially growing cells liberate $Mg^{2+}$ when transferred to $Mg^{2+}$-free medium [100,102]. The released $Mg^{2+}$ was less effectively reaccumulated by Gram-positive organisms than by Gram-negative ones [100,102]. Webb showed that the reaccumulation of $Mg^{2+}$ by A. aerogenes (Gram-negative) and by B. megaterium (Gram-positive) was inhibited at $0°$ C or in the presence of 2,4-dinitrophenol, and thus was not due to surface binding of $Mg^{2+}$ [103]. Increasing concentrations of $Mg^{2+}$ (but not of $Mn^{2+}$, $Ca^{2+}$, or $K^+$) inhibited the uptake of $Ni^{2+}$, $Co^{2+}$, and $Zn^{2+}$ by A. aerogenes; and increasing concentrations of $Ni^{2+}$, $Co^{2+}$, and $Zn^{2+}$ caused both an increase in the amount of $Mg^{2+}$ liberated by A. aerogenes and a decrease in the $Mg^{2+}$ reaccumulation [103]. These findings led Webb to conclude that $Ni^{2+}$, $Co^{2+}$, and $Zn^{2+}$ are transported by the same system that transports $Mg^{2+}$ in A. aerogenes and also in E. coli and B. megaterium. Bucheder and Broda [16] have shown energy-dependent uptake of $Zn^{2+}$ by E. coli. Since their assays were run

in the absence of magnesium and at relatively high (10 $\mu$M) concentrations
of $Zn^{2+}$, it is not clear whether such uptake occurred via the system concerned with magnesium accumulation or was due to a separate micronutrient $Zn^{2+}$ transport system (see Chapter 4).


B.  Gram-Positive Bacteria

In addition to Webb's [100,102,103] atomic absorption studies of the reaccumulation of $Mg^{2+}$ liberated by Gram-positive bacilli, there have been
several investigations of radioactive magnesium accumulation with
Bacillus subtilis.  Work in our laboratory [82,83,88] has demonstrated
that magnesium uptake by B. subtilis occurs by a specific transport system
that is energy dependent.  The magnesium transport system of log phase
cells of B. subtilis shows a lower affinity for magnesium than was found
for the comparable transport systems of either E. coli or R. capsulata
(Table 1-3).  As in other bacteria, manganese was a competitive inhibitor
of magnesium accumulation by B. subtilis.
    During sporulation of B. subtilis, the rate of magnesium accumulation
declined, although the manganese and calcium transport rates increased
[82,83,88].  At 4.5 or 8 hr after the end of log-phase growth (and into the
sporulation process), the residual magnesium transport function appeared
to be saturated even at the lowest tested concentration of 34 $\mu$M $Mg^{2+}$; the
$K_m$ was probably less than 10 $\mu$M.  Stationary-phase B. subtilis, grown 9
hr beyond log phase but not sporulating, also had a low rate of $Mg^{2+}$ uptake comparable to that found for late sporulating cells [82].  These data
were thus suggestive of two magnesium transport systems — with only a
high-affinity system functioning in late sporulation.  Determinations of $K_m$
and $V_{max}$ of the major system during growth or early sporulation would
not be influenced by this high-affinity system because of its low $K_m$ and
$V_{max}$.  However, lack of additivity of two $Mg^{2+}$ systems was noted with
studies of E. coli mutants [68], and interpretations of the B. subtilis $Mg^{2+}$
transport as being due to two systems with regulatory control or one system with variable kinetic parameters are therefore premature.  In contrast
to B. subtilis, there was little change in the rate of magnesium accumulation during the growth cycle of R. capsulata grown phototrophically (although the rate did change for R. capsulata grown aerobically in the dark
[38]).
    Manganese induced the loss of $^{28}Mg^{2+}$ from B. subtilis.  There was a
difference, however, between E. coli and B. subtilis with respect to the
action of cobalt on magnesium exchange.  Although $Co^{2+}$ inhibited $Mg^{2+}$
accumulation in both bacteria, cobalt did not stimulate the loss of cellular
$^{28}Mg^{2+}$ from B. subtilis (or R. capsulata) as it did from E. coli [82]; see
Figure 1-4.  In fact, $Co^{2+}$ actually inhibited $^{28}Mg^{2+}$ loss from B. subtilis.
This difference between E. coli and B. subtilis must be interpreted in
terms of our model for magnesium metabolism given in Figure 1-2.  If

TABLE 1-3   Kinetic Parameters of Magnesium Transport in Various Organisms [a]

| Species | $K_m$ ($\mu$M) | $V_{max}$ (nmol/min per mg) | Competitive inhibitors |
|---|---|---|---|
| E. coli | 31 (broth) | 8.0 | $K_i(Mn^{2+}) = 2$ mM |
| | 18 (Tris) | | $K_i(Co^{2+}) = 0.4$ mM |
| R. capsulata | 55 | 1.8 | $Co^{2+}$, $Mn^{2+}$, $Fe^{2+}$ |
| B. subtilis | 250 | 4.4 | $K_i(Mn^{2+}) = 0.5$ mM |
| B. subtilis (citrate system)[b] | 450 | 123. | $Mn^{2+}$ |
| S. aureus | 70 | 9.6 | |
| Euglena gracilis | 300 | 1.0 nmol/hr per $10^5$ cells | |
| Human KB cells | 100 | 0.4 nmol/hr per $10^5$ cells | |

[a] E. coli data from Ref. 85; R. capsulata, cf. Table 1-2; B. subtilis, Refs. 82 and 105; S. aureus, Ref. 104; E. Gracilis, D. Kohl and S. Silver, unpublished data; and KB cells, Ref. 6.

[b] Citrate inducible [Ref. 105].

$Co^{2+}$ bound competitively to the $Mg^{2+}$ transport carrier but was not itself transported in B. subtilis and was both bound and transported in E. coli, then $Co^{2+}$-inhibited $^{28}Mg^{2+}$ egress would be due to the inhibition of $^{24}Mg^{2+}$-$^{28}Mg^{2+}$ exchange.

In addition to the aforementioned two magnesium transport systems of B. subtilis, there exists another, apparently separate system which mediates the cotransport of magnesium (or other divalent cations) and citrate [105]. Citrate transport was dependent upon the presence of $Mg^{2+}$ (or $Mn^{2+}$, $Co^{2+}$, or $Ni^{2+}$), and the divalent cations were accumulated by the cells simultaneously with citrate. The system was inducible by citrate and was energy dependent. Cells must be induced for citrate transport in order to carry out citrate–dependent accumulation of divalent cations. The kinetic constants of citrate accumulation and citrate-dependent $Mg^{2+}$ accumulation were very similar, specifically, $K_m = 0.55$ mM citrate and $0.45$ mM $Mg^{2+}$,

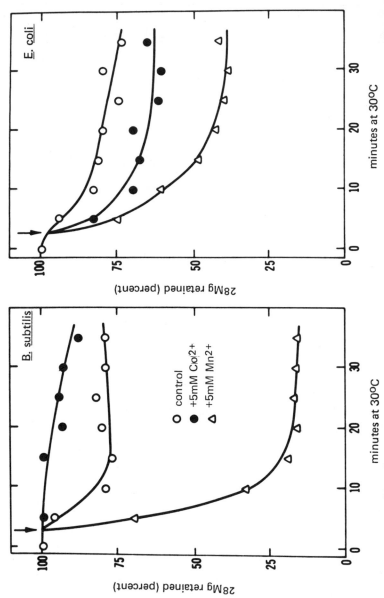

FIG. 1-4  Exchange of $^{28}Mg^{2+}$ by B. subtilis and E. coli. Cells were grown in supplemented tryptone broth with added $^{28}Mg^{2+}$. After centrifugation and resuspension in supernatant fluids from equivalent nonradioactive cultures, aliquots were distributed; $Mn^{2+}$ or $Co^{2+}$ were added where indicated. Samples were Millipore filtered and washed, and the radio-activity was normalized to that in three initial samples (per strain). (From Ref. 82 with permission.)

and $V_{max}$ = 145 nmol citrate/min per milligram and 123 nmol $Mg^{2+}$/min per gram, respectively, suggesting that both substrates may be bound as a complex to the same "transporter" molecule (see also Section III. C).

The accumulation of magnesium early in germination of B. subtilis spores has been reported [25]. Magnesium (and also potassium) active transport were "turned on" at the start of spore germination, prior to RNA and protein synthesis, and therefore prior to the biochemical events usually associated with the initial outgrowth stage of germination. This was shown by the uptake of $^{28}Mg^{2+}$ and $^{42}K^+$ before detectable incorporation of [$^{14}C$]uracil into ribonucleic acid and [$^{14}C$]leucine into protein. The uptake of cations was furthermore unaffected by actinomycin D, rifamycin SV, and chloramphenicol [25]; therefore, latent transport systems must exist in dormant spores.

One additional study of magnesium accumulation by a Gram-positive bacillus should be related because of its bearing on previous findings of magnesium accumulation by E. coli. Schmidt et al. [81] reported that Bacillus cereus T was able to accumulate $Mg^{2+}$ in its cell "sap" against a concentration gradient and maintained an internal free concentration of about 6 mM $Mg^{2+}$. Their method involved the fluorometric determination of magnesium in the high-speed supernatant (from a 100,000g centrifugation of French-Press disrupted cells, after the cells had been collected by centrifugation through silicone oil. Using this same method, however, Hurwitz and Rosano [35] previously reported that the intracellular free magnesium concentration of E. coli was the same as the extracellular magnesium concentration, which led them to postulate free permeability and the absence of magnesium active transport in E. coli [35]. The finding that B. cereus does accumulate $Mg^{2+}$ against a concentration gradient, coupled with the results with $^{28}Mg^{2+}$ from other laboratories, calls into question the technical aspects of the results of Hurwitz and Rosano with E. coli [35]. Their method, in any case, does not allow for rapid measurements of directional $Mg^{2+}$ movement and thus is inherently less appropriate for the detection of active magnesium transport.

Recent experiments in our laboratory [A. Weiss, unpublished data; see also Ref. 104] indicate as expected that magnesium accumulation by S. aureus, another Gram-positive bacterium, is energy dependent, i.e., magnesium accumulation was inhibited by the uncoupler CCCP, by NaCN, and at $0°C$. The $K_m$ for uptake was approximately 70 μM, only slightly higher than the $K_m$ values determined for E. coli and R. capsulata, which are Gram-negative, and lower than the $K_m$ for Gram-positive B. subtilis (Table 1-3). The $V_{max}$ for accumulation of $Mg^{2+}$ by S. aureus was approximately 9.6 nmol/min per milligram. Magnesium uptake by S. aureus was inhibited by $Zn^{2+}$ and $Mn^{2+}$, and possibly by $Co^{2+}$. Nonradioactive $Mg^{2+}$ and $Mn^{2+}$, but neither $Co^{2+}$ nor $Zn^{2+}$, promoted the release of previously accumulated $^{28}Mg^{2+}$ from S. aureus. In this respect, S. aureus was similar to B. subtilis (Fig. 1-4) and R. capsulata. Intracellular $^{28}Mg^{2+}$ did not exchange with $Co^{2+}$ in any of these three bacteria.

## C.  Mutants

Several kinds of mutants altered in magnesium transport have already been mentioned.  We will now proceed to discuss these and other mutants in greater detail.

Lusk [51] and Lusk et al. [55] isolated two mutants of E. coli which required high levels of magnesium for growth.  The magnesium requirement for 50% of the maximum growth rate was 1-3 $\mu$M for wild-type E. coli, but 300 $\mu$M or 600 $\mu$M $Mg^{2+}$ for the two mutants.  Supplementation of the medium with high levels of calcium or strontium reduced the magnesium requirement to that of wild-type cells.  The magnesium content of the mutant cells, when they were grown either on 10 mM $Mg^{2+}$ or on 0.01 mM $Mg^{2+}$ plus 1 mM $Ca^{2+}$, was similar to that of the parental type.  In addition, the mutant cells grown in a medium containing high levels of calcium and limiting magnesium stopped growing when the magnesium was exhausted from the medium.  These findings showed that calcium could not substitute for magnesium and suggested that magnesium transport was not directly affected in these mutants.  More recently, Lusk and Kennedy [53] reported that the growth of one of the mutants (A324-1) was inhibited by low concentrations of sodium ions and that it was this inhibition that was relieved by added magnesium, calcium, or strontium.  The uptake of $^{28}Mg^{2+}$ by mutant A324-1 (in the absence of $Na^+$) was as rapid as with the wild-type strain.  Furthermore, growth-inhibitory concentrations of sodium did not affect $^{28}Mg^{2+}$ transport [53].  More cardiolipin and less phosphatidyl ethanolamine were incorporated in the membranes of mutant cells after addition of sodium than in the absence of sodium [53]; but rates of synthesis of protein, RNA, DNA, and total lipid did not.change.  It is still premature to conclude that altered phospholipids are the immediate cause of inhibition by sodium or to speculate as to how magnesium and calcium relieve this inhibition, but it is clear that magnesium transport is unaltered in this mutant.  The growth inhibition of mutant A324-1 by sodium and the relief of this inhibition by divalent cations is reminiscent of the growth inhibition of bacteriophage T4 rII mutants on E. coli K-12 strains lysogenic for bacteriophage lambda.  Such rII mutant growth was also $Na^+$ inhibited and relieved by divalent cations or polyamines [30].  The basic rII defect is still unknown; but rII of phage T4, like the mutation in E. coli A324-1, is thought to involve a change in membrane structure.

The defect in the other magnesium-requiring mutant isolated by Lusk [51] and Lusk et al. [55], A324-2, is still unknown.  Strain A324-2 required high concentrations of magnesium when grown on amino acids as carbon source but not when grown on glucose, galactose, lactose, or glycerol [51].  Unlike A324-1, growth of strain A324-2 in low-magnesium medium containing any of these four carbon sources was not inhibited by sodium.

Nelson and Kennedy [62,63] isolated mutants of E. coli with altered magnesium transport.  These mutants were selected for their ability to survive exposure to cobalt in the absence of magnesium (cor-4 and cor-53) or by their ability to grow in the presence of very high concentrations (0.2 M) of magnesium (mgr-19). The uptake of cobalt was very low in each of the three mutants as compared with the wild-type, regardless of whether the cells were grown on high (10 mM) or low (50 μM) magnesium concentrations. The rate of magnesium uptake in the mutants was similar to that of the wild-type when the mutants were grown on low concentrations of magnesium; but when grown on high concentrations of magnesium, the rate of magnesium uptake was much lower in the mutants than in the wild-type. It was concluded that there are two systems for the uptake of magnesium by E. coli (Table 1-1), only one of which remained in these mutants. System II, present in both mutant strains and wild-type, was repressible and had little affinity for cobalt. System I, which was missing in the corA mutants, was nonrepressible and could catalyze the uptake of $Co^{2+}$ as well as of $Mg^{2+}$.

Mutants with altered magnesium transport properties were also isolated as manganese resistant [86]. Experiments with $^{28}Mg^{2+}$ led to the conclusion that the manganese-resistant mutants were altered in the inhibition constant ($K_i$) for manganese as a competitive inhibitor of magnesium accumulation. As with the other E. coli magnesium transport mutants, the molecular basis leading to the manganese-resistant phenotype is unknown. It is interesting to note that the gene determining resistance to manganese (mng) in E. coli is chromosomal (mapping at about 40 min on the genetic map [2]), whereas genes conferring resistance to heavy metals are usually plasmid borne.

Recently, Park et al. [68] isolated new mutants missing the repressible $Mg^{2+}$ transport system (mgt) as well as double mutants (corA mgt) that lacked both systems and therefore accumulated magnesium only by means of non-energy-dependent nonsaturable passive diffusion. The study of these mutants led to the formulation of Table 1-1. Still a third class of mutants (corB) carried out $Co^{2+}$ transport only when grown on high $Mg^{2+}$ concentrations. CorB is likely a regulatory gene that causes System I to be inducible, instead of constitutive as in wild-type strains [68]. Both CorA and CorB of mutations affected $Co^{2+}$ and $Mn^{2+}$ transport in parallel [68]. The corA mutation mapped at 83.5 min on the new E. coli genetic map, cotransducible with ilv and with uncA [2,68]. This means that corA is not identical with the $Mn^{2+}$-resistance locus mng, which maps at 39 min [2,86]. In addition, mng mutants were $Co^{2+}$ sensitive [86], and corA mutants were $Mn^{2+}$ sensitive [68]. Nevertheless, corA and mng are both thought to affect the function of $Mg^{2+}$ System I. The regulatory gene corB was not cotransducible with either ilv or corA; corB maps at 93 min [2,68]. CorB must form a regulatory protein that can function at a distance rather than being part of a hypothetical corA operon. The other locus affecting $Mg^{2+}$ transport, mgt, was 5-10% cotransducible with malB at about 91 min on the map (still a fourth map locus!). Clearly, even the number of genetic loci affecting magnesium transport is yet to be fixed. The recent studies of Park et al. [68] have also unearthed a $Ca^{2+}$ sensitivity with the corA mutants, but it seems premature to ascribe this to $Ca^{2+}$/movement coupled with $Mg^{2+}$ uptake in the opposite direction.

Cobalt-resistant mutants of B. subtilis were unaltered in their citrate-dependent magnesium transport system [105]. However, one such mutant exhibited only 33% of the initial magnesium uptake rate shown by the parental strain, when assayed for magnesium transport by the high-affinity citrate-independent magnesium transport system. Since the growth of B. subtilis was retarded in ribose minimal media by the presence of the citrate analog DL-erythro-fluorocitrate (probably by inhibition of aconitase activity [65]), mutants were isolated that were resistant to 2-fluoro-DL-erythro-citrate. These mutants were defective in coupled citrate-magnesium transport, as indicated by a decreased ability to transport citrate; no direct measurements of magnesium transport were made [65]. Because the fluorocitrate-resistant mutants were able to grow, although slowly, on minimal medium containing citrate as sole carbon source, it was suggested that B. subtilis has a second system for the transport of citrate [65]. This second system would transport citrate slowly and with lower affinity than the described citrate-magnesium transport system.

D.  Transport by Subcellular Vesicles

Isolated membrane vesicle "ghosts" from E. coli and other bacteria actively accumulate sugars and amino acids. Membrane vesicles also actively take up cations. Vesicles accumulate the monovalent cations $K^+$ and $Rb^+$, in the presence of the ionophore valinomycin [9], and $Mn^{2+}$ (the presence of 10 mM calcium is needed for E. coli [7] but not for B. subtilis [8]); $Ca^{2+}$ has been shown to be accumulated by inside-out vesicles [75, 95 (see also Chapter 2 in this volume)].

D. L. Nelson (personal communication) found that vesicles of E. coli carried out energy-dependent accumulation of magnesium. The rate of uptake of magnesium followed Michaelis-Menten saturation kinetics, with a $K_m$ of about 13 μM and a $V_{max}$ of 0.6 nmol/min per milligram of protein. The vesicles also took up $^{60}Co^{2+}$; and $Co^{2+}$, $Ni^{2+}$, $Mn^{2+}$, and nonradioactive $Mg^{2+}$ inhibited $^{28}Mg^{2+}$ accumulation. Vesicles which had previously accumulated $^{28}Mg^{2+}$ released $^{28}Mg^{2+}$ in the presence of $^{24}Mg^{2+}$, $Co^{2+}$, or $Mn^{2+}$. Stimulated egress was also found to be saturable with respect to external $Mg^{2+}$, with a $K_m$ between 10 and 50 μM $Mg^{2+}$. There was great variability between batches of membranes with regard to magnesium transport; the accumulation properties with $^{60}Co^{2+}$ appeared more reproducible. It is unfortunate that these results have not been published and have not been further pursued by any of the laboratories known to be interested in magnesium transport.

E.  Chelates and Ionophores

Ionophores are small membrane-soluble molecules which transport ions
across otherwise impermeable membranes.  Additional information on
divalent cation ionophores appears in Chapters 2 and 3 on calcium and
manganese transport.  Therefore we will limit our comments here to currently available suggestions of a relationship of ionophores to microbial
magnesium utilization.

Two antibiotic ionophores, X537A and A23187 (known to transport
calcium and discussed in some detail in Chapter 2), are also known to have
an affinity for magnesium: X537A has a greater affinity for $Ca^{2+}$ than for
$Mg^{2+}$, but the affinity of A23187 is greater for $Mg^{2+}$ than $Ca^{2+}$ [69,71].
To our knowledge, no use has been made of either of these ionophores in
studies of microbial magnesium transport.  In the fission yeast Schizosaccharomyces pombe, cell division is associated with a doubling of magnesium
concentration.  A23187 promotes the passive transport of divalent cations,
so the effects of A23187 on division were studied [22].  A23187 blocked
cell division very close to the end of the cell cycle [22].  Furthermore,
cells treated with A23187 had lower $Ca^{2+}$ and $Mg^{2+}$ contents than did untreated cells [22].

We recently suggested [83] that manganese accumulation by B. subtilis
involves a low-molecular-weight extracellular component (referred to as
the "stimulator") synthesized by the cells; $^{54}Mn^{2+}$ was shown to bind
directly to the stimulator, and a similar stimulator for magnesium accumulation may also exist.  Suggestive evidence for such a magnesium
stimulator was found by Meers and Tempest [60] from the behavior of
mixed microbial populations in magnesium-limited chémostat cultures.
Under magnesium-limited conditions Gram-negative organisms invariably
outgrew Gram-positive ones, but the ability of B. subtilis and B. megaterium to outgrow each other or Torula utilis, a yeast, was dependent upon
the size of the inoculum.  Batch cultures of the three organisms were
grown in low-magnesium medium which was supplemented with equal volumes of supernatant fluids from magnesium-limited cultures of the three
species.  The resulting growth suggested the presence of growth-promoting
substances in the extracellular fluids.  The concentration of these substances in magnesium-limited B. subtilis cultures varied with population
density.  The growth-promoting substances were somewhat species specific
in that the growth of each of the Bacillus species was promoted to the
greatest degree by extracellular fluid from a culture of the same species.
It was suggested that the effect of inoculum size on the growth of Gram-
positive organisms in magnesium-limited mixed cultures was due to the
presence of a substance(s) secreted into the extracellular fluid that
facilitated magnesium accumulation.

## IV.  TRANSPORT OF MAGNESIUM BY OTHER MICROBES

### A.  Fungi

Until recently the only fungal cells studied to any extent with respect to transport were the yeasts.  Even today, only members of one class of fungi — the Ascomyces — have been utilized.  The Ascomycetes are the largest class of fungi, with the characteristic feature of sexually produced spores that are contained in a sac or ascus.  The vegetative growing structure may consist of either single cells (e.g., yeasts) or septate filaments, each segment of which may contain several nuclei.

Most cation transport studies of bakers' yeast (Saccharomyces cerevisiae) have concentrated on alkali metal cations.  Reports from Rothstein's laboratory have also described the transport of divalent metal cations into yeast [29, 76], with results somewhat similar to those described above for the $Mg^{2+}$ system I of E. coli, namely, a high affinity for $Mg^{2+}$ but also lesser affinities for other divalent cations including $Co^{2+}$, $Mn^{2+}$, and $Ni^{2+}$. With the rapid rate of changing ideas about cellular transport, it is no criticism to state that the yeast experiments were conducted in earlier times and with less sophisticated techniques than those currently being employed utilizing mutants as primary experimental material.  There was a rapid binding to the anionic groups of the surface of the cells upon addition of cations to yeast cell suspensions.  This binding exhibited little discrimination between many divalent cations (e.g., $Mn^{2+}$, $Mg^{2+}$, $Ca^{2+}$, $Sr^{2+}$), and the bound cations were exchangeable.  In addition to binding, some divalent cations were transported into the cells and were then no longer exchangeable with extracellular cations.  With energy-starved cells, little uptake was observed; but in the presence of, or after previous exposure to, glucose and phosphate, $Mg^{2+}$ and $Mn^{2+}$ were rapidly accumulated.  It was proposed that $Ni^{2+}$, $Co^{2+}$, and $Zn^{2+}$, which were also absorbed by the yeast into a nonexchangeable state, were accumulated by the same system that transported $Mg^{2+}$ and $Mn^{2+}$.  This conclusion was made on the basis of the following kinds of data: (a) direct comparisons of $Mn^{2+}$, $Co^{2+}$, $Ni^{2+}$, and $Zn^{2+}$ transport; (b) competition experiments between pairs of cations; (c) demonstration of the dependence on a common energy source; and (d) dependence of transport of all these cations on pretreatment with phosphate and the requirement for concomitant phosphate accumulation.  The affinity series was found to be $Mg^{2+} > Co^{2+} > Zn^{2+} > Mn^{2+} > Ni^{2+} > Ca^{2+} > Sr^{2+}$. The affinities of $Mn^{2+}$, $Zn^{2+}$, and $Co^{2+}$ were only roughly determined, and the $K_m$ values for the first four cations were apparently less than $10\,\mu M$. The $K_m$ for $Ni^{2+}$ was higher, probably about 0.5 mM.

It was clear from Rothstein's data that the divalent cations were actively accumulated by S. cerevisiae.  What was not abundantly clear was that they were all transported by the same system.  Although some pairs of divalent cations showed mutual inhibition of uptake [29], no direct studies of the uptake of magnesium, the divalent cation presumably with the highest affinity for the accumulation system, were carried out.

It was also suggested that $Mg^{2+}$ could be transported by the yeast monovalent cation accumulation system [1,19]. A number of cations (including $H^+$, $Rb^+$, $Cs^+$, $Na^+$, $Li^+$, $Mg^{2+}$, $Ca^{2+}$, and $NH_4^+$) appeared to share the $K^+$ transport carrier; $Mg^{2+}$ and $Ca^{2+}$ markedly inhibited $K^+$ uptake. Although the affinity of $Mg^{2+}$ for this system was very low, $Mg^{2+}$ at a relatively low concentration (2 mM) inhibited $K^+$ uptake by 25-30%. The noncompetitive inhibition of $K^+$ accumulation by low concentrations of $Ca^{2+}$ and $Mg^{2+}$ led to the proposal that two sites were involved in the uptake process [1]. The site with the higher relative affinity for $K^+$ was termed the transporting or carrier ("C") site. The second (modifier or "M") site was thought to influence the maximal rate of uptake of $K^+$ and other cations. We prefer to attribute effects of divalent cations on the $K^+$ transport system to more general and nonspecific effects on the yeast cell surface.

$Mg^{2+}$ counteracts both the mutagenic and growth-inhibitory actions of $Mn^{2+}$ in S. cerevisiae [70]. However, it was reported that manganese-resistant strains accumulate $Mn^{2+}$ equally as well as do $Mn^{2+}$-sensitive (wild-type) strains [70]. We suggest that in S. cerevisiae, as in E. coli, B. subtilis, and R. capsulata, $Mn^{2+}$ is a low-affinity substrate of the $Mg^{2+}$ accumulation system. It is likely that the $Mg^{2+}$ accumulation system is altered in $Mn^{2+}$-resistant mutants of S. cerevisiae.

Rapid accumulation of magnesium has been associated with cell division in Schizosaccharomyces pombe, a "fission yeast" [21]. During or immediately following cell division in S. pombe, there was a large uptake of magnesium, maintaining a constant total magnesium level per cell [21]. Since this doubling in magnesium was associated with division, it was suggested that magnesium (and also calcium) is essential for division, possibly in order to promote the breakdown of microtubules involved in nuclear division and cell plate formation [21,22].

Investigations with other fungi have yielded information on intracellular magnesium concentrations and localization [66]. The total content of $Mg^{2+}$ in Penicillium chrysogenum was found to be dependent on the concentrations of $Mg^{2+}$, phosphate, and other ions (including $Fe^{3+}$) in the growth medium, the stage of development of the fungus, and its growth rate. Magnesium apparently exists in several states within the fungal cell. Bound $Mg^{2+}$ is either bound to ribosomes or exists as a polymeric magnesium orthophosphate $[PO_4 : 1.5Mg]_n$. $[PO_4 : 1.5Mg]_n$ was extracted from the fungi with a solution of 7.5 M urea in 0.3 N $NH_4OH$ and sedimented on centrifugation (45,000g for 30 min) [66]. This high-molecular weight material was acid-labile and therefore orthophosphate rather than polyphosphate. An X-ray diffraction study showed that polymeric magnesium orthophosphate existed in fungal cells in an amorphous form which crystallized during extraction and purification. Polymeric magnesium orthophosphate was found in a number of other fungi in addition to P. chrysogenum, including Penicillium notatum, Aspergillus niger, Aspergillus oryzae, Phycomyces blakesleanus, and <u>Endomyces magnusii</u>. It was suggested that formation and breakdown of intracellular polymeric magnesium orthophosphate is one of the mechanisms of regulation of the level of free cellular $Mg^{2+}$. In addition to bound $Mg^{2+}$, some of the cellular $Mg^{2+}$ existed in an osmotically free state; this consisted of free $Mg^{2+}$ ions and low-molecular weight complexes of $Mg^{2+}$ [66]. Determinations of the concentrations of "free" $Mg^{2+}$ varied from 3.3 to 20 mM during growth and development of <u>P. chrysogenum.</u>

The accumulation of $Mg^{2+}$ by <u>P. chrysogenum</u> was shown to occur against a concentration gradient, indicating active transport. Phosphate was essential for the accumulation of $Mg^{2+}$ [66,67], as was the case for $Mg^{2+}$ accumulation by the yeast <u>S. cerevisiae</u> [19]. Phosphate was also necessary for the retention of $Mg^{2+}$ by <u>P. chrysogenum</u> in the presence of exogenous $Fe^{3+}$ [66]. Transfer of <u>P. chrysogenum</u> from normal medium to one deprived of phosphate resulted in an increase in the level of osmotically free magnesium compared to the amount of bound magnesium (as magnesium orthophosphate) [67]. Magnesium in <u>E. magnusii</u> and <u>S. cerevisiae</u> was localized in intracellular vacuoles. Biochemical studies with <u>Saccharomyces carlsbergensis</u> supported this conclusion. The vacuoles contained about 30 times as much osmotically free magnesium as the cytosole, and there was more than 100 times as much orthophosphate in the vacuoles as in the cytosole [50]. Thus, it has been suggested that there is an intracellular magnesium gradient and that a magnesium transport system is present in the tonoplast membrane that surrounds the vacuole [50]. This intracellular transport system would regulate the free $Mg^{2+}$ in the cytosole as diagramed in Figure 1-5. Furthermore, the tonoplast membrane transport system for $Mg^{2+}$ would provide a neutralizing cation for the high levels of inorganic polyphosphates found in the vacuoles of yeast cells and of higher plants [37,58]. Whereas vacuole $Mg^{2+}$ is one or two orders of magnitude higher than that in the cytoplasm, vacuole and cytoplasmic potassium concentrations are about equal [58], leaving no need to postulate an intracellular potassium transport system. Lutoids, polydisperse vacuole-like organelles of <u>Hevea</u> (a latex-producing plant), also accumulate $Mg^{2+}$ to a higher concentration than is found in the cytoplasm [72].

Transport studies with isolated vacuoles [13,36] should yield rich rewards of new insights into ionic and osmotic regulation in fungi and higher plants. Although no such vacuole cation or anion transport studies have been reported, recent success in characterizing arginine transport in isolated vacuoles [13] — the first reported vacuole transport studies — points the way. Arginine constitutes some 16% of the 0.65 M free amino acid pool found in vacuoles, which contain most of the free amino acids of yeast cells [13]. Transport assays on yeast spheroplasts and on vacuoles

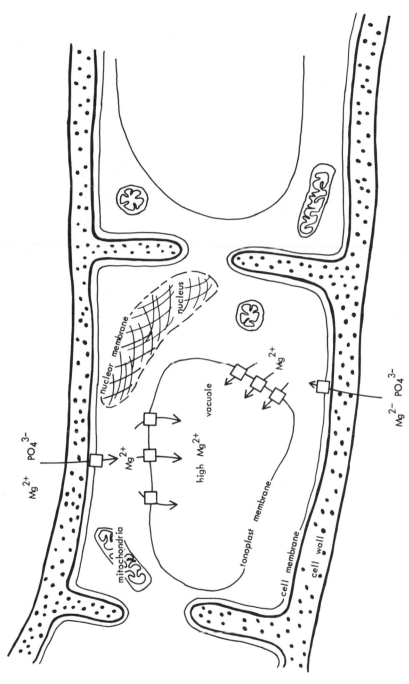

FIG. 1-5 Magnesium transport and accumulation in fungi. The nucleus, mitochondria, and vacuoles are the major "compartments" within the cytoplasm. The concentration of $Mg^{2+}$ in the cytoplasm is higher than in the surrounding medium. Since $Mg^{2+}$ is concentrated to even higher levels in the vacuoles, the presence of two (hypothetical) $Mg^{2+}$ transport systems is suggested: a transport system across the cellular membrane, and a tonoplast membrane system with different properties [66, 67].

isolated from them showed energy-dependent uptake in both cases, but with differing physiological and kinetic properties.  The existence of both high-affinity and low-affinity systems was indicated both for the cytoplasmic and the tonoplast membranes.  The vacuole high-affinity system of S. cerevisiae had a $K_m$ of 30 μM, whereas that for the intact spheroplasts showed a higher affinity ($K_m = 1.5$ μM).  These two systems also differed in their sensitivity to inhibition by other amino acids and by arginine analogs [13].  The most striking difference was that the cytoplasmic membrane system appeared to carry out concentrative, unidirectional active transport of arginine that was inhibitable by dinitrophenol and azide, whereas vacuole arginine transport was by exchange of extravesicular arginine for intramembrane arginine and was insensitive to dinitrophenol and azide [13].  No net accumulation occurred in these in vitro vacuole assays, although presumably concentrative uptake must have occurred within the cells in order to establish the 100 mM arginine level found in vacuoles.  This arginine for arginine exchange process in vacuoles showed the same temperature dependence as did arginine uptake by yeast spheroplasts.  With another yeast, Candida utilis, the vacuole arginine system showed a somewhat lower affinity ($K_m = 120$ μM) and a $V_{max}$ of 1.6 pmol/min per $10^6$ vacuoles [23].  Compared with vacuoles from the same organism, the C. utilis spheroplasts had a lower $K_m$ of 13 μM and a higher $V_{max}$ of 9.7 pmol/min per $10^6$ spheroplasts [23].  Since each isolated vacuole had essentially the same amino acid content as that of the spheroplast from which it was isolated [13] and presumably the vacuoles are somewhat smaller, the $V_{max}$ values per surface area unit were perhaps similar for vacuoles and spheroplasts.  We look forward to analogous vacuole $Mg^{2+}$ studies and then to studies of the regulation of cellular magnesium.

Cobalt transport has been studied in intact Neurospora crassa [96].  Since cobalt has been shown to be a competitive inhibitor of, and alternate substrate for, magnesium transport in several bacteria [38, 62, 82], it is not surprising that $Mg^{2+}$ severely depressed $Co^{2+}$ uptake by N. crassa.  $Mg^{2+}$ interfered with both surface binding and intracellular uptake of $Co^{2+}$.  It was suggested [96] that because $Co^{2+}$ was not significantly concentrated within the mycelia (11-13 mM internal $Co^{2+}$ at 10.2 mM external $Co^{2+}$) $Co^{2+}$ uptake by N. crassa is a facilitated diffusion process, not active transport.  However, we would suggest that $Co^{2+}$ is actively accumulated in N. crassa by a system which is primarily responsible for $Mg^{2+}$ accumulation.  The lack of $Co^{2+}$ concentration would be due to the low affinity of the $Mg^{2+}$ accumulation system for $Co^{2+}$ and the high external concentrations tested.  Furthermore, the inhibition of $Co^{2+}$ uptake by DNP and azide suggests an energy-dependent process.

B.  Algae

Algae require magnesium for normal growth.  Chlorella formed abnormally
large cells, and division ceased in magnesium-deficient media [27].
Euglena cells grown in a medium deficient in magnesium and other divalent
cations ($Mn^{2+}$, $Zn^{2+}$, and $Co^{2+}$) also formed enlarged cells and cell division was retarded [106].  Wolken [107] studied the effects of magnesium
and manganese deficiency on the growth of Euglena gracilis.  He also
measured $^{28}Mg^{2+}$ incorporation into Euglena, although these results were
undoubtedly preliminary since no data or conclusions were stated [107].
     To our knowledge, there is no published data on magnesium accumula-
tion by algal cells.  There are only the report (without data) by Wolken
[107] and an unpublished study of $^{28}Mg^{2+}$ accumulation by Euglena gracilis
(S. Silver and D. Kohn, unpublished data) which demonstrated magnesium
transport similar in many ways to the $Mg^{2+}$ transport systems of bacteria:
(a) $Mg^{2+}$ was concentrated by the cells to an extent about 100 times the
extracellular $Mg^{2+}$ concentration; (b) $Mg^{2+}$ accumulation was energy
dependent; and (c) $Mn^{2+}$, $Co^{2+}$, and $^{24}Mg^{2+}$ inhibited $^{28}Mg^{2+}$ accumulation.
E. gracilis may be grown on a light-dark schedule (green cells), in dark-
ness (bleached white cells), or in the dark followed by exposure to light for
various times ("greening" cells).  No significant difference in the rate of
uptake of $Mg^{2+}$ was found among green, greening, or white cells, which is
not surprising considering the subtle differences between light- and dark-
grown photosynthetic bacteria (cf. Section III. A).  The $K_m$ and $V_{max}$ for
$^{28}Mg^{2+}$ uptake in Euglena are found in Table 1-3 along with the similar
values for bacteria.  One difference between algal and bacterial cells was
that, under a variety of conditions, no efflux of accumulated $^{28}Mg^{2+}$ from
Euglena could be demonstrated.  Detailed studies of $Mg^{2+}$ transport by
algal cells and chloroplasts derived from such cells are bound to be of
interest, especially in light of recent reports that $Mg^{2+}$ may exchange for
chloroplast $H^+$ during light-induced proton pumping with spinach chloroplasts [4, 5].


V.  NONMICROBIAL SYSTEMS

Studies of $Mg^{2+}$ in whole animals have been concerned primarily with the
$Mg^{2+}$ content of the serum or organs or with effects of $Mg^{2+}$ deficiency on
the growth and development of the animal [97].  This can be seen in the
Proceedings of the Second International Symposium on Magnesium [17],
held in Montreal in May 1976.  Although a Society for Development of

Research on Magnesium exists as an offshoot of the American College of Nutrition, the emphasis of magnesium research in animals has been strictly nutritional and clinical. Few investigations have been carried out with the intention of studying the "uptake of magnesium" by whole animals. Typical is that by Sabbot and Costin [79], who used $^{28}Mg^{2+}$ to show that stress conditions (produced by drenching rats in cold water) resulted in statistically significant increases of $Mg^{2+}$ uptake in certain brain areas and in the pituitary gland.

A few experiments have been done with $Mg^{2+}$ in animal cells, and these will now be discussed.

## A. KB Cells

Suspension cultures of human KB cells accumulate $^{28}Mg^{2+}$ by a specific transport system [6]. The uptake of $^{28}Mg^{2+}$ followed saturation kinetics with a $K_m$ and $V_{max}$ rather similar to bacterial values (Table 1-3). Inhibitors of energy-dependent processes, such as CCCP, cyanide, and ouabain (a cardiac glycoside that inhibits the $Na^+$, $K^+$-ATPase responsible for alkali cation transport) all reduced the uptake of $Mg^{2+}$ by 50% or more. $Mn^{2+}$ at low concentrations (0.1 mM) and $Ca^{2+}$ (1.0 mM) did not affect $Mg^{2+}$ accumulation; but 1 mM $Mn^{2+}$ inhibited $^{28}Mg^{2+}$ uptake by 80% (although this level of $Mn^{2+}$ was cytotoxic, so that the direct effect of $Mn^{2+}$ on $Mg^{2+}$ transport was uncertain). $^{28}Mg^{2+}$ accumulated by KB cells was exchangeable with $Mg^{2+}$ added after 5 hr of exposure and also was released by agents that caused nonspecific membrane leakiness. Similar $^{28}Mg^{2+}$ uptake was seen with mouse fibroblasts [89].

## B. Adipocytes

Magnesium accumulation by membrane vesicles derived from rat fat cells was measured by a method which utilized the ligand 8-hydroxyquinoline-5-sulfonic acid, whose magnesium complex fluoresces distinctly [26]. In the presence of the hormones epinephrine, norepinephrine, and adrenocorticotropic hormone, magnesium accumulation by these vesicles was stimulated. $Mg^{2+}$, $Ca^{2+}$, and ATP were required for active $Mg^{2+}$ accumulation by vesicles. Calcium may potentiate the magnesium accumulation by contributing to the integrity of the plasma membrane vesicles. Although the hormones utilized in this study stimulated cyclic AMP production, no effect of added cyclic AMP on magnesium uptake was demonstrable.

An increase in the magnesium content of intact adipocytes following hormone stimulation was also shown by use of $^{28}Mg^{2+}$ and atomic absorption spectroscopy [26]. In these studies, an increase in magnesium content due to the addition of hormones was measured after a 30-min incubation. No determinations of the rate of $Mg^{2+}$ uptake in the presence or absence of hormones by intact adipocytes or plasma membrane vesicles were made.

C.   Squid Axons

The animal cells just discussed were concentrating magnesium against a
chemical gradient generally from about 1 mM external $Mg^{2+}$ to perhaps
3 mM internal within the cells [6] (see also Table 2-1 in Chapter 2, Section I).   A rather different situation exists with the squid axon bathed in
sea water, which contains 55 mM $Mg^{2+}$.   Squid axons excrete $^{28}Mg^{2+}$ by
an energy-dependent process [2a,3].   The internal concentration of $Mg^{2+}$
in squid axons was about 6 mM and about half of this magnesium was
judged "free" from measurements of diffusional mobility and movement in
a potential field [2a,3].   This approximately 20:1 $Mg^{2+}$ ratio was maintained by an efflux system that was inhibited more than 80% by cyanide and
by uncouplers.   The inhibited efflux was restored by intraaxonal injection
of ATP, suggesting the possibility that a membrane-imbedded $Mg^{2+}$-ATPase
is involved in $Mg^{2+}$ efflux.   $Mg^{2+}$ efflux required external sodium (half
maximum at 100 mM) for activity.   The rate of $Mg^{2+}$ uptake by squid axons,
which in steady state equaled the rate of efflux at about 0.6 pmol/sec per
square centimeter of surface area, was unaffected by cyanide.   Magnesium
influx was accelerated almost 10 times by the removal of $Na^+$; and since
$Na^+$ removal from the bathing solution reduced the efflux rate 80% as well,
the net effect was an increase in intraaxonal $Mg^{2+}$ [2a,3].
        Squid axons "pump" $Mg^{2+}$ out against a concentration gradient and
against a potential gradient (internal negative); this clearly represents
"active transport" in the strict sense of being against an electrochemical
gradient.   In contrast, energy-dependent uptake of $Mg^{2+}$ is carried out by
KB cells and adipocytes, as already described.   KB cells and adipocytes
establish and maintain a concentration gradient of perhaps 3:1 against the
serum or media level (see Table 2-1 in Chapter 2, Section I), and with
special low $Mg^{2+}$ media this gradient could be at most 50:1.   The Nernst
potential equilibrium (see Chapter 2, Section VIII.C) would maintain a
gradient of 3:1 with a potential of -14 mV (internal negative and within the
normal range found with animal cells), obviating the need for $Mg^{2+}$ active
transport in these cases.   However, the Nernst equation equilibrium will
only be established if there is ready permeability across the membrane;
and it appears that most membranes are not readily permeable to $Mg^{2+}$.
Certainly the high degree of specificity of the $Mg^{2+}$ uptake by KB cells
[6] suggests specific membrane carriers.   Furthermore, in preliminary
experiments with $^{28}Mg^{2+}$ and human red blood cells, we could not demonstrate any accumulation of magnesium by the cells — either in the presence of glucose at 37° C, or at 4° C over a 12-hr period (S. Silver and H.
Lubowitz, unpublished data).   The indications were that the human red
blood cell, at least, completely lacks $Mg^{2+}$ transport function and is effectively impermeable to $Mg^{2+}$ as well.   If red blood cell precursors are
like tissue culture cells and adipocytes, then an inactivation of transport
must occur during cell maturation.   The level of passive permeability

would be negligible, and therefore the Nernst equation prediction would be inappropriate. The extreme case of the inappropriateness of the free-permeability potential-equilibrium model is found with the bacterial transport systems that we discussed earlier. Here there is strong evidence for energy-dependent accumulation systems of high specificity. However, the accumulation ratio of inside concentration of magnesium to outside rarely exceeded 10- to 50-fold (only an approximation made by using poorly-based estimates of the "free" internal concentration, already described), although a Nernst potential equilibrium with a typical value of -180 mV internal negative and free permeability would give rise to an inside-to-outside ratio of $10^6 : 1$ (far beyond the experimental range of possible values). The free-permeability-equilibrium model requires one to postulate a $Mg^{2+}$ efflux system for energy-dependent $Mg^{2+}$ egress so as to keep down the magnesium gradient, while in fact in bacterial systems it is magnesium uptake that is dependent on metabolic energy. Essential impermeability of the membrane explains the failure to reach Nernst potential equilibrium.

### D.   Chick Embryo Fibroblasts

$Mg^{2+}$ accumulation per se has not been studied in cultures of chick embryo fibroblasts; however, effects of $Mg^{2+}$ deprivation on growth of these cells have been noted. The rate of DNA synthesis in cultures of chick fibroblasts was reduced when the concentration of $Mg^{2+}$ added to the medium was lowered [78]. Because the direct lowering of medium $Mg^{2+}$ had erratic results, lowering of extracellular $Mg^{2+}$ was also accomplished by use of phosphorylated compounds that bind $Mg^{2+}$ preferentially to $Ca^{2+}$. Adenosine triphosphate and adenosine diphosphate at concentrations below that of $Mg^{2+}$ stimulate DNA synthesis; but at higher concentrations, ATP and ADP inhibited DNA synthesis by reducing the proportion of cells in the S (DNA synthesis) phase. Sodium pyrophosphate, which strongly complexes with $Mg^{2+}$, caused a striking decrease in DNA synthesis when added at concentrations greater than that of the available $Mg^{2+}$ [78]. The inhibition by sodium pyrophosphate was reversible by excess $Mg^{2+}$. Pyrophosphate-induced $Mg^{2+}$ limitation also resulted in the reduction in the rates of RNA and protein synthesis, 2-deoxy-D-glucose uptake, and lactic acid formation to similar extents as the reduction caused by removal of serum from the medium. Thus $Mg^{2+}$ deprivation inhibited the same intracellular metabolic pathways as did serum deprivation and cell density-dependent growth inhibition. This is consistent with the known requirement for $Mg^{2+}$ for transphosphorylation reactions. From these results, Rubin [78] suggested that compartmentalization of $Mg^{2+}$ within the cell (and hence effects on processes whose rate-limiting steps are transphosphorylation reactions) is the key element in the coordinate control of metabolism, differentiated function, and growth.

E.  Mitochondria

Mitochondria accumulate $Mg^{2+}$ in vitro in an energy-dependent manner.
An early report of inorganic phosphate accumulation by beef heart mitochondria indicated that magnesium was required for phosphate translocation and that $Mg^{2+}$ was only actively accumulated when it accompanied
phosphate [14].  Later it was shown that $Mg^{2+}$ and phosphate may be taken
up independently but that the presence of both ions results in higher rates
of uptake [40].  The observed rates of $Mg^{2+}$ accumulation by rat liver
mitochondria were similar to rates of $K^+$ influx measured under comparable
conditions [39].  $Mg^{2+}$ influx by mitochondria was inhibited by rotenone (a
respiratory inhibitor) and by the uncoupler p-trifluoromethoxy(carbonylcyanide)phenylhydrazone (FCCP).  Rat liver mitochondrial $Mg^{2+}$ was found
to exchange only slowly with external $Mg^{2+}$ [39].  As was true of $Mg^{2+}$
accumulation by intact adipocyte cells, $Mg^{2+}$ accumulation in rat liver
mitochondria could be hormonally stimulated, in this case by parathyroid
hormone [39].
    A factor has been isolated from animal tissues that acts on mitochon-
drial magnesium movements.  This molecule, called cytoplasmic or
cytosol metabolic factor (CMF), was partially purified and is thought to be
a cyclic peptide of about 2,200-daltons molecular weight [10].  Kun and
coworkers [47] have shown that CMF can prevent the loss of $Mg^{2+}$ from
mitochondria usually caused by the addition of DNP and ADP.  It was suggested that CMF is either a $Mg^{2+}$ carrier or else an activator of a mitochondrial site for $Mg^{2+}$ that plays a rate-limiting role in mitochondrial
metabolism [47].  The factor seems to act specifically on $Mg^{2+}$ movements in various biological membranes, i.e., those of plant chloroplasts
and bacteria [11] as well as of mitochondria [10].  ADP was required for
the action of CMF in preventing the efflux of $Ca^{2+}$ and $Mg^{2+}$ from mitochondria [12].  These mitochondrial magnesium studies, both those on the
cytoplasmic factor and those concentrating on the nature of the magnesium
transport system itself, are very preliminary.  There is nowhere near the
understanding of magnesium transport in mitochondria that there is of the
calcium transport system (see Chapter 2).  We are even uncertain as to
whether separate divalent cation systems function in mitochondria as they
do in bacterial cells.


VI.  SUMMARY

All cells require magnesium for growth and normal metabolism.  Microorganisms satisfy their magnesium requirement by actively accumulating
magnesium by specific energy-dependent magnesium transport systems.
In some microbes it has been found that cobalt and manganese are inhibitors
of and alternate substrates for the magnesium transport system.  As with
other essential nutrients, the transport systems for magnesium are under

careful regulation and change quantitatively in response to growth conditions. Studies of bacteria which possess mutations affecting magnesium accumulation are leading to an increased understanding of the means by which organisms accumulate and utilize this vital metal.

## ACKNOWLEDGMENTS

The research in this laboratory and the writing of this chapter were supported by National Science Foundation grant BMS71-01456 and National Institutes of Health grant AI08062. We are grateful to E. P. Kennedy, J. E. Lusk, and D. L. Nelson, with whom we exchanged materials and data during the times of initial characterization of magnesium transport systems, and who in addition provided unpublished data and manuscripts to aid in writing this review. L. Okorokov and I. Kulaev have added ideas on vacuole magnesium metabolism. Kathleen Farrelly drew the figures and watched over the bibliography.

## REFERENCES

1.  W. McD. Armstrong and A. Rothstein, Discrimination between alkali metal cations by yeast. II. Cation interactions in transport. J. Gen. Physiol., 50, 967-988 (1967).
2.  B. J. Bachmann, K. B. Low, and A. L. Taylor, Recalibrated linkage map of Escherichia coli K-12. Bacteriol. Rev., 40, 116-167 (1976).
2a. P. F. Baker, Regulation of intracellular Ca and Mg in squid axons. Fed. Proc., 35, 2589-2595 (1976).
3.  P. F. Baker and A. C. Crawford, Mobility and transport of magnesium in squid giant axons. J. Physiol., 227, 855-874 (1972).
4.  J. Barber, Cation control in photosynthesis. Trends in Biochem. Sci., 1, 33-36 (1976).
5.  J. Barber, J. Mills, and J. Nicolson, Studies with cation specific ionophores show that within the intact chloroplast $Mg^{++}$ acts as the main exchange cation for $H^+$ pumping. FEBS Letters, 49, 106-110 (1974).
6.  R. S. Beauchamp, S. Silver, and J. W. Hopkins, Uptake of $Mg^{2+}$ by KB cells. Biochim. Biophys. Acta, 225, 71-76 (1971).
7.  P. Bhattacharyya, Active transport of manganese in isolated membranes of Escherichia coli. J. Bacteriol., 104, 1307-1311 (1970).
8.  P. Bhattacharyya, Active transport of manganese in isolated membrane vesicles of Bacillus subtilis. J. Bacteriol., 123, 123-127 (1975).

9.   P. Bhattacharyya, W. Epstein, and S. Silver, Valinomycin-induced
     uptake of potassium in membrane vesicles from Escherichia coli.
     Proc. Nat. Acad. Sci. U.S., 68, 1488-1492 (1971).

10.  A. Binet, C. Gros, and P. Volfin, A cytoplasmic molecule active on
     membranar $Mg^{2+}$ movements. I. Isolation and properties. FEBS
     Letters, 17, 193-196 (1971).

11.  A. Binet and P. Volfin, A cytoplasmic molecule active on mem-
     branar $Mg^{2+}$ movements. II. Its role as a factor of the membrane
     integrity in mitochondria, chloroplasts and bacteria. FEBS Letters,
     17, 197-202 (1971).

12.  A. Binet and P. Volfin, ADP requirement for prevention by a cyto-
     solic factor of $Mg^{2+}$ and $Ca^{2+}$ release from rat liver mitochondria.
     Arch. Biochem. Biophys., 164, 756-764 (1974).

13.  T. Boller, M. Dürr, and A. Wiemken, Characterization of a specific
     transport system for arginine in isolated yeast vacuoles. Eur. J.
     Biochem., 54, 81-91 (1975).

14.  G. P. Brierley, E. Bachmann, and D. E. Green, Active transport
     of inorganic phosphate and magnesium ions by beef heart mitochondria.
     Proc. Natl. Acad. Sci. U.S., 48, 1928-1935 (1962).

15.  T. D. Brock, Effects of magnesium ion deficiency on Escherichia
     coli and possible relation to the mode of action of novobiocin.
     J. Bacteriol., 84, 679-682 (1962).

16.  F. Bucheder and E. Broda, Energy-dependent zinc transport by
     Escherichia coli. Eur. J. Biochem., 45, 555-559 (1974).

17.  M. Cantin (ed.), Proceedings of the Second International Symposium
     on Magnesium, Spectrum Publ., Holliswood, New York, 1977, in
     press.

18.  P. A. Cohen and H. L. Ennis, Amino acid regulation of RNA synthe-
     sis during recovery of Escherichia coli from $Mg^{2+}$ starvation.
     Biochim. Biophys. Acta, 145, 300-309 (1967).

19.  E. J. Conway and M. E. Beary, Active transport of magnesium
     across the yeast cell membrane. Biochem. J., 69, 275-280 (1958).

20.  M. A. Devynck, P. L. Boquet, P. Fromageot, and E. J. Simon,
     On the mode of action of levallorphan on Escherichia coli: effects on
     cellular magnesium. Mol. Pharmacol., 7, 605-610 (1971).

21.  J. H. Duffus and L. J. Paterson, The cell cycle in the fission
     yeast Schizosaccharomyces pombe: changes in activity of magnesium
     dependent ATP'ase and in total internal magnesium in relation to cell
     division. Z. Allg. Mikrobiol., 14, 727-729 (1974).

22.  J. H. Duffus and L. J. Paterson, Control of cell division in yeast
     using the ionophore A23187 with calcium and magnesium. Nature,
     251, 626-627 (1974).

23.  M. Dürr, T. Boller, and A. Wiemken, Polybase induced lysis of
     yeast spheroplasts: A new gentle method for preparation of vacuoles.
     Arch. Microbiol., 105, 319-327 (1975).

24. R. G. Eagon, G. P. Simmons, and K. J. Carson, Evidence for the presence of ash and divalent metals in the cell wall of Pseudomonas aeruginosa. Can. J. Microbiol., 11, 1041-1042 (1965).

25. E. Eisenstadt and S. Silver, Restoration of cation transport during germination. In Spores V (H. O. Halvorson, R. Hanson, and L. L. Campbell, eds.), American Society for Microbiology, Washington, D. C., 1972, pp. 443-448.

26. D. A. Elliott and M. A. Rizack, Epinephrine and adrenocorticotropic hormone-stimulated magnesium accumulation in adipocytes and their plasma membranes. J. Biol. Chem., 249, 3985-3990 (1974).

27. B. J. Finkle and D. Appleman, The effect of magnesium concentration on growth of Chlorella. Plant Physiol., 28, 664-673 (1953).

28. E. Frieden, The chemical elements of life. Sci. Amer., 227, 52-60 (1972).

29. G.-F. Fuhrmann and A. Rothstein, The transport of $Zn^{2+}$, $Co^{2+}$ and $Ni^{2+}$ into yeast cells. Biochim. Biophys. Acta, 163, 325-330 (1968).

30. A. Garen, Physiological effects of rII mutations in bacteriophage T4. Virology, 14, 151-163 (1961).

31. A. Gloe, N. Pfennig, H. Brockman, Jr., and W. Trowtizsch, A new bacteriochlorophyll from brown-colored chlorobacteriaceae. Arch. Microbiol., 102, 103-109 (1975).

32. Th. Günther and F. Dorn, Über die intrazelluläre Mg-Ionenaktivität von E. coli-Zellen. Z. Naturforsch., 24B, 713-717 (1969).

32a. Th. Günther and C. F. Hoffman, Zum Magnesium-Stoffwechsel von E. coli. Z. Klin. Chem. Klin. Biochem., 11, 237-242 (1973).

33. A. Hurst, A. Hughes, D. L. Collins-Thompson, and B. G. Shah, Relationship between loss of magnesium and loss of salt tolerance after sublethal heating of Staphylococcus aureus. Can. J. Microbiol., 20, 1153-1158 (1974).

34. C. Hurwitz and C. L. Rosano, The intracellular concentration of bound and unbound magnesium ions in Escherichia coli. J. Biol. Chem., 242, 3719-3722 (1967).

35. C. Hurwitz, and C. L. Rosano, Is there an active transport system for $Mg^{2+}$ in Escherichia coli? Unpublished manuscript (1974).

36. K. J. Indge, The isolation and properties of the yeast cell vacuole. J. Gen. Microbiol., 51, 441-446 (1968).

37. K. J. Indge, Polyphosphates of the yeast cell vacuole. J. Gen. Microbiol., 51, 447-455 (1968).

38. P. L. P. Jasper, Cation transport systems of Rhodopseudomonas capsulata. Ph.D. Thesis, Washington University, St. Louis, Missouri, 1975.

39. J. H. Johnson and B. C. Pressman, Rates of exchange of mitochondrial $Mg^{2+}$ determined from $^{28}Mg$ flux measurements. Arch. Biochem. Biophys., 132, 139-145 (1969).

40. J. D. Judah, K. Ahmed, A. E. M. McLean, and G. S. Christie, Uptake of magnesium and calcium by mitochondria in exchange for hydrogen ions. Biochim. Biophys. Acta, 94, 452-460 (1965).

41. D. Kennell and A. Kotoulas, Magnesium starvation of Aerobacter aerogenes. I. Changes in nucleic acid composition. J. Bacteriol., 93, 334-344 (1967).

42. D. Kennell and A. Kotoulas, Magnesium starvation of Aerobacter aerogenes. II. Rates of nucleic acid synthesis and methods for their measurement. J. Bacteriol., 93, 345-356 (1967).

43. D. Kennell and A. Kotoulas, Magnesium starvation of Aerobacter aerogenes. IV. Cytochemical changes. J. Bacteriol., 93, 367-378 (1967).

44. D. Kennell and B. Magasanik, The relation of ribosome content to the rate of enzyme synthesis in Aerobacter aerogenes. Biochim. Biophys. Acta, 55, 139-151 (1962).

45. P. Kent, S. C. Bunce, R. A. Bailey, D. A. Aikens, and C. Hurwitz, Interference by polyamines in the measurement of magnesium ion at physiological pH with the divalent cation-selective electrode. Anal. Biochem., 62, 75-80 (1974).

46. A. L. Kopecky, D. P. Copeland, and J. E. Lusk, Viability of Escherichia coli treated with colicin K. Proc. Nat. Acad. Sci. U.S., 72, 4631-4634 (1975).

47. E. Kun, E. B. Kearney, I. Wiedemann, and N. M. Lee, Regulation of mitochondrial metabolism by specific cellular substances. II. The nature of stimulation of mitochondrial glutamate metabolism by a cytoplasmic component. Biochemistry, 8, 4443-4449 (1969).

48. J. Lederberg, Bacterial protoplasts induced by penicillin. Proc. Nat. Acad. Sci. U.S., 42, 574-577 (1956).

49. L. Leive, Release of lipopolysaccharide by EDTA treatment of E. coli. Biochem. Biophys. Res. Commun., 21, 290-296 (1965).

50. L. P. Lichko and L. A. Okorokov, The compartmentalization of magnesium and phosphate ions in Saccharomyces carlsbergensis cells. Dokl. Akad. Nauk SSSR, in press (1976).

51. J. E. Lusk, Transport of magnesium in Escherichia coli. Ph.D. thesis, Harvard University, Cambridge, Massachusetts, 1970.

52. J. E. Lusk and E. P. Kennedy, Magnesium transport in Escherichia coli. J. Biol. Chem., 244, 1653-1655 (1969).

53. J. E. Lusk and E. P. Kennedy, Altered phospholipid metabolism in a sodium-sensitive mutant of Escherichia coli. J. Bacteriol., 109, 1034-1046 (1972).

54. J. E. Lusk and D. L. Nelson, Effects of colicins El and K on permeability to magnesium and cobaltous ions. J. Bacteriol., 112, 148-160 (1972).

55. J. E. Lusk, R. J. P. Williams, and E. P. Kennedy, Magnesium and the growth of Escherichia coli. J. Biol. Chem., 243, 2618-2624 (1968).

56. R. A. MacLeod and E. Onofrey, Nutrition and metabolism of marine bacteria. III. The relation of sodium and potassium to growth. J. Cell. Comp. Physiol., 50, 389–401 (1957).

57. S. L. Marchesi and D. Kennell, Magnesium starvation of Aerobacter aerogenes. III. Protein metabolism. J. Bacteriol., 93, 357–366 (1967).

58. Ph. Matile, Vacuoles. In Plant Biochemistry, 3rd ed. (J. Bonner and J. E. Varner, eds.), Academic Press, New York, 1976, pp. 189–224.

59. B. J. McCarthy, The effects of magnesium starvation on the ribosome content of Escherichia coli. Biochim. Biophys. Acta, 55, 880–888 (1962).

60. J. L. Meers and D. W. Tempest, The influence of extracellular products on the behaviour of mixed microbial populations in magnesium-limited chemostat cultures. J. Gen. Microbiol., 52, 309–317 (1968).

61. S. Natori, R. Nozawa, and D. Mizuno, The turnover of ribosomal RNA of Escherichia coli in a magnesium-deficient stage. Biochim. Biophys. Acta, 114, 245–253 (1966).

62. D. L. Nelson and E. P. Kennedy, Magnesium transport in Escherichia coli. Inhibition by cobaltous ion. J. Biol. Chem., 246, 3042–3049 (1971).

63. D. L. Nelson and E. P. Kennedy, Transport of magnesium by a repressible and a nonrepressible system in Escherichia coli. Proc. Natl. Acad. Sci. U.S., 69, 1091–1093 (1972).

64. H. C. Neu and L. A. Heppel, The release of enzymes from Escherichia coli by osmotic shock and during the formation of spheroplasts. J. Biol. Chem., 240, 3685–3692 (1965).

65. P. Oehr and K. Willecke, Citrate-$Mg^{2+}$ transport in Bacillus subtilis. Studies with 2-fluoro-L-erythro-citrate as a substrate. J. Biol. Chem., 249, 2037–2042 (1974).

66. L. A. Okorokov, L. P. Lichko, V. M. Kadomtseva, V. P. Kholodenko, and I. S. Kulaev, Metabolism and physicochemical state of $Mg^{2+}$ ions in fungi. Mikrobiologiya, 43, 410–416 (1974).

67. L. A. Okorokov, L. P. Lichko, V. P. Kholodenko, V. M. Kadomtseva, S. B. Petrikevich, E. I. Zaichkin, and A. M. Karimova, Free and bound magnesium in fungi and yeasts. Folia Microbiol., 20, 460–466 (1975).

67a. G. W. Ordal, Control of tumbling in bacterial chemotaxis by divalent cation. J. Bacteriol., 126, 706–711 (1976).

68. M. H. Park, B. B. Wong, and J. E. Lusk, Mutants in three genes affecting transport of magnesium in Escherichia coli: genetics and physiology. J. Bacteriol., 126, 1096–1103 (1976).

69. B. C. Pressman, Properties of ionophores with broad range cation selectivity. Fed. Proc., 32, 1698–1703 (1973).

70. P. P. Puglisi, G. Lucchini, and A. Vecli, Resistenza allo ione manganoso ($Mn^{2+}$) in Saccharomyces cerevisiae. Atti Assoc. Genet. Ital., 15, 159–170 (1970).

71. P. W. Reed and H. A. Lardy, A23187: a divalent cation ionophore. J. Biol. Chem., 247, 6970-6977 (1972).
72. D. Ribaillier, J.-L. Jacob, and J. d'Auzac, Sur certains caracteres vacuolaires des lutoïdes du latex d'Hevea brasiliensis Mull. Arg. Physiol. Veg., 9, 423-437 (1971).
73. A. Rodgers, The exchange properties of magnesium in Escherichia coli ribosomes. Biochem. J., 90, 548-555. (1964).
74. C. L. Rosano and C. Hurwitz, Interrelationship between magnesium and polyamines in a pseudomonad lacking spermidine. Biochem. Biophys. Res. Commun., 37, 677-683 (1969).
75. B. P. Rosen and J. S. McClees, Active transport of calcium by inverted membrane vesicles of Escherichia coli. Proc. Nat. Acad. Sci. U.S., 71, 5042-5046 (1974).
76. A. Rothstein, A. Hayes, D. Jennings, and D. Hooper, The active transport of $Mg^{++}$ and $Mn^{++}$ into the yeast cell. J. Gen. Physiol., 41, 585-594 (1958).
77. M. A. Rouf, Spectrochemical analysis of inorganic elements in bacteria. J. Bacteriol., 88, 1545-1549 (1964).
78. H. Rubin, Central role for magnesium in coordinate control of metabolism and growth in animal cells. Proc. Nat. Acad. Sci. U.S., 72, 3551-3555 (1975).
79. I. Sabbot and A. Costin, Cold stress induced changes in the uptake and distribution of radiolabelled magnesium in the brain and pituitary of the rat. Experientia, 30, 905-906 (1974).
80. R. Scherrer and P. Gerhardt, Influence of magnesium ions on porosity of the Bacillus megaterium cell wall and membrane. J. Bacteriol., 114, 888-890 (1973).
81. G. B. Schmidt, C. L. Rosano, and C. Hurwitz, Evidence for a magnesium pump in Bacillus cereus T. J. Bacteriol., 105, 150-155 (1971).
82. H. Scribner, E. Eisenstadt, and S. Silver, Magnesium transport in Bacillus subtilis W23 during growth and sporulation. J. Bacteriol., 117, 1224-1230 (1974).
83. H. E. Scribner, J. Mogelson, E. Eisenstadt, and S. Silver, Regulation of cation transport during bacterial sporulation. In Spores VI (P. Gerhardt, R. N. Costilow, and H. L. Sadoff, eds.), American Society for Microbiology, Washington, D.C., 1975, pp. 346-355.
84. S. Silver, Active transport of magnesium on Escherichia coli. Proc. Nat. Acad. Sci. U.S., 62, 764-771 (1969).
85. S. Silver and D. Clark, Magnesium transport in Escherichia coli: interference by manganese with magnesium metabolism. J. Biol. Chem., 246, 569-576 (1971).
86. S. Silver, P. Johnseine, E. Whitney, and D. Clark, Manganese-resistant mutants of Escherichia coli: physiological and genetic studies. J. Bacteriol., 110, 186-195 (1972).

87.  S. Silver, E. Levine, and P. M. Spielman, Cation fluxes and perme-
     ability changes accompanying bacteriophage infection of Escherichia
     coli. J. Virol., 2, 763-771 (1968).
88.  S. Silver, K. Toth, P. Bhattacharyya, E. Eisenstadt, and H.
     Scribner, Changes and regulation of cation transport during bacterial
     sporulation. In Comparative Biochemistry and Physiology of Trans-
     port: Proceedings of the 5th International Conference on Biological
     Membranes (L. Bolis, K. Bloch, S. E. Luria, and F. Lynen eds.),
     North-Holland Publ., Amsterdam, 1974, pp. 393-408.
89.  S. Silver, L. Wendt, P. Bhattacharyya, and R. S. Beauchamp,
     Effects of polyamines on membrane permeability. Ann. N.Y. Acad.
     Sci., 171, 838-862 (1970).
90.  D. W. Tempest, Quantitative relationships between inorganic cations
     and anionic polymers in growing bacteria, In Microbial Growth:
     Symposium of the Society for General Microbiology, Vol. 19 (P. M.
     Meadow and S. J. Pirt, eds.), Cambridge University Press, New
     York, 1969, pp. 87-111.
91.  D. W. Tempest, J. W. Dicks, and J. L. Meers, Magnesium-limited
     growth of Bacillus subtilis in pure and mixed cultures, in a chemostat.
     J. Gen. Microbiol., 49, 139-147 (1967).
92.  D. W. Tempest, J. R. Hunter, and J. Sykes, Magnesium-limited
     growth of Aerobacter aerogenes in a chemostat. J. Gen. Microbiol.,
     39, 355-366 (1965).
93.  D. W. Tempest and R. E. Strange, Variation in content and distribu-
     tion of magnesium, and its influence on survival, in Aerobacter aero-
     genes grown in a chemostat. J. Gen. Microbiol., 44, 273-279 (1966).
94.  A. Tissières, J. D. Watson, D. Schlessinger, and B. R. Hollingworth,
     Ribonucleoprotein particles from Escherichia coli. J. Mol. Biol.,
     1, 221-233 (1959).
95.  T. Tsuchiya and B. P. Rosen, Characterization of an active trans-
     port system for calcium in inverted membrane vesicles of Escherichia
     coli. J. Biol. Chem., 250, 7687-7692 (1975).
96.  G. Venkateswerlu and K. Sivarama Sastry, The mechanism of uptake
     of cobalt ions by Neurospore crassa. Biochem. J., 118, 497-503
     (1970).
97.  W. E. C. Wacker and B. L. Vallee, Magnesium. In Mineral
     Metabolism (C. L. Comar and F. Bronner, eds.), Vol. 2, Pt. A,
     Academic Press, New York, 1964, pp. 483-521.
98.  M. Webb, The influence of magnesium on cell division. 2. The effect
     of magnesium on the growth and cell division of various bacterial
     species in complex media. J. Gen. Microbiol., 3, 410-417 (1949).
99.  M. Webb, The influence of magnesium on cell division. 3. The effect
     of magnesium on the growth of bacteria in simple chemically defined
     media. J. Gen. Microbiol., 3, 418-424 (1949).

100. M. Webb, The utilization of magnesium by certain gram-positive and gram-negative bacteria. J. Gen. Microbiol., 43, 401-409 (1966).
101. M. Webb, The influence of certain trace metals on bacterial growth and magnesium utilization. J. Gen. Microbiol., 51, 325-335 (1968).
102. M. Webb, Effects of magnesium deficiency on ribosomal structure and function in certain gram-positive and gram-negative bacteria. Biochim. Biophys. Acta, 222, 416-427 (1970).
103. M. Webb, Interrelationships between the utilization of magnesium and the uptake of other bivalent cations by bacteria. Biochim. Biophys. Acta, 222, 428-439 (1970).
104. A. Weiss and S. Silver, Plasmid-determined cadmium resistance in Staphylococcus aureus: cadmium, manganese and zinc transport systems. J. Bacteriol., in preparation (1977).
105. K. Willecke, E.-M. Gries, and P. Oehr, Coupled transport of citrate and magnesium in Bacillus subtilis. J. Biol. Chem., 248, 807-814 (1973).
106. J. J. Wolken, A molecular morphology of Euglena gracilis var. bacillaris. J. Protozool., 3, 211-221 (1956).
107. J. J. Wolken, Euglena: An Experimental Organism for Biochemical and Biophysical Studies, 2nd ed., Appleton-Century-Crofts, New York, 1967.
108. N. D. Zinder and W. F. Arndt, Production of protoplasts of Escherichia coli by lysozyme treatment. Proc. Nat. Acad. Sci. U.S., 42, 586-590 (1956).

Chapter 2

CALCIUM TRANSPORT IN MICROORGANISMS

Simon Silver

Biology Department
Washington University
St. Louis, Missouri

I.  INTRODUCTION

Calcium is the fifth most abundant element by weight found in the crust of
the earth (following only oxygen, silicon, aluminum, and iron) [45,158].
In typical microbial growth media, calcium is the most abundant of the
divalent cations, occurring at much higher concentrations than those of
magnesium, for example (Table 2-1).  Yet, within microbial cells the
calcium level is low when compared to the external concentration; magnesium is the predominant intracellular divalent cation.  This chapter is
concerned with the mechanism by which the lowered intracellular calcium
level is maintained and with the few known exceptional cases where high
intracellular calcium levels are found.

      The situation in microorganisms is not unique: all living cells main-
tain low cytoplasmic calcium ion levels and high intracellular magnesium
levels in spite of the reverse concentration ratio in the media bathing the
cells.  Data on this point tend to be inadequate and confusing, because
massive amounts of calcium may be bound to the cell wall of plant cells
and microorganisms.  Massive amounts of calcium may be compartmentalized intracellularly in such structures as the mitochondria [11,22,85] and
the sarcoplasmic reticulum surrounding muscle fibers [90,92].  In order
to maintain differential cation gradients across semipermeable cell and
organelle membranes, highly specialized and highly specific calcium transport systems have evolved which bring about the energy-dependent movement of calcium across the membranes.  We shall be concerned here primarily with the calcium system in the bacterial cytoplasmic membrane,
which functions to lower intracellular calcium.  This system has been
studied to a limited extent with intact bacterial cells.  Studies with isolated
cytoplasmic membranes have the advantage of eliminating complications
due to binding to intracellular components.  In addition, it is possible to
prepare both right-side-out and inside-out membranes from the same
organism by varying the isolation procedure, thus allowing studies of the
polarity of the calcium transport system in an experimentally controlled
manner.  Furthermore, the movement of calcium across the intracellular
chlorophyll-containing chromatophore membranes isolated from photosynthetic bacteria allows questions pertaining to the polarity of the chromatophore membrane in contrast to the normal cellular membrane.
These studies lead to the unified hypothesis that all microbial cells actively
secrete calcium by a highly specific, metabolically active process.  Since
most microbial active transport systems are oriented inwardly to concentrate materials needed for growth [70-72,138], this polarity is unusual.
Outwardly oriented systems are not without precedent, however.  The
active outward secretion of protons was hypothesized by Mitchell [98] as
the primary coupling system for providing energy to mobilize inwardly
oriented systems.  Evidence for proton extrusion has been rapidly accumulating and is reviewed by Harold [57] and Harold and Altendorf [58].

TABLE 2-1  Typical Concentrations of Calcium, Magnesium, and Manganese

| Fluid | $Ca^{2+}$ (mM) | $Mg^{2+}$ (mM) | $Mn^{2+}$ (mM) | Ratio $Ca^{2+}/Mg^{2+}$ | References |
|---|---|---|---|---|---|
| Human plasma | 2.5 | 0.95 | 0.00005 | 2.6:1 | b |
| Human milk | 8.6 | 1.4 | 0.0001 | 6.1:1 | b, c |
| Cow's milk | 34.2 | 5.3 | 0.0004 | 5.9:1 | b, c |
| Human feces | [0.64 | 0.20 | 0.004][a] | 3.2:1 | b |
| Tryptone broth | 0.15 | 0.05 | <0.00002 | 3:1 | d |
| Sea water | 10.0 | 55.0 | 0.0003 | 1:5.5 | e |
| River water (typical) | 0.4 | 0.2 | 0.001 | 2:1 | e |
| River water (Missouri River at St. Louis) | 1.4 | 0.71 | <0.0005 | 2:1 | f |
| Bacterial cell water | 0.025 | 25.0 | 0.01 | 1:1,000 | d |
| Animal cell water | 0.001 | 3.3 | — | 1:3,300 | g |

[a] Grams per day.

[b] From K. Diem and C. Lentner (eds.), Documenta Geigy: Scientific Tables, 7th ed., Geigy Pharmaceuticals, Ardsley, New York, 1970.

[c] From K. Simkiss, Calcium in Reproductive Physiology, Chapman & Hall, London, 1967.

[d] Data from our laboratory by atomic absorption spectroscopy.

[e] From H. J. M. Bowen, Trace Elements in Biochemistry, Academic Press, New York, 1966.

[f] 1974 yearly averages from the St. Louis County Water Company.

[g] From Refs. 11 and 62.

Cyclic AMP (adenosine 3':5'-cyclic monophosphate) appears to be extruded from Escherichia coli cells by an energy-dependent process during regulation of carbohydrate metabolism [124]. Another outwardly oriented transport system is that for lactic acid in Streptococcus faecalis [60]. Glycolysis in these cells produces massive amounts of lactic acid which must be excreted. Still another example is the outwardly oriented movement of $Na^+$ from bacterial cells [59,157], which is associated with cellular accumulation of $K^+$ (much as accumulation of $Mg^{2+}$ is at least coincidentally associated with $Ca^{2+}$ excretion). $K^+$ accumulation can be associated with either $Na^+$ or $H^+$ extrusion [58,59], and $Na^+$ extrusion appears to be primarily paired with $H^+$ uptake [157].

## II.   CALCIUM TRANSPORT IN INTACT
##        BACTERIAL CELLS

That bacterial cells might have a transport system for the metabolically active extrusion of calcium was first suggested by Silver and Kralovic [134] on the basis of experiments showing three to four times greater accumulation of $^{45}Ca$ by E. coli incubated below $5°C$, as compared with cells in the more physiologically normal ranges of $15-37°C$. Heat-killed cells bound still more calcium. This suggested that the cell membrane barriers limited the access of $Ca^{2+}$ to intracellular binding sites and that metabolically active processes resulted in further lowering of intracellular calcium, presumedly by moving the calcium across the cell membrane. More recently these experiments have been expanded [136] not only in detail but also with additional microbial cells such as Bacillus subtilis and Bacillus megaterium. B. subtilis shows the same patterns as E. coli, but this Gram positive microbe accumulates quantitatively 6-8 times more $^{45}Ca$ below $5°C$ (on a μmoles $Ca^{2+}$ per mg cell mass basis), facilitating such studies. The accumulation at low temperature shows substrate specificity for $Ca^{2+}$ greater than for $Sr^{2+}$ or $Mn^{2+}$. $Mg^{2+}$ and monovalent cations were without effect. Such specificity is generally considered diagnostic for "carrier mediated" transport across membranes as opposed to passive leakage processes. Calcium accumulation also shows "saturation kinetics" (Michaelis-Menten kinetics), which is again characteristic of carrier mediated as compared with passive diffusion processes. The accumulation is reversible, in that more than 80% of the accumulated calcium is extruded upon warming the cells from $0°C$ to $20°C$; this extrusion process is energy-dependent in that it is inhibited by cyanide [136].

Low temperature was not the only means of inducing calcium accumulation by intact bacterial cells. The addition of energy poisons of the uncoupler or proton conducting class stimulated calcium accumulation by E. coli and B. subtilis cells. The optimum concentration of uncoupler was proportional to its potency in affecting other energy-dependent processes:

p-trifluoromethoxyphenyl carbonylcyanide hydrazone (FCCP) stimulated
calcium accumulation at lower concentrations than m-chlorophenyl car-
bonyl cyanide hydrazone (CCCP) which was in turn more potent than penta-
chlorophenol, tetrachlorosalicylanilide or dinitrophenol [136]. Again, the
uptake stimulated by uncouplers displayed substrate specificity and saturation kinetics. Low temperature- and uncoupler-stimulated calcium accumulation has been found in still another type of microbe, the photosynthetic bacterium Rhodopseudomonas capsulata [69]. As with E. coli, the
level of calcium accumulation is rather low with this Gram negative
organism.

Bronner et al. [15] and Golub et al. [50] found an alternative method
of studying calcium accumulation in intact cells of B. megaterium.
Washed cells grown on limiting glucose accumulate $^{45}$Ca. The addition of
an electron-transport substrate, reduced phenazine methosulfate, inhibited
the accumulation of $^{45}$Ca and caused the release of previously accumulated
calcium [15,50]. Similarly, reduced phenazine methosulfate caused the
release of some of the calcium accumulated by B. megaterium during
sporulation [15,50]. Although the relationship between calcium uptake in
energy-starved cells and that in sporulating bacteria requires much further work, this encouraging new approach allows direct experimentation
on the question. In all, studies with a variety of intact cells provide
plausible evidence for a calcium extrusion system as a general characteristic of bacterial cells.


III.  STUDIES WITH SUBCELLULAR
      MEMBRANES

A.  Right-side-out and Inside-out
    Membranes of Escherichia coli

By far the most clear and elegant evidence on calcium accumulation by
E. coli comes from the work of Rosen and McClees [121] and Tsuchiya
and Rosen [146], who measured calcium and amino acid accumulation by
membranes prepared by two methods and inferred that the membranes had
two orientations. Membranes prepared by osmotic lysis of E. coli cells
by methods devised by Kaback [70-72] have the normal right-side-out
orientation of the original cells [5,78] and have been extensively used for
studies of inwardly directed transport systems for amino acids, sugars,
and cations [72]. These right-side-out membranes accumulate amino
acids in an energy-dependent process (for example, proline with energy
from reduced phenazine methosulfate in Fig. 2-1d). Right-side-out membranes from E. coli do not accumulate calcium with or without an energy
source (Fig. 2-1c, but see the following discussion). The alternative
method of preparing cell membranes from E. coli cells, by explosive

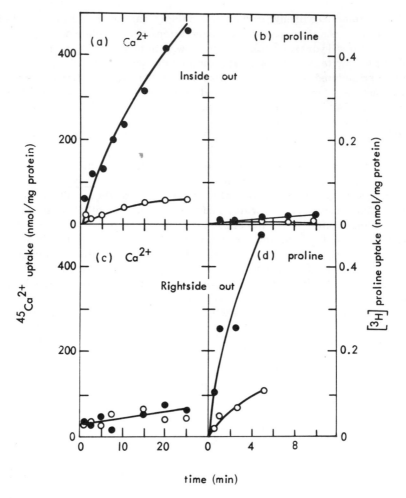

FIG. 2-1   Calcium uptake by everted membranes and proline transport by
right-side-out membranes: O, no energy source; ●, energy from reduced
phenazine methosulfate. (Redrawn from Rosen and McClees [121], with
permission.)

decompression in a French Pressure chamber, yields membranes apparently with the opposite "inside-out" orientation [47]. The physical forces
bringing about these two orientations are not known; one hypothesis centers
on the importance of physical connections between the outer and inner
membranes which would remain during disruption in the French Pressure
chamber but would have been removed by lysozyme action before osmotic
lysis. In any case, the French-Press membranes show the opposite

transport characteristics from osmotic lysis membranes. They accumulate calcium in an energy-dependent process (Fig. 2-1a) and are unable to accumulate proline under any conditions (Fig. 2-1b). This experiment was designed in such a manner as to demonstrate an energy-dependent calcium transport system with an orientation in the cell membrane opposite to that of the proline transport system [121].

In addition to determining the polarity of $Ca^{2+}$ transport in E. coli membranes, Rosen and McClees [121] demonstrated that calcium uptake by membranes was energy dependent. Respiratory substrates [e.g., NADH (reduced nicotinamide adenine dinucleotide), D-lactate, or succinate] or adenosine triphosphate (ATP) was required. Uptake was subject to inhibition by uncouplers, by cyanide when respiration driven [146], and by dicyclohexylcarbodiimide (DCCD) when ATP driven [121]. Since either respiratory substrates or ATP can drive calcium accumulation, this places the energy coupling for the calcium transport system in E. coli in the class of proton-motive force driven systems [58,98] and not in the class of transport systems that appear to be more directly coupled to ATP as an energy source [8]. This latter class includes at least one for phosphate [114] and two for potassium [43].

Tsuchiya and Rosen [146] measured the basic characteristics of calcium uptake by everted E. coli membranes in considerable detail. Coaccumulation of phosphate was required for continued calcium accumulation beyond a minute or so [121]. The stoichiometry of 1.5 $Ca^{2+}$ per $PO_4^{3-}$ [121,149] suggested accumulation of insoluble $Ca_3(PO_4)_2$ that is consistent with the lack of exchangeability of the accumulated calcium and its failure to be released by uncouplers. (Note that with presumably similar vesicles prepared from B. subtilis, $^{45}Ca$ was readily exchangeable with nonradioactive calcium [136], although these assays were run in the presence of 10 mM $PO_4^{3-}$. However, the B. subtilis experiments measured facilitated diffusion occurring at 1,000 times lower rates than the active $Ca^{2+}$ uptake by E. coli vesicles. It is likely that the solubility limit of $[Ca^{2+}] \times [PO_4^{3-}]$ was never exceeded in these experiments.) A careful consideration of the kinetics of calcium accumulation by E. coli membranes suggested that two systems were operating: (1) a high-affinity system with a $K_m$ of 4.5 μM (and a $V_{max}$ of 2 nmol/min per milligram of protein); and (2) a low-affinity system with a $K_m$ of 340 μM (and a $V_{max}$ of 85 nmol/min per milligram of protein).[1] The low-affinity system has a similar $K_m$ to those reported for energy-inhibited (360 μM) and sporulating (380 μM) B. subtilis, whereas the high-affinity system might be more closely related to the 9 μM $K_m$ system of calcium uptake found with everted membranes from B. megaterium [49, 50]. Monovalent cations, especially $K^+$, stimulated calcium uptake when added above 100 mM. Divalent cations were without effect at 0.1 mM; but at 1 mM levels, $Mg^{2+}$ and $Sr^{2+}$ inhibited calcium accumulation by about

---

[1] $K_m$ is the concentration for half-saturation of rate and $V_{max}$ the maximum rate of uptake.

half and inexplicably $Mn^{2+}$ stimulated calcium accumulation.  Neither $La^{3+}$ nor ruthenium red at $10 \mu M$ had any effect on $^{45}Ca$ accumulation by E. coli membranes [146], although these materials specifically and competitively inhibit calcium accumulation by mitochondria (cf. Section VIII. C) and also inhibit calcium extrusion by red blood cells (cf. Section VIII. A).

Given that the normal orientation of both the proton and the calcium transport systems are outward, this raises the interesting questions of coupling between the two and whether electroneutrality is preserved (two $H^+$ for one $Ca^{2+}$ antiport[2]) or whether the movement of calcium is electrogenic (one $H^+$ in for one $Ca^{2+}$ out).  In its simplest form the Mitchell hypothesis [98; cf. 57] proposes that only proton extrusion will be outwardly electrogenic and will generate the transmembrane potential, internal negative.  All other transport systems are predicted to be either electroneutral or to run down the potential gradient.  However, Tsuchiya and Rosen [149] favor electrogenic one-for-one calcium/proton antiport on the basis of the following indirect evidence:  (a) coaccumulation of the neutralizing anion phosphate [149]; (b) the lack of effect or small stimulation of calcium accumulation by valinomycin (which would be expected to dissipate the unfavorable internal positive potential built up by electrogenic calcium accumulation) [149]; and (c) an experiment showing that valinomycin-induced potassium efflux, which generates an internal negative potential, can drive calcium uptake (B. P. Rosen and T. Tsuchiya, personal communication).  The actual calcium/proton stoichiometry has not been measured, and we consider the question of electrogenic or electroneutral calcium egress from cells unanswered and of prime importance.  Two equally noncompelling arguments against electrogenic calcium movement in intact cells are the following:  (a) such movement would have to work against the normal membrane potential that is highly internal negative [57, 58]; and (b) the comovement of phosphate is not the overall situation in intact cells which accumulate phosphate [e.g., 114] while expelling calcium — both by energy-requiring mechanisms.  It is likely that the $Ca^{2+}/H^+$ antiport mechanism explains the unusual requirement for an alkaline external pH for calcium accumulation both with everted membranes [121, 146, 149] and with uncoupler-treated intact cells (H. E. Scribner, unpublished data).  A pH difference of 1.0 across the membrane would be able to maintain a calcium gradient of 100:1.

Tsuchiya and Rosen [147, 148] have also used the calcium transport system in studies demonstrating the need for proton impermeability for the functioning of proton-driven transport systems.  With membranes that were proton permeable because of loss of the $Mg^{2+}$ ATPase [either by elution with ethylenediaminetetraacetate (EDTA) or because of a mutational defect], calcium transport and other energy-coupled phenomena could not

---

[2]Antiport, symport, and uniport are terms introduced by Mitchell [98] to characterize different "coupling" mechanisms between substrates transported across the cell membrane either alone (uniport) or obligatorally coupled with another substrate, often protons, moving either in the same direction (symport) or in the opposite direction (antiport) across the membrane.

be demonstrated, even when the energy source was from respiration of NADH and not hydrolysis of ATP. Upon rebinding of the ATPase to the membrane, a lowering of membrane proton permeability was shown [147] concomitant with a reestablishment of a functioning calcium transport system [148].

B.  Membranes from Bacillus subtilis, Bacillus megaterium, and Azotobacter vinelandii

Three other laboratories have reported experiments with calcium uptake in subcellular bacterial membranes. Silver and associates [136] studied calcium uptake by aged (more than 1 year at $-70^\circ$ C) right-side-out vesicles of E. coli and B. subtilis prepared by Kaback's osmotic lysis procedures. With everted E. coli membranes, Rosen and McClees [121] found calcium transport activity was lost within a few hours, even at $4^\circ$ C, although activity was stable upon storage at $-70^\circ$ C with 50% glycerol present [146]. With aged right-side-out membranes, highly specific calcium uptake was found. Calcium accumulation was not energy dependent although these membranes had retained the energy-dependent proline transport also found in fresh membranes. The relationship between the energy-dependent uptake of calcium by inside-out membranes [121,146] and the non-energy-dependent uptake by aged right-side-out membranes [136] remains in doubt. However, one can hypothesize [49] that aging (even at $-70^\circ$ C) causes a loss of the energy coupling to the calcium efflux system, which subsequently functions in non-energy-requiring facilitated carrier-mediated transport in the opposite direction. Further work is required to reconcile the different results from different laboratories. Silver et al. [136] found the rate of calcium uptake by aged B. subtilis membranes was about eight times that with aged E. coli membranes. A similar ratio of relative rates was obtained with intact cells with proton-conducting uncouplers (see Section II).

Golub and Bronner [49] and Golub et al. [50] found energy-dependent calcium accumulation with fresh membrane vesicles from B. megaterium, even though the membranes had been prepared by osmotic lysis procedures. However, by varying the buffer in which osmotic lysis occurred, Bronner et al. [15] and Golub et al. [50] found that the relative rate of calcium uptake by the membranes was inversely related to the relative rate of uptake by a "normally" oriented transport system, in this case for glutamate. Thus they were able to conclude that the calcium uptake by freshly prepared osmotically lysed membranes was due to a variable fraction of membranes with everted orientation. French-Press prepared membranes of B. megaterium accumulated calcium but not glutamate [50]. Golub and Bronner [49] observed a loss of energy-dependent calcium uptake by B. megaterium membranes over a few days storage and a concomitant rise in the rate of non-energy-dependent calcium accumulation. Whether a single system is changing from active transport in one orientation to facilitated diffusion in the other direction remains an open question.

Bhattacharyya and Barnes [9] recently reported the energy-dependent uptake of $^{45}$Ca by membrane vesicles from A. vinelandii prepared by osmotic lysis procedures. These membranes have many properties in common with those produced by Bronner et al. [15] and Golub et al. [50] and probably are a mixture of membranes some with normal outward orientation and some with everted orientation. The Azotobacter membranes concentrated calcium with substrate specificity, Michaelis-Menten kinetics, and strict dependence upon respiratory energy sources. Unlike the E. coli membranes of Rosen and McClees [121], the Azotobacter membranes did not respond to added ATP as an energy source unless pretreated with trypsin. The calcium accumulated by the Azotobacter membranes in the absence of phosphate was rapidly released when either an uncoupler or EGTA$^3$ was added [9], unlike calcium accumulated by everted E. coli membranes in the presence of phosphate buffer. In the absence of high phosphate, the small amount of calcium accumulated by E. coli membranes was also readily released and exchanged [121]. These results are consistent with the calcium being retained within the membrane by continued calcium/proton exchange rather than being bound in insoluble phosphate salts. Calcium/proton exchange in everted membranes would also be consistent with the inhibition of calcium uptake by nigericin and the lack of an effect of valinomycin with Azotobacter membranes [9]. Azotobacter differs from the bacteria with which we have previously dealt with in that it has a calcium requirement for growth (see Section VII. A). Barnes found that intact cells of A. vinelandii also accumulate calcium by an energy-independent process that is apparently missing in spheroplasts prepared from the cells. We will later ascribe this whole cell calcium binding to cell walls (Section VII. A), but here we are dealing only with membrane transport.

C.   Intracellular Membranes

The membranes prepared by osmotic lysis and explosive decompression in all cases just described are derived from the cellular "cytoplasmic" membrane since they are prepared from cells having only this membrane (Fig. 2-2a). Photosynthetic bacteria such as Rhodopseudomonas capsulata and nitrogen-fixing bacteria such as A. vinelandii also contain small intracellular extensions of the cell membrane containing the photosynthetic or respiratory apparatus of the cells, respectively (Fig. 2-2b). These intracellular membranes are formed by invagination from the cellular membranes. They are therefore thought to have the reversed polarity or sidedness of the cytoplasmic membrane. The photophosphorylating membrane vesicles (chromatophores) do accumulate protons, showing the opposite polarity to the cellular membranes in this respect [129]. Jasper [69] in our laboratory has found energy-dependent uptake of $^{45}$Ca by chromatophore membranes from R. capsulata, consistent with the model

---

$^3$ EGTA is ethyleneglycoltetraacetic acid.

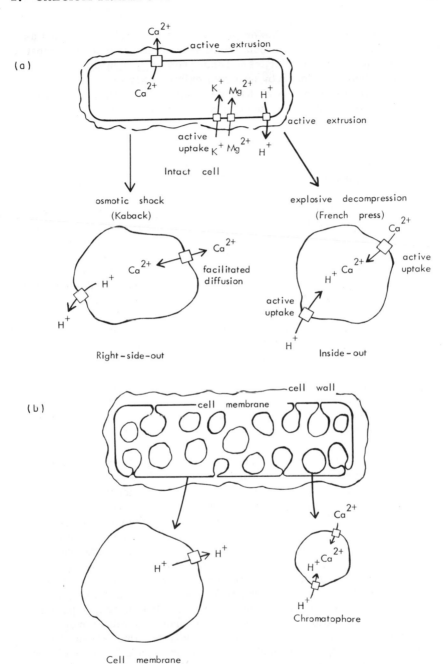

FIG. 2-2 Orientation of membrane transport systems. (a) Orientation of cell membrane transport systems. (b) Orientation of transport systems in subcellular membranes.

we have been developing in this chapter of oriented calcium transport as a
general characteristic of microbial membranes. It would be of interest to
isolate the small intracellular membranes from A. vinelandii and directly
to measure their orientation by means of calcium (and proton) translocation. All of the experiments we have described so far support the concept
of an outwardly oriented calcium permease as a universal constituent of
bacterial membranes.

IV.  CALCIUM ACCUMULATION
     DURING SPORULATION

Massive calcium accumulation occurs during sporulation in members of
the genus Bacillus, along with the accumulation of other divalent cations
[32,151]. Generally, calcium is found associated with dipicolinic acid
(DPA) in a 1:1 complex that can constitute 20% of the dry weight of the
eventual spore [103,151]. Calcium dipicolinate had been considered
responsible for the extraordinary heat resistance of bacterial spores [53].
That conclusion is probably erroneous, however, since some mutants are
devoid of DPA and yet completely heat resistant [56,159]. Since calcium
accumulation begins an hour or two earlier in sporulation than does DPA
synthesis [151] and can be separated from DPA synthesis by use of
actinomycin D [108], the question arose as to whether calcium was pas-
sively flowing into the cell where it was bound by a chelate-"sink" such as
DPA or whether the entry to the cell was via a transmembrane transport
system. The relative rate of calcium uptake increases from 0 about 3 hr
after sporulation begins, peaks about 6 hr later, and then rapidly declines
[130,135], suggesting the synthesis and regulation of a specific transport
system.
     More direct studies have shown that calcium uptake during sporulation
is due to the functioning of a membrane-associated carrier-mediated transport system. Bronner et al. [13] and Bronner and Freund [14] suggested
the existence of a calcium "pump" based on studies of the timing of calcium
accumulation and attempted to isolate the cellular calcium-binding component that will be described later (Section V). Eisenstadt and Silver [39]
studied the specificity and saturation kinetics of calcium uptake during
sporulation, both of which were indicative of carrier-mediated uptake.
Our current picture of the uptake and sequestering of calcium is summarized in Figure 2-3. After consideration of components of this model, we
will develop the experimental results upon which it is based. The first
step is the accumulation of calcium by the sporulating cell by means of a
membrane-located transport system [39]. Whether this transport is by a
calcium/proton antiport mechanism as shown in Figure 2-3 is not known.
We also do not know whether calcium accumulation during sporulation is
mediated by the same system that carried out calcium egress during

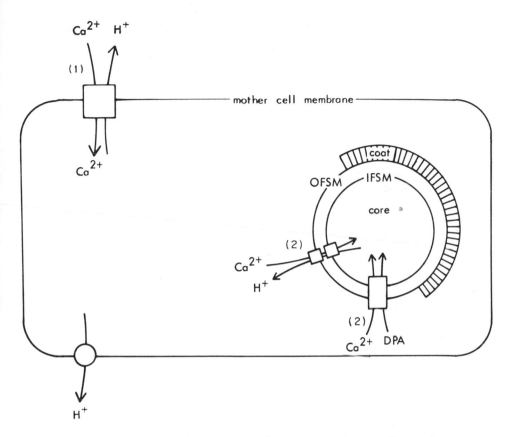

FIG. 2-3 Scheme for calcium accumulation during bacterial sporulation. Steps: (1) Calcium uptake by the mother cell membrane. Whether antiport for protons and/or by a system related to the extrusion system of log phase cells is unknown. (2) Movement of $Ca^{2+}$ from the mother-cell cytoplasm into the developing spore core. Currently, we cannot say whether this movement is active or passive; whether the movement is antiport with protons or symport with the anion DPA; and what is the relationship between movement across the outer and inner forespore membranes (OSFM and ISFM), which arise with opposite polarities.

earlier growth phases (cf. Section III). It is conceivable that this $Ca^{2+}$ efflux system functions in an opposite polarity. Since data are not readily available about the energy requirement for calcium accumulation during sporulation [39], uptake during sporulation may be either facilitated diffusion or by active transport. Alternatively, the egress system may be inactivated and a new sporulation-specific calcium uptake system synthesized. Calcium transport mutants should provide the distinction between these alternatives. A common system used with different polarities during different stages of growth and sporulation would lead to the prediction that mutants defective in calcium efflux during growth (perhaps calcium sensitive because of this) would be unable to accumulate calcium during sporulation. These might then produce light-density heat-sensitive spores. If totally different systems were involved in calcium movement during growth and during sporulation, mutants might be isolated defective in only one of the two processes. Once within the mother-cell cytoplasm the calcium must move across two additional membranes to reach its ultimate location within the developing spore cytoplasm, or "core" (Fig. 2-3). These are the outer forespore membrane (that originated by invagination from the mother cell cytoplasmic membrane and therefore has the opposite polarity of the mother-cell membrane) and the inner forespore membrane (that originated as the cellular membrane of the small "protospore" and therefore has the normal polarity). Several hypothetical alternatives are available and experiments to test these models have just begun.

Electroneutrality could be maintained either by a calcium/proton antiport mechanism or by symport with the dicarboxylic acid DPA. Both alternatives are diagramed in Fig. 2-3. There is also the question of whether movement across the outer and inner forespore membranes involves membrane transport systems (as we have diagramed) or might occur by passive diffusion through a more-or-less freely permeable membrane. At least one membrane transport system is likely to be required in order to establish the partitioning of components and the formation of the unique internal structure of the developing spore.

Given the model, what are the data, however incomplete, that led us to construct it? By measuring exchangeability with added nonradioactive calcium and release by agents such as toluene and lysozyme, we were able to distinguish four positions or stages of calcium movement during sporulation [39]: (1) $Ca^{2+}$ in solution outside the cells; (2) $Ca^{2+}$ in solution within the mother-cell cytoplasm (exchangeable and releasable by toluene, which disrupts membrane permeability barriers); (3) $Ca^{2+}$ bound within the forespore with DPA and no longer subject to release by toluene but still extractable by lysozyme (which disrupts the cell wall leading to osmotic lysis); and (4) $Ca^{2+}$ bound in a stable completed spore so that it could no longer be released by either toluene or lysozyme treatment. The conversion from stage 3 to stage 4 involved synthesis of the spore coat that protects the otherwise lysozyme-sensitive structure of the spore cell wall, or "cortex." Both the DPA and the $Ca^{2+}$ are found distributed within the spore cytoplasm, or core (Fig. 2-3). This only became clear during the last few years [84,128], and previously alternative locations for the $Ca^{2+}$-DPA complex were favored [103,151]. Although calcium movement across the mother-cell membrane is carrier mediated [39], the question of facilitated transport followed by chelation versus energy-dependent active

transport across the membrane remains open.   Experiments with energy
inhibitors gave equivocal results [39].

The mechanism of calcium movement from the mother-cell compartment into the forespore has been approached by isolating intact forespores
from sporulation stages during which active calcium accumulation occurs
[41,81].   Calcium movement is interconnected with the question of the
site of synthesis and movement of DPA, since movement of the two might
be directly or indirectly coupled.   Apparently DPA is synthesized exclusively in the mother-cell compartment and then rapidly moved into the
forespore (Fig. 2-3).   The enzymes for DPA biosynthesis are found in the
mother-cell cytoplasm [6,159], but the newly synthesized DPA is found
exclusively in the developing forespore [41,81].   Protoplasts of B. megaterium have a $K_m$ for calcium uptake below 50 μM [41], 10 times less than
the comparable value for B. subtilis [39], whereas the forespores of
B. megaterium accumulate appreciable calcium only when the level is
raised to 10 mM or so [41].   With 10 mM external calcium the forespores
accumulated as much calcium in 2 hr as the protoplasts from which they
had been isolated did from 0.1 mM $Ca^{2+}$ [41].   At 0.1 mM $Ca^{2+}$, the forespores showed no detectable uptake [41,81].   As with Rosen and McClees's
[121] studies with everted E. coli membranes, accumulation of calcium by
forespores of B. megaterium required a counterion for continued accumulation but not for the initial transport event.   In this case, some 20 to 30
times more $^{45}$Ca was accumulated by forespores in the presence of 2 mM
$PO_4^{3-}$ buffer in addition to the 10 mM $Ca^{2+}$ [41].   Addition of DPA was
ineffective in these experiments.   The calcium accumulated by forespores
in the presence of $PO_4^{3-}$ differed from that accumulated within the mother-
cell protoplast in that the "in vitro"-accumulated calcium would exchange
with added nonradioactive calcium (only at 30°C; not at 0°C), whereas
"in vivo"-accumulated calcium was not exchangeable, presumedly because
it was sequestered with DPA.   Clearly, the mechanism of calcium movement across the forespore membrane is now ready for direct experimental
attack.

In summary, the current data are most easily explained as support
for a calcium transport system in the mother-cell membrane followed by
another accumulation system for calcium in the forespore.   Forespore
transport may be coupled to the movement of newly synthesized DPA
(Fig. 2-3).

Another question is that of the relationship of the sporulating cell's
inwardly directed calcium transport system to the growing cell's outwardly
directed system: Is a new sporulation-specific system synthesized by the
cells after the inactivation of the previously outwardly oriented calcium
system [121,136], or does the same system function in different orientations during different stages of the sporulation cycle?   If the latter case
proves correct, then the question of energy coupling will be particularly
interesting.   According to the chemiosmotic hypothesis [57,58,98], the

outwardly oriented calcium movement is likely to be coupled antiport with
the inward movement of protons, down the pH gradient (and electroneutral),
whereas the coupling for inwardly directed calcium movement would be
electrogenic and a form of carrier-mediated diffusion driven by the potential gradient (internal negative). It should now be possible to test these
predictions of the chemiosmotic coupling hypothesis directly. Still another
question with developing spores is that of the orientation of the three membranes involved. The mother-cell membrane has the normal membrane
orientation. The early stages of spore formation involve the envelopment
of the small cell that will become the spore by the larger ultimate mother
cell. This yields two membrane coverings to the spore: the inner forespore membrane, which originates from the cytoplasmic membrane of the
small protospore cell and therefore has the normal right-side-out orientation; and the outer forespore membrane (Fig. 2-3), which arises from
invagination of the mother-cell membrane and therefore has an inside-out
orientation. What mechanisms are involved in moving calcium (and other
materials) across both the outer and inner forespore membranes?

Since spores contain as much as 20% of their dry weight as $Ca^{2+}$-DPA,
it is reasonable to ask what the function of this complex might be. The
early postulate [53,103,151] of a primary role for $Ca^{2+}$-DPA in maintaining heat stability was eliminated by the isolation of heat-stable spores of
mutants unable to synthesize DPA [56,159]. Nevertheless, there is much
data showing a correlation of heat resistance with $Ca^{2+}$-DPA content [103]
or with the ratio of $Ca^{2+}/Mg^{2+}$ in the spores. The more $Ca^{2+}$-DPA and
the higher the $Ca^{2+}/Mg^{2+}$ ratio, the more heat resistant are the spores.
Dehydration of the spore core by mechanical compression due to the expansion of overlying peptidoglycan cortex layers is now considered the
likely basis for the extreme heat resistance of spores [52]. The expanded
electronegative peptidoglycan will be largely neutralized by monovalent
cations, and calcium as a bridging cation would be expected to cause heat
sensitization — and indeed under some conditions it does [52]. Movement
of the calcium from the core into the cortex would facilitate the contraction
of the cortex layers [52], causing rehydration of the core and concomitant
heat sensitization [35]. Fitz-James [44] suggested an alternative function
for calcium in stabilizing the spore membranes during the transition from
an anhydrous state (the dormant spore) into a normally hydrated cell form.
In studies of the requirements for converting dormant spores into intact
protoplasts, calcium was needed during an early stage of protoplast
formation. Although $Sr^{2+}$ could replace $Ca^{2+}$ in this role, neither $Mg^{2+}$
nor $Mn^{2+}$ provided the needed stability. After hydration and protoplast
formation were complete, calcium and DPA were released and only $Mg^{2+}$
was needed for protoplast stability. Fitz-James [44] proposed that the
$Ca^{2+}$ is not totally bound to DPA but that appreciable amounts of calcium
are bound to the head groups of membrane phospholipids. During the dehydration process associated with contraction of the spore cortex [35,52,

103], this calcium would serve to stabilize [38] the membrane structure. During early stages of rehydration during spore germination, calcium would be required to stabilize the membrane and to prevent lysis. A combination of both displacement of the membrane calcium by magnesium that is taken up early in germination [40] and chelation of the displaced calcium by excess DPA would mark the intermediate stage in rehydration, after which calcium is no longer required for membrane stability — only magnesium [44]. Calcium and DPA then flow out of the germinating spore [32,34], roughly in parallel but by separate mechanisms.


V.  CALCIUM-BINDING FACTORS

A primary goal in studies of microbial cation transport is the isolation and identification of the chemical nature of the "carrier" which is initially hypothesized on the basis of substrate specificity and saturation kinetics. With calcium and other cations, two models are available: the calcium-binding proteins of vertebrates found and studied by Huang et al. [65] and Wassermann et al. [153], and the low-molecular-weight chelates or ionophores of microbial origin, discussed in Section VIII.D.

The isolation and preliminary characterization of a low-molecular-weight calcium-binding factor was reported by Bronner and Freund [14]. This factor was assayed using the chelex binding assay developed for the intestinal binding protein (see Section VIII). It is synthesized by B. megaterium after the end of log-phase growth and early in the sporulation cycle. In fact, the calcium-binding activity peaks during sporulation just before net calcium accumulation begins, and the level of the chelate then declines during the period of massive calcium accumulation [14]. This material could be precipitated by ethanol and run on a Sephadex G25 or Biogel P2 column at a position indicative of a molecular weight of a few hundred daltons for the major activity peak. The calcium-binding factor from B. megaterium was clearly not DPA, however, as determined by comparing chromatographic mobilities and by direct chemical analysis of the active fractions for DPA [14,15].


VI.  EUCARYOTIC MICROBES

Most monographs on the growth and physiology of yeasts and other fungi that have appeared during the last decade have a paragraph or two stating that calcium is required for growth, or for maximum growth yield, or at a specific developmental stage during the growth of many fungi. This is much like the situation with most calcium-requiring bacterial species (see Section VII.A). The one exception comes from the work of Cameron and LéJohn [20,21] and LéJohn et al. [88], who have studied calcium accumulation and function in the water mold Achlya.

Achyla develops from a single-cell spore, through a coenocytic hyphal growth stage, and then through septation of the hyphal tips into mononucleate cells that develop into spores [21]. Calcium was required at all stages — during both proliferative growth and during sporulation. The requirement for calcium was highly specific in that $Mg^{2+}$, $Mn^{2+}$, $Co^{2+}$, and $Fe^{2+}$ could not replace $Ca^{2+}$. Barium competitively inhibited the calcium growth response and also inhibited calcium accumulation; apparently strontium was not tested [21].

The basis of the calcium requirement was studied by adding chelates to remove available extracellular calcium. The accumulation of six amino acids and thymidine was shown to be $Ca^{2+}$ dependent whereas the accumulation of uracil was not [21]. Calcium is accumulated by Achyla during growth [21,88], while hyphae whose growth has been inhibited by other means and dormant spores do not accumulate $Ca^{2+}$[21]. Calcium accumulation in Achyla is by a process not dependent on metabolic energy and therefore not by an active transport system [88]. Two classes of fixed calcium were found. In the first class, about 40% of the accumulated calcium was extracellular, removable by chelates (citrate, EDTA, EGTA), and bound to a low-molecular-weight glycopeptide that could be removed from Achyla by osmotic shock [88]. Binding to this glycopeptide was greatly stimulated by mercury and mercurials [88]. The second class consisted of the remaining 60% of the accumulated calcium and was apparently intracellular. This calcium was not associated with structures such as the nucleus, mitochondria, or ribosomes, but rather is found in the soluble cytoplasmic fraction [21]. Although this intracellular calcium is quantitatively low, it presents a problem, because evidence was presented earlier that cells do not accumulate calcium (Section II). We will present in Section VII.B the reasons why intracellular calcium should be toxic! The most plausible explanation is that this cellular calcium is largely compartmentalized in the intracellular vacuole, which is known to contain divalent cations in concentrations above those in the cytosol [94]. There is currently no reason to believe that fungal cells — or even those of higher plants — have cytosolic calcium levels appreciably above those listed in Table 2-1 for bacterial and animal cells.

Although it did not require energy, calcium accumulation by osmotically shocked Achyla followed classical saturation kinetics ($K_m$ about $40\,\mu M$) and was inhibited by both mercurials and by ruthenium red (a specific inhibitor of the mitochondrial calcium system; cf. Section VIII. C). Cytokinins appeared to stimulate calcium accumulation by osmotically shocked Achyla [88]. Apparently this calcium is in an intracellular compartment.

We suggest that further studies with Achyla will demonstrate that the nonexchangeable cell calcium is not really free intracellularly during growth but is associated with the vacuole and/or other membranous structures and that the uptake by osmotically shocked cells is indicative of their generally inhibited state, i.e., time is required for recovery before

further growth can proceed [88]. Studies of calcium movement during the
recovery stage should show that Achyla is like other microbes with regard
to the general properties of calcium metabolism.

Calcium fluxes are critical in the early developmental stages of the
brown marine algae Fucus and Pelvetia. Unidirectional light exposure
induces germination of the round algal egg into a polarized structure with
a lighted thallus (or growth) end and a shaded rhizoid (or holdfast) end
[119]. During early germination the rate of calcium influx through the
rhizoid end of the egg was five times that through the thallus end. The
ratio of efflux rates for $Ca^{2+}$ was reversed and 2.5 times greater through
the thallus end. This resulted in a net calcium current from rhizoid to
thallus of approximately 2 pA per egg, and a total current of this magnitude
was measured [119]. Between 6 and 12 hr into germination with Pelvetia
eggs, this differential calcium current disappeared and equal influx and
egress rates were measured at the rhizoid and thallus ends of the now
completely polarized eggs. The mechanisms considered to govern calcium
fluxes were a regulated rate of passive leakage (for influx) and an equally
controlled rate of energy-dependent efflux by $Ca^{2+}$ "pumps" [119]. More
and more, it appears that all living cells share the same fundamental
calcium transport mechanisms that can be regulated and adjusted for control of development and growth.

There is still another cellular role of calcium that may be important
in microorganisms: calcium is required for aggregation of the cellular
slime mold, Dictyostelium discoideum [93], and is involved in intracellular
microfilament aggregation. Since eucaryotic microbes including algae
[106] and amoeba [144,155] contain actin-like proteins that are involved
in cellular motility [144], one may ascribe a regulatory role to calcium in
these processes too. Cyclic AMP is the trigger for Dictyostelium aggregation; and the process may involve an increase in membrane permeability
to calcium [93,116], followed by calcium flux into the cell and the ATP-
dependent polymerization of globular actin monomers into actin filaments.
The calcium content and distribution have been followed during development of D. discoideum [91]. Total calcium content decreases from 3.5
to below 2.0 µg per $10^8$ cells during the early aggregation stages governed
by cyclic AMP [91]. These total calcium measurements therefore cannot
be following an increase in intracellular cytosolic calcium. During slug
migration and stalk and spore formation, the total calcium increases
dramatically. Radioautography of $^{45}Ca$ showed a preponderance of the
calcium toward the anterior ends of the slugs, which are destined to become stalk cells, and not in the posterior prespore cells [91]. We assume
that most of the total calcium is extracellular and associated with the stalk
cell-wall components. Vacuoles may also be involved in calcium compartmentalization. The finding of calcium and cyclic AMP involvement in
slime mold aggregation [93] has suggested that this system would be a
useful model for $Ca^{2+}$-cyclic AMP governed hormonal regulation of animal

cells [116]. To complete this speculative note, we may also ask whether actin-like proteins [97] — and roles for calcium — should be sought in procaryotic bacterial cells.

There is a microbial example of behavioral control by calcium, namely, that of ciliary activity in protozoa [37, 79]. Paramecium aurelia responds to external stimuli by reversing the direction of ciliary orientation and beat in order to reverse the direction of swimming. Ciliary reversal is completely dependent upon calcium flux through the cellular membrane. When swimming forward and running into an obstacle or toxic substance, P. aurelia increases its membrane conductance to calcium, leading to an influx of calcium (if there is external calcium available) and an increase in internal concentration. This was tested with Paramecia which were made permeable to calcium with mild detergent treatment. These demonstrated the ciliary reversal when exposed both to adequate calcium and to $Mg^{2+}$-ATP to drive the ciliary movement [37]. When the internal concentration rose above $1 \mu M$, the cilia reversed their orientation from posterior to anterior, reversed the power stroke of their beat — and the paramecia swam backwards. As the calcium was pumped from the cells by a membrane active transport system [17] and as the internal calcium level dropped below $1 \mu M$, the ciliary orientation and beat were reversed once again so that the cells swam in a forward direction. The studies with P. aurelia have included internal recordings of membrane potential and internal excitation and electrical depolarization, as well as control of external calcium concentrations.

Over 300 mutants mapping at 20 genetic loci are defective in stimulus response in Paramecium [79]. The "pawn" mutants (in three separate genetic loci and so called because, like the chess piece, they can move only forward) have lost the ability to reverse their ciliary orientation and have lost the calcium action potential as measured with microelectrodes [79, 127a]. Some pawn mutations appear to affect the calcium channel itself and others to affect the voltage-dependent "gating" mechanism that governs the normal inward flux of calcium. These two classes fall in different genes and can be distinguished by the degree of resistance to the toxic $Ca^{2+}$ analog $Ba^{2+}$ [127a]. Further evidence that the pawn mutants are defective in calcium transport is the finding that detergent-treated cells of the pawn mutants could be induced to swim backwards by adding calcium [79]. The cilia of these cells are normal and capable of backward strokes, but the calcium flux that triggers the swing in ciliary direction does not occur in the pawn mutants. Recently, direct measurements of calcium movements using $^{45}Ca$ have supplemented the microelectrode studies of a calcium potential [16, 17]. Stimulating the avoiding response resulted in a five- to tenfold increase in the rate of $^{45}Ca$ influx concomitant with increased backward swimming [16]. Pawn mutants unable to generate a calcium action potential did not show this increase in $^{45}Ca$ uptake [16]. The gated influx of calcium was "downhill" (from a higher external concentration to a lower internal concentration) and was countered by a system carrying out energy-dependent calcium efflux [17]. Pawn mutants did not show the stimulated increase in $^{45}$Ca uptake, and temperature-sensitive pawn mutants [79] showed the increased $Ca^{2+}$ uptake only when grown at lower (25° C) temperatures [16]. With another class of mutant called "fast-2" (because it swims more rapidly and turns less often when stimulated than wild-type [79]), increased $Ca^{2+}$ influx was stimulated by high $K^+$ but not by $Na^+$, in direct correlation with stimulation of backward swimming by $K^+$ but not by $Na^+$ with this mutant strain [16]. The gated $^{45}$Ca influx showed a half-saturation value below 10 $\mu$M and several divalent cations ($Mg^{2+}$, $Sr^{2+}$, $Ba^{2+}$, $Ni^{2+}$, $Mn^{2+}$, and $Co^{2+}$) all inhibited calcium uptake when added at 10- to 500-times higher levels than calcium [17]. Detailed kinetic studies analogous to those with E. coli are still required to define the the specificity of the gated $Ca^{2+}$ uptake. Calcium efflux was measured with cells "loaded" with $^{45}$Ca at 0° C [17]. Although efflux occurred at 23° C but not at 0° C, the inhibitors cyanide and azide were without effect on $^{45}$Ca efflux — under conditions where cellular ATP levels were reduced 95% by the inhibitors; $Ca^{2+}$ efflux was also unaffected by external calcium or by influx-stimulating levels of $K^+$. At physiological temperatures (23° C), the calcium efflux system maintained an internal $Ca^{2+}$ level at least 10 times lower than the external concentration [17].

Again, the protozoan cells show properties in common with those we have described for bacterial cells in Section II, although there is no evidence for a role for $Ca^{2+}$ in bacterial motility or chemotaxis [3].


VII.   BACTERIAL FUNCTIONS OF CALCIUM

Calcium is required for the growth of many bacterial species, including photosynthetic bacteria [48] and Azotobacter [104]. Calcium is also required for the adsorption of many bacteriophages [2]. As a result of numerous observations of this nature, calcium is often added to standard bacteriological media without consideration as to whether it is truly required for growth, or in what manner it may function where required. We have proposed earlier that calcium is normally transported out of bacterial cells and that the intracellular level of calcium is very low. In this section, we will develop the thesis that all microbial calcium functions are at the cellular membrane or external to the membrane. There are good data for calcium involvement in the synthesis and stability of cell wall structures. Calcium may be a specific activator of extracellular enzymes. But, as far as we are aware, there is no required intracellular role for calcium in bacterial cells; muscle cells are, of course, another matter.

A.   Calcium as a Growth Factor

The timing and effects of calcium deprivation have been studied only in a
few cases.  Usually only the need to add calcium to the medium is known.
Calcium deficiency causes growth defects and development of pleomorphic
forms including Y-shaped bifid forms in Lactobacillus bifidus, probably
from alteration of cell wall formation [77].  High (2.5 mM) $Ca^{2+}$ is required for growth of the plague bacterium Yersinia pestis at $37°$ C, but
there is no $Ca^{2+}$ requirement at $26°$ C [54].  Upon shifting calcium-
deprived cells from $26°$ C to $37°$ C, DNA synthesis appeared to be one of
the first growth processes affected [54].  Protein accumulation and RNA
synthesis continued for a longer time (R. R. Brubaker, personal communication).  The synthesis of extracellular V and W antigens responsible
for plague virulence was repressed by added calcium [18].  Avirulent
yersiniae mutants did not synthesize the antigens and were able to grow at
$37°$ C without added $Ca^{2+}$ [18,54].  Therefore, the difference between
intracellular and extracellular calcium levels in host animal cells has been
considered an important factor in regulating the growth and virulence of
this intracellular parasite [18,54].


B.   Cell Wall Roles

The relatively high calcium requirement (0.4 mM) for optimum encystment of A. vinelandii has been attributed to specific $Ca^{2+}$-alginic acid
complexes formed by and contributing to the intine and exine layers of the
cyst coat [105,123].  However, X-ray microanalysis has shown that the
major location of calcium in the cyst is in the central body and not the
outer coat layers [123].  Without high medium calcium, abnormal slime
is synthesized as the cells attempt to produce the outer layer materials.
With adequate calcium, uronic acids (polymannuronic acid and polyguluronic
acid) form specific alginate gels with calcium interchain linkages [105,
123].  Detailed physicochemical studies of cation binding by alginic acid
gels [64,83,99,139] have shown highly selective, very tight, binding of
calcium.  Alginate in solution discriminates moderately (about 7:1) in
favor of calcium over magnesium, but alginate gels can show discrimination ratio of about 40:1 to 50:1 and sometimes nearly 100:1 in favor of
calcium binding [139].  This gel selectivity occurs as the alginate fibers
align side by side and divalent $Ca^{2+}$ functions to provide interstrand salt
bridges that stabilize the gel [99,139].  The degree of calcium selectivity
of alginate gels varies with the proportion of polyguluronic acid as opposed
to polymannuronic acid.  (Alginic acids contain runs of "pure" chains of
one or the other uronic acid, in addition to mixed stretches with both [64,
83,123].)  Both the exine and intine layers contain about 50% heteropolymeric
polymannuronic-gluronic acid [123].  Most of the remaining uronic acid in
the intine layer is polymannuronic acid, and most in the exine layer is

polyguluronic acid. The extracellular slime produced during abortive encystment is predominantly polymannuronic acid and has a ratio of mannuronic acid to guluronic acid residues of nearly 6:1 [123]. Conversion of polymannuronic acid to hetero-polymannuronic-guluronic acid and then to homopolyguluronic acid (which shows greater selectivity in binding calcium [139]) is carried out by an extracellular Azotobacter enzyme that requires calcium for functioning [63]. In those few cases involving alginic acid, the known locus of the cell surface calcium goes well beyond the less specific "cell wall" calcium of earlier studies [24,66,74]. Calcium can also stabilize cell wall structure by bridging carboxyl groups in peptidoglycan chains [89]. Moreover, anionic compounds such as alginic acid and peptidoglycan are not the exclusive candidates for cell wall calcium binding. $Ca^{2+}$ binding to uncharged sugar moieties such as fucose may also function in carbohydrate bridge formation [28].


C.  Calcium in Enzymes

There are only a small number of hypotheses for functional role(s) of calcium in bacterial cells and, therefore, few reasons why calcium levels are kept low within cellular cytoplasm. First and foremost is the proposal that calcium may play a regulatory role by maintaining exoenzymes in an inactive form until the time that they are secreted from bacterial cells. This would serve to protect the cytoplasmic nucleic acids from nuclease destruction [31], the cytoplasmic proteins from protease attack, and so forth. A more subtle variant of this hypothesis suggests that cells might control cellular calcium very tightly so as to activate internal calcium-requiring enzymes at times when their enzymatic activity is of value. An alternative hypothesis for calcium action concerns questions of membrane potential, motility, and chemotaxis. This is related to the mode of action of calcium in regulating the direction of ciliary beat and therefore motility of Paramecium [16,37]. Both roles for calcium may indeed be used in nature.

At present, we have data on the role(s) of calcium with specific enzymes. The primary literature is spread among the numerous papers on enzyme purification and function and has never been compiled (to our knowledge) to form a general theory of calcium-activated enzymes. We have not examined the literature on the 840 enzymes listed by Dixon and Webb [33] and counted by Mildvan [96], nor the additional literature on enzymes newly described during the last 10 years. The only compilation on exoenzymes with which we are familiar was made by Pollock [110], who listed about 40 types of exoenzymes including 19 cases of specific activation or stabilization of exoenzymes by $Ca^{2+}$. Even 15 years ago, it was clear that calcium played a special role with some, but not all, exoenzymes [110]. A perusal of the 13 volumes of The Enzymes (edited by Boyer [12]) supports the argument that calcium specific enzymes are always extracellular in microbes and generally so in mammalian tissues as well. Internal enzymes are either calcium indifferent, calcium inhibited, or require any of several divalent cations. Usually $Mg^{2+}$ or $Mn^{2+}$ are alternatives to calcium.

In general, the literature on cation interactions with enzymes follows a few straightforward rules which fit the functions of calcium in enzymes quite well. Three basic types of complex are recognized [96]: E-S-M, E-M-S, and E$\lessgtr^M_S$ (with E representing the enzyme; S, the substrate; and M, the metal cation).

E-S-M complexes are often less specific in their cation requirement, and these may cover the large variety of enzymes with which $Ca^{2+}$, $Mg^{2+}$, $Mn^{2+}$, or other cations will stimulate activity. Especially with charged (e.g., phosphorylated) substrates, the divalent cation may function to neutralize the substrate and sometimes to fold it into an appropriate conformation compatible with the enzyme active site [96]. Among the many enzymes for which calcium can function in vitro in place of $Mg^{2+}$ are the numerous ATP-kinases where the $Ca^{2+}$-ATP complex can function as substrate instead of the $Mg^{2+}$-ATP complex. These include nucleotide kinases and sugar kinases. Not all kinases will utilize $Ca^{2+}$; for some $Ca^{2+}$ is inhibitory, since the $Ca^{2+}$-ATP complex is not acceptable to the enzyme. In general, these enzymes show a range of effectiveness of the ATP complexes, with the $Mg^{2+}$ complex showing highest activity, followed by $Mn^{2+}$ and then $Ca^{2+}$ (with sometimes $Ba^{2+}$, $Zn^{2+}$, and $Co^{2+}$ complexes also slightly active). An example of how this specificity can vary with related enzymes is that of the arginase kinase of northern hemisphere lobster, which will use $Ca^{2+}$-ATP, whereas that of the southern hemisphere crayfish will not [101].

The membrane-bound $Ca^{2+}/Mg^{2+}$ ATPase found in bacteria [1] is unlikely to be directly associated with the calcium extrusion system already described since it requires either $Ca^{2+}$ or $Mg^{2+}$ for maximum activity in vitro. Also, the evidence is clear that this enzyme functions during oxidative phosphorylation in aerobic organisms [1] and occurs on the inner surface of the cell membrane [47]. Nevertheless, this is a much studied $Ca^{2+}/Mg^{2+}$ enzyme and its cation-specific roles are of further interest. The ATPase binds to the cell membrane, using either $Mg^{2+}$ or $Ca^{2+}$. Consequently washing in low $Mg^{2+}$-$Ca^{2+}$ buffer or with EDTA releases the soluble ATPase (bacterial F1). This complex of several hundred thousand daltons molecular weight (differing somewhat from species to species [1]) can reattach specifically to the original sites on the "stripped" membranes by a process requiring either $Mg^{2+}$ or $Ca^{2+}$. The reassociated ATPase reconstitutes oxidative phosphorylation and other related functions of the enzyme-membrane complex. The solubilized bacterial coupling factor (BF1) in the case of <u>Micrococcus lysodeikticus</u> functions as a $Ca^{2+}$-requiring ATPase; $Mg^{2+}$ is inhibitory with this solubilized unattached enzyme. Of course, the ATP hydrolysis reaction is the reverse of physiologically important ATP synthesis. Nevertheless, this change in specificity for the

ATP hydrolysis reaction is striking.  Similar results are found with the
ATPase from chloroplasts (chloroplast-coupling factor, CF1, involved in
photophosphorylation) from both spinach plants and from the alga Euglena
gracilis.  The membrane-attached ATPase utilizes $Mg^{2+}$ for photophosphorylation and for ATP hydrolysis.  The solubilized, released, ATPase
requires $Ca^{2+}$ for hydrolysis and is in fact inhibited (spinach) or 10 times
less active (Euglena) with $Mg^{2+}$.  The spinach ATPase can function with
$Mg^{2+}$ only in the presence of high levels of carboxylic acids (e.g., 60 mM
maleate).

E-M-S and $E{<}^{S}_{M}$ complexes are more cation specific and include those
enzymes that require calcium uniquely and those $Mg^{2+}$-specific enzymes
for which calcium is strongly inhibitory.  A single example of this latter
class is the phosphoribosylpyrophosphate synthetase, which is a calcium-
sensitive magnesium enzyme that forms a complex of the form M-E-M-S
[142].  The known calcium-specific enzymes are all exoenzymes and, in
addition to mammalian trypsin and fibrin, include many microbial hydro-
lytic enzymes such as proteases, nucleases, and lipases [110].

A thoroughly studied class of $Ca^{2+}$-activated enzymes are the microbial α-amylases.  The B. subtilis enzyme has six $Ca^{2+}$ binding sites per
enzyme molecule [143].  Loosely bound $Ca^{2+}$ (which can be removed by
dialysis against EDTA) does not affect enzyme activity; but removal of
this $Ca^{2+}$ makes the enzyme less stable, so that changes in conformation
(as followed by ORD[4] spectra as well as activity measurements) occur at
pH values above 6 and at high temperatures.  Under these conditions the
enzyme with a full complement of $Ca^{2+}$ is completely stable.  A difference
of 7-8 kcal/mole has been calculated for the calcium stabilization against
heat inactivation [143].  The remaining tightly bound calcium, which is
required for enzyme function, can only be removed with high temperature
and EDTA.  Partial reactivation of the enzyme can be achieved by adding
back $Ca^{2+}$.  The related α-amylase from Bacillus amyloliquefaciens also
utilizes $Ca^{2+}$ for activity and for stabilization.  With B. amyloliquefaciens,
a crystalline $Sr^{2+}$-containing enzyme has been obtained from cells grown
with $Sr^{2+}$ and without $Ca^{2+}$.  This $Sr^{2+}$ enzyme has full activity but is more
pH- and temperature-sensitive [143] than is the $Ca^{2+}$ enzyme.

Aspergillus niger makes two $Ca^{2+}$-α-amylases.  One contains one
$Ca^{2+}$/molecule that can be removed by EDTA with loss of activity.  Partial
restoration of activity is achieved by adding $Ca^{2+}$ back again.  The fungal
Taka-amylase A has 10 g-atoms $Ca^{2+}$ per mole of enzyme [143].  Nine g-
atoms participate in stabilizing the enzyme and are loosely bound and removable by dialysis with EDTA.  The tenth $Ca^{2+}$ is stable to dialysis
against EDTA at $0°$C for 150 hr but is removed during dialysis against
EDTA at $33°$C.  Enzyme activity is regained upon readding $Ca^{2+}$.

There has been a suggestion that calcium may regulate messenger
RNA degradation in vivo as it can in vitro [31].  The intracellular RNAse(s)
that are involved in messenger breakdown are still unknown.  However, there

---

[4] ORD is optical rotary dispersion.

is one extracellular RNAse that requires calcium and that has been studied in such great detail that it needs two separate chapters in The Enzymes [7,30]: the staphylococcal nuclease. This 3'-nucleotidohydrolyase attacks both DNA and RNA and requires 10 mM $Ca^{2+}$ for full activity. Extensive X-ray crystallographic data are available [7] marking the exact position of $Ca^{2+}$ in the ternary complex of $Ca^{2+}$, the enzyme, and a $Ca^{2+}$-requiring inhibitor thymidine-3':5'-diphosphate (pTp). The inhibited complex (although inactive) is stabilized against low pH (below 3.7), proteolytic attack by trypsin, high temperatures ($55°$-$65°C$), and reagents such as urea and guanidinium chloride. In "exchange" reactions, the inhibited $Ca^{2+}$-pTp-enzyme complex has 35 fewer hydrogen atoms exchangeable with solvent hydrogens than were available with the uninhibited enzyme [7]. Calcium thus helps "rigidify" the enzyme [7]. Other interactions with $Ca^{2+}$ are rather complicated, as the enzyme can hydrolyze DNA and RNA and also can function in two modes, i.e., endonucleolytic and exonucleolytic, depending upon the reaction conditions, including pH and $Ca^{2+}$ concentration. Less calcium is required for activity at high pH. The staphylococoal nuclease with $Sr^{2+}$ replacing $Ca^{2+}$ has appreciable DNAse activity but no RNAse activity [7]. Although $Ca^{2+}$ functions both to help bind the substrate to the enzyme, and in a catalytic role in nuclease activity [7], the binding function is not $Ca^{2+}$ specific but the catalytic role is. Calcium affects both the $V_{max}$ (absolutely required) and the $K_m$ (highest affinity at about 10 mM Ca) of the enzyme [7]. At this stage, we should warn the reader that not all nucleases require calcium: The Enzymes, Vol. 4, p. 211, has a table of five fungal RNAses that are neither stimulated nor inhibited by $Ca^{2+}$. The same is true of numerous bacterial nucleases. Our conclusion is not that all extracellular enzymes are calcium enzymes but the converse: that all calcium-requiring enzymes are extracellular.

With regard to lipases, there is no general pattern of responses to calcium. Phospholipase A1 is a $Ca^{2+}$-stimulated enzyme that has been purified from the outer membrane of E. coli [125]. Phospholipase A2 (from snake venom or hog pancreas) requires $Ca^{2+}$ for maximum activity; $Ca^{2+}$ apparently affects enzyme conformation rather than substrate [55]. $Mg^{2+}$, $Mn^{2+}$, or $Cd^{2+}$ can also function here. Phospholipase C from Bacillus cereus and from Clostridium welchii differ in substrate range and are extracellular enzymes, but both are apparently $Mg^{2+}$ activated and $Ca^{2+}$ inhibited. Neuraminidases can be of viral, bacterial, or animal origin. Influenza virus enzyme is unaffected by $Ca^{2+}$; Vibrio cholerae and human plasma enzymes are $Ca^{2+}$ stimulated, but Diplococcus pneumoniae and Clostridum perfringens neuraminidases are neither stimulated nor inhibited by $Ca^{2+}$ [51].

D.  Other Roles of Calcium

There are undoubtedly other specific functions of calcium at the cell surface, and we expect the list of known calcium functions to increase with time.  Three additional bacterial roles for calcium come to mind:  (a) Calcium is known to stabilize or rigidify the membranes of B. subtilis [38] and artificial membranes [69a].  Magnesium can play this role but to a lesser extent, and the monovalent cations not at all [38,69a].  This probably is the basis for use of divalent cations to stabilize osmotically labile protoplast membranes.  By maintaining magnesium internally and calcium externally, cells should introduce a difference in rigidity of the two sides of the membrane lipid bilayer.  Transient calcium fluxes would cause transient changes in membrane fluidity.  (b) A rather different role for calcium is suggested by the requirement for high calcium (30 mM) for DNA uptake during genetic transformation of E. coli [29].  It was postulated that $Ca^{2+}$ is needed to increase cell permeability to DNA [29], and one possible means for this would be localized phospholipase digestion [125] of the outer membrane permeability barrier.  (c) The third additional role for calcium comes from its requirement for bacteriophage growth [2]. This too is an external requirement, and Lanni [82] has shown that it occurs early during the infection cycle.  Penetration of T5 viral DNA into the infected bacterial cell occurs in the absence of calcium, but one of the early subsequent stages, that of stabilization of the infected cells to the physical trauma of blending in a Waring blender, occurs only in the presence of high calcium.  This stabilization process is expected to result from calcium-dependent changes in the structure or composition of the cell membrane.  Although macromolecular synthesis studies defined the limited nature of intracellular RNA and protein synthesis in T5-infected cells with low calcium [102], these studies have not pinned down the basis for the $Ca^{2+}$ requirement.

The isolation of calcium mutants should clarify the functions of calcium in bacterial cells.  Calcium-requiring, calcium-sensitive, and calcium-resistant mutants may be imagined:  the calcium-requiring mutants might point to a critical cellular role for calcium; calcium-sensitive mutants might be defective in the calcium extrusion system; and calcium-resistant mutants might have either high capacity for calcium egress or altered forms of calcium-sensitive components of the cell.  Isolation of mutants has just begun, and establishment of the physiological bases of their altered calcium properties has not been accomplished.  Park et al. [107] found that mutants defective in one of the magnesium transport systems discussed in Chapter 1 were sensitive to calcium.  Since these corA mutants were originally selected for resistance to $Co^{2+}$, a toxic alternative substrate for the magnesium transport system, there was the possibility of a $Mg^{2+}/Ca^{2+}$ antiport mechanism in magnesium uptake [107].  However, direct $Ca^{2+}$ transport studies with these corA mutants have not been done. Furthermore, B. P. Rosen (personal communication) has also isolated calcium-sensitive mutants of E. coli.  Although it is not known whether his mutants are related to those of Park and associates, everted membranes from Rosen's calcium-sensitive strains showed normal $Ca^{2+}$ uptake, suggesting some other basis for calcium sensitivity. Rosen (personal communication) did isolate a strain with unusual calcium transport properties. Everted membranes from this strain showed two to three times higher calcium transport rates than those of wild-type membranes, but the cells were neither calcium sensitive nor calcium resistant. This strain was found during a screening of various mutant stocks for calcium transport activity. The original hypothetical basis for the search was irrelevant since the mutant was found as a "hidden" second mutation in a double-mutant stock and, once separated from the primary mutation by transduction, intact cells of the calcium mutant were phenotypically indistinguishable from wild type. Therefore, as of this writing (1977), calcium mutants are only a promise for the future.

There is another aspect of calcium metabolism of microbes to consider: intracellular calcification by a wide variety of microbial strains [42,141]. Intracellular crystals of hydroxyapatite, $3[Ca_3(PO_4)_2] \cdot Ca(OH)_2$, have been detected by electron microscopy and by X-ray diffraction. The crystals of hydroxyapatite were formed only after prolonged incubation (for example 14 days) with cells that previously grew rapidly for a few hours. The crystals were found only in cells in the process of disintegration [42,141]. It is clear that the calcification process is of little interest to the dead bacterial cells. It is of interest, however, to workers in oral biology, since (a) primarily microbes of oral origin have been studied, and (b) the process is sometimes considered to be related to calcification mechanisms in vertebrate animals [131].

We have been concerned with cellular and subcellular aspects of microbial calcium metabolism. Another level of calcium metabolism by microbes that we have not really considered at all is microbial influences in calcium geochemistry (see Zajic [158]).

## VIII.  NONMICROBIAL MODEL SYSTEMS

Studies of calcium transport and function provide a situation where those of us who work with E. coli must turn to mammalian systems for models to guide our studies. Several of the mammalian systems have advanced to a remarkable state of detailed biochemical understanding. The vitamin-D-dependent calcium-binding protein of the vertebrate intestinal mucosa was implicated in $Ca^{2+}$ transport across the intestinal wall [153]. Its entire amino acid sequence of 85 amino acids including 15 lysine residues has now been determined [65]. Nevertheless, the precise role that the calcium-binding protein plays in calcium movement and even which side of the intestinal cell (mucosal or serosal) the binding protein occurs are still unresolved questions.

There is the calcium transport system in mitochondria (Section VIII. C), which is driven by a proton-motive force, as we believe to be the case in bacteria. There are also directly ATP-coupled $Ca^{2+}$-ATPase systems in mammalian cell membranes (Section VIII.A) and in sarcoplasmic reticulum membranes (Section VIII.B). Although the mechanisms are different from those of microbes, the state of understanding is such that the protein components of the sarcoplasmic reticulum have been isolated and put back into membranes to reconstitute activities of the original membranes. Since we believe that the roles of calcium in and out of cells are generally the same in all cell types, we can explain the purpose of this section of our chapter on microbial calcium by reversing the famous adage and stating, "The elephant, E. coli!"

A.  Red Blood Cells and Membrane Ghosts

From the time of the initial report of energy-dependent calcium extrusion from red blood cells a decade ago [126], it has become clear that such calcium extrusion systems are common to all animal cell plasma membranes [11] and function to maintain a very low intracellular calcium level — perhaps $10^{-6}$ M $Ca^{2+}$ or less [11,62]. The red blood cell calcium transport system has recently been reviewed by Schatzmann, its discoverer, and the following brief description comes from this review [127] and a few other sources. No claim for complete coverage of the nonmicrobial models is intended.

With a mammalian plasma $Ca^{2+}$ level of greater than 1 mM and a cellular calcium level of less than 1 μM (Table 2-1), the problem of an energy-dependent mechanism to maintain the gradient of more than 1,000 arises immediately. A tightly sealed membrane barrier might help; but over the lifetime of a red blood cell, energy input is required for $Ca^{2+}$ egress. The transmembrane potential gradient of only about 10 mV and in the "wrong" direction eliminates a potential source for energy input.

Study of the red blood cell calcium efflux system involved use of human red blood cells lysed in a hypotonic solution containing $^{45}Ca$, a buffer, and Mg-ATP, and then "sealed" by raising the tonicity with KCl or another salt. Surprisingly, about 80% of the lysed cells rapidly became sealed for calcium movements; and the subsequent loss of $^{45}Ca$ from the resealed membranes was shown to be strictly ATP dependent. A $Mg^{2+}$- and $Ca^{2+}$-dependent ATPase activity was implicated in this uphill calcium transport. This ATPase was shown to differ from the better studied $Na^+/K^+$ ATPase both in that it did not require monovalent cations and also in that it was insensitive to cardiac glycosides such as ouabain [127]. Intracellular $Mg^{2+}$ was required as well as intracellular $Ca^{2+}$ for calcium efflux to occur, but $Mg^{2+}$ and $Mn^{2+}$ were not transported out of the membrane ghosts; $Sr^{2+}$ was, however, an alternative substrate for the calcium efflux system in red cell ghosts. Parallel $Ca^{2+}$ and $Mg^{2+}$ measurements ruled out a $Ca^{2+}/Mg^{2+}$ exchange reaction. Intracellular magnesium did not increase during calcium efflux [127]. Complications of membrane binding of $Ca^{2+}$

and $Ca^{2+}$-ATP complex formation prevented determinations of the affinity constant for calcium efflux except with the use of the chelate EGTA. With $Ca^{2+}$-EGTA buffers, the $K_m$ for calcium efflux appeared to be about 4 μM $Ca^{2+}$ [127]. The maximum rate of calcium efflux was 0.15 μmol per liter of packed cell volume per hour, and the maximum rate of efflux was maintained even at outside/inside $Ca^{2+}$ ratios of 700:1, demonstrating the uphill nature of the transport and placing a thermodynamic energy requirement on the system of about 4 kcal/mole. ATP hydrolysis can yield up to 13 kcal per mole of usable free energy [127], giving a maximum possible stoichiometry for calcium efflux of more than 3 $Ca^{2+}$ moved per ATP hydrolysed. Here, the question of stoichiometry is as central to understanding direct coupling ATPases as that of proton/calcium ion ratios are to systems with proton-motive coupling. Although a value of 1 $Ca^{2+}$/ATP had been previously suggested [127], a stoichiometry of 2 $Ca^{2+}$/ATP was determined using $La^{3+}$-inhibited ATPase activity as a measure of the transport ATPase [113]. The $Ca^{2+}$/ATP ratio of 2 would correspond with the similar ratio for sarcoplasmic reticulum calcium transport (Section VIII. B). There have been no studies of the electrogenic or electroneutral nature of red blood cell $Ca^{2+}$ egress.

The kinetic parameters of the $Ca^{2+}$-specific ATPase have been measured to determine whether they are similar to those for calcium movement and again to provide a $Ca^{2+}$/ATP stoichiometry. A $K_m$ of $5 \times 10^{-5}$M for $Mg^{2+}$-ATP was measured for $Ca^{2+}$-dependent ATP hydrolysis. The $K_m$ of $Ca^{2+}$ for ATPase activity was between 1 and 2 μM; $Sr^{2+}$ was about equally as effective as $Ca^{2+}$ [127]. Calcium and magnesium were both shown to function on the same side (i.e., inside) of the membrane as ATP. Inorganic $PO_4^{3-}$ is released intracellularly [127]. $La^{3+}$, used as a specific inhibitor of mitochondrial calcium uptake, inhibited the $Ca^{2+}$-ATPase activity only by half when added at 100 μM externally to resealed red blood cell ghosts; however, $La^{3+}$ completely eliminated calcium efflux [113]. These results suggest that another enzyme with $Ca^{2+}$-ATPase activity is found on the red cell membranes in addition to the $Ca^{2+}$ efflux ATPase and the $Na^+/K^+$ exchange ATPase. The inhibition of calcium efflux by $La^{3+}$ (and similarly by the lanthanides $Ho^{3+}$ and $Pr^{3+}$) occurred from the outside of right-side-out membrane ghosts, that is, on the opposite membrane surface from the high affinity side for $Ca^{2+}$ [127].

The most recent addition in knowledge of the $Ca^{2+}$ efflux ATPase has been the tentative identification of a phosphorylated membrane protein as part of the $Ca^{2+}$ ATPase [75]. This protein has an apparent molecular weight of 150,000 daltons in SDS[5] acrylamide gel electrophoresis, and hydrolysis of bound phosphate by hydroxylamine indicated the presence of an acyl phosphate group. By assuming the addition of 1 $^{32}PO_4^{3-}$ per protein molecule, it was estimated that this protein represented 0.02% of the total red cell ghost protein [75].

---

[5] SDS is sodium dodecylsulfate.

ATP-dependent $Ca^{2+}$ uptake was studied with specially prepared inside-out membrane fragments or vesicles from red blood cells [19, 154, 156]. These inside-out membranes showed a $K_m$ for $^{45}Ca$ uptake of about $100\,\mu M$ and a similar $K_m$ for $Ca^{2+}$ for the activity of the ouabain-resistant $Mg^{2+}/Ca^{2+}$-ATPase thought to be responsible for the calcium transport. Only ATP among purine triphosphates supported $Ca^{2+}$ uptake by inside-out membranes. An advantage of the inside-out vesicles, of course, is that one can vary ATP concentrations and other components at will throughout an experiment, rather than having all internal components beforehand in the sealing solution. The calcium accumulated by everted membranes appeared to be "internal" in that it was not removed by subsequent incubation in the presence of EGTA [19]. Apparent concentration gradients of more than 200 to 1, inside/outside, were established [19]. However, because of the possibility of internal membrane binding [127], one cannot really state the free internal calcium concentration. Ruthenium red inhibited the erythrocyte vesicle calcium uptake system only by about half, as compared to the complete inhibition found with mitochondrial calcium transport (Section VIII. C). Ruthenium red did not inhibit the $Ca^{2+}$-ATPase activity at all with inside-out membrane vesicles [19]. However, ruthenium red was reported completely to inhibit the $Ca^{2+}$-ATPase activity of red cell membranes in earlier studies [154]. The sidedness of the ruthenium red target may account for the different results with different membranes, since a molecule as large and highly charged as ruthenium red, $[(NH_3)_5Ru\text{-}O\text{-}Ru(NH_3)_4\text{-}O\text{-}Ru(NH_3)_5]^{6+}$, would not be expected to pass across the membrane. $La^{3+}$ (100 $\mu M$) slowed the efflux of previously accumulated $^{45}Ca$ and inhibited the $Ca^{2+}$-ATPase activity by about 70% [154]. Concomitant phosphate uptake was not required for prolonged (over an hour) uptake of calcium by red cell membranes [19, 156], and accumulated calcium was gradually lost from membranes when ATP was removed from the system [156]. Presumably internal calcium was free in solution and not in a calcium phosphate precipitate.

Red blood cell calcium efflux has been studied in considerable detail, and the calcium transport system seems characteristic of that found in many other types of intact mammalian cells. Similar calcium efflux is responsible for maintaining low internal calcium in isolated kidney cells [11] and tissue culture cells [11, 80]. Although there are quantitative questions [11] on the relative importance of cell membrane efflux and mitochondrial uptake (see Section VIII. C) in maintaining the low intracellular calcium level, mitochondrial calcium storage can regulate cellular calcium only over brief periods, and there is no obvious mechanism for emptying the cell of its mitochondrial calcium. It remains for the cell membrane $Ca^{2+}$-ATPase system to maintain low intracellular calcium over the long run, just as the cell membrane $Ca^{2+}$ efflux system plays this role in bacterial cells (cf. Sections II and III). The calcium efflux system of mammalian cells functions not only to protect the intracellular milieu by

maintaining a low calcium level, but also in intercellular hormonal communication [116]. Transient increased calcium uptake is always associated with those cellular responses which also involve changes in cyclic
AMP levels. Microtubule contraction can be stimulated by increased
intracellular calcium and "coupled" to serve a variety of intracellular
functions [116] that we will not discuss in this review of microbial calcium
functions.


B.  Sarcoplasmic Reticulum

The sarcoplasmic reticulum is the membranous structure that accumulates
calcium from the proximity of muscle fibers following contraction — resulting in muscle fiber relaxation [46]. The calcium is later released
rapidly in response to a neuromuscular stimulus, resulting in the next
muscle contraction. Aside from its interest in studies of muscle physiology [46], the sarcoplasmic reticulum calcium pump is the best understood
of any calcium transport system in terms of reaction sequences and
molecular components. The sarcoplasmic reticulum contains only three
major membrane proteins; the ATPase constitutes two-thirds of the total
membrane protein. This very simplicity strikes envy in the hearts of
those of us who have looked at complicated inner membrane gels from
bacterial cells! Reviews have appeared frequently on the structure and
function of the sarcoplasmic reticulum [e.g., 90,92]. The following brief
treatment of the mass of available information comes primarily from a
very lucid review [90] by one of the major contributors to our knowledge
of sarcoplasmic reticulum structure and from a few selected additional
and recent papers.
        The isolation of these proteins began only a few years ago [92], but
the proteins of the sarcoplasmic reticulum have already been purified,
characterized, and reconstituted into membrane structures possessing the
calcium transport function [90,115]. The major protein of the membrane
is the ATPase molecule itself. About 65% of the total membrane protein
is found as a polypeptide chain of molecular weight 102,000 daltons, which
can be phosphorylated with $\gamma$-$^{32}$P-labeled ATP to the extent of one $^{32}$P
per polypeptide chain [90]. The phosphate is added to an aspartyl residue.
The purified ATPase binds tightly two calcium ions per polypeptide chain
with a binding constant of about 2 $\mu$M [67], which is similar to the half-
saturation constant for calcium transport. Limited tryptic digestion of
the purified ATPase created two similar sized half molecules that did not
show immunological cross-reactivity. The slightly larger polypeptide retained ATPase activity and also the $^{32}$P of the phosphorylated protein.
This portion of the ATPase appeared to extend out from the surface of the
membrane, as it could be cleaved from intact membranes with trypsin.
"Knob"-like projections seen in electron micrographs of sarcoplasmic
reticulum membranes probably consist of one (or perhaps several, from

size considerations) copies of this half-protein. Such knobs were removed
from the membranes by limited trypsin digestion. The other half of the
ATPase is embedded in the membrane itself. Amino acid analysis showed
a preponderance of hydrophobic amino acid residues in this half-molecule
[90]. The overall composition of the sarcoplasmic reticulum membrane
is 62% protein and 38% lipid by weight. In addition to loosely associated
lipid, the ATPase is surrounded with a "shell" of tightly bound phospholipid
that is required for ATPase activity. This phospholipid can be removed
only with difficulty [104], and it must be reestablished for reconstitution of
membranes with calcium transport function [115].

The two other major proteins of the sarcoplasmic reticulum are known
as the "high-affinity calcium-binding protein" (molecular weight 56,000
daltons) and "calsequestin" (another calcium-binding protein, the molecular
weight of which is somewhat in question but about 50,000 daltons) [90].
Since 12% of the sarcoplasmic reticulum protein is the high-affinity
calcium-binding protein and 19% is calsequestin, the three major proteins
comprise well over 90% of the total protein. There are also small amounts
of low molecular weight acidic proteins and a proteolipid (molecular weight
of only 6,000 daltons, with one or two covalently bound fatty acids). If
present to the extent of 6% of the total protein, this molecule would exist
in a one-to-one ratio with the ATPase. The compositional data suggests
that less than 6% of the protein is present in proteolipid form, but uncer-
tainties associated with membrane proteins require caution. Limited access
to trypsin suggests that both the high-affinity calcium-binding protein and
calsequestin are to be found on the inside of the sarcoplasmic reticulum
membranes, where some granular matrix materials have been visualized
in thin-section electron micrographs. Fractionation of sarcoplasmic
reticulum membranes by density centrifugation results in a preparation of
membranes that microscopy reveals to be empty and that lack the two binding proteins. The high-affinity calcium-binding protein is named for its
tight (3-4 $\mu$M $Ca^{2+}$ dissociation constant) binding of a single $Ca^{2+}$ per polypeptide chain. However, this highly acidic protein also binds an additional
25 $Ca^{2+}$ per molecule with lower affinity. With its presumed internal site
where the ATPase would maintain a high calcium level (estimated at over
10 mM, if it were free $Ca^{2+}$), this protein could function as a calcium
"sink" during the uptake-release cycle [90].

Calsequestin, the third of the major sarcoplasmic reticulum proteins,
is also assigned the role of intramembrane calcium "sink" [90]. Calsequestin binds 43 mol of $Ca^{2+}$ per mole of protein with a dissociation constant that is 50 $\mu$M $Ca^{2+}$ or higher depending upon the ionic conditions.
The calcium is bound relatively specifically, since $Mn^{2+}$ is unable to displace calcium, $Mg^{2+}$ is only partially bound, and $Sr^{2+}$ is bound similarly
to calcium [90]. The binding of calcium is due to the very acidic nature of
the protein, which contains approximately 19% each glutamic and aspartic
acid residues. The protein appears largely as a random coil in the absence

of calcium. Careful measurements have shown increased helical structure and burying of hydrophobic tyrosines as $Ca^{2+}$ associates with calsequestin. The calcium accumulated by sarcoplasmic reticulum is not irreversibly bound, and the addition of the calcium ionophore X537A caused rapid release of accumulated calcium [23, 90].

Not only is the structure of the sarcoplasmic reticulum known in great detail, but also the functioning of the $Ca^{2+}$ pump ATPase has been thoroughly dissected. In this case, the biochemical evidence goes a long way beyond that which loosely defined a "pump." A large series of experiments including $^{18}O$ exchange experiments between HOH and $PO_4^{3-}$ and $^{32}P$ labeling and exchange experiments were carried out by Kanazawa and Boyer [73]. These led to a model for the calcium-ATPase cycle shown in Figure 2-4, and similar models have appeared from other laboratories. Figure 2-4, as we have drawn it, emphasizes the calcium uptake process as a directional sequence. In the original formulation [73], the emphasis was on the reversibility of each of the steps, since $HOH-PO_4^{3-}$ exchange of $^{18}O$ could be measured in both directions in the absence of ATP or calcium. The ATPase function is also reversible: ATP can drive calcium accumulation, or a preformed calcium gradient can drive ATP synthesis [90]. The Kanazawa-Boyer [73] model contains eight distinct steps. Experimental evidence exists for each of these. The first step is the association of two $Ca^{2+}$ ions with the outer surface of the ATPase molecule along with the binding of an ATP molecule (Fig. 2-4). Although $Ca^{2+}$ binding is not required for ATP binding [90, 145], $Ca^{2+}$ is required for ATPase activity and for phosphorylation of the ATPase from ATP [67, 68, 73, 90]. The $Mg^{2+}$ and ADP are released at the exterior of the membrane. The protein remains phosphorylated, and the calcium moves across the membrane or through the embedded portion of the membrane ATPase. The two $Ca^{2+}$ ions are released into the sarcoplasmic reticulum interior [67]. Kanazawa and Boyer [73], suggest that the next step, the conversion of the ATPase from a high-$Ca^{2+}$ affinity/low-$Mg^{2+}$ affinity form to a high-$Mg^{2+}$ affinity/low-$Ca^{2+}$ affinity form is the essential step where the ATP-energized protein drops into an unenergized form by a spontaneous conformational change without dephosphorylation. Prior to this step, the reversibility of the reactions allows a $Ca^{2+}$ gradient to drive ATP synthesis from ADP and $PO_4^{3-}$. Phosphorylation of the enzyme with $^{32}P$-ATP required $Ca^{2+}$ and showed a half-saturation constant of 0.35 $\mu M$ $Ca^{2+}$; $Mg^{2+}$ could replace $Ca^{2+}$, but the $Mg^{2+}$ half-saturation constant was 10.6 mM (giving a calcium-specificity ratio of 30,000:1) [145]. The maximum rate of phosphorylation of the ATPase was about as high with $Sr^{2+}$ as with $Ca^{2+}$, but the $K_m$ for $Sr^{2+}$ in this reaction was 27.5 $\mu M$ $Sr^{2+}$, nearly 100 times higher than that for $Ca^{2+}$ [145]. In measurements of subsequent dephosphorylation of the phosphorylated protein, $Mg^{2+}$ was more effective than calcium with a ratio of 1:2.5 for the half-saturation concentrations [145]. This magnesium is later deposited on the exterior of the membrane along with the $^{32}PO_4^{3-}$ that originated from the ATP [90]. A portion of the reaction cycle was studied as a magnesium-

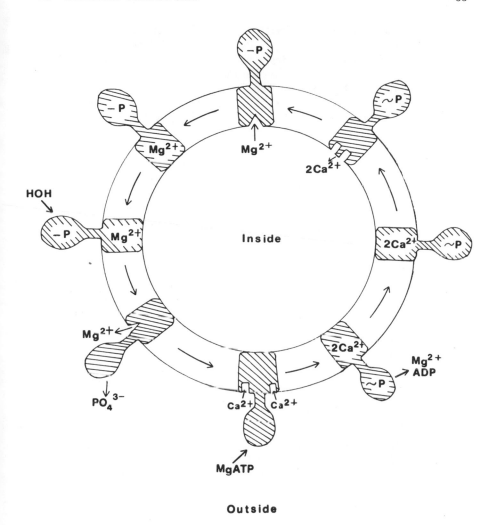

FIG. 2-4  The calcium uptake cycle of the sarcoplasmic reticulum: $\sim P$ represents a high-energy form of the phosphorylated ATPase molecule, and -P represents the comparable low-energy phosphorylated form. (Modified from Kanazawa and Boyer [73] with permission.)

dependent $^{18}O$-labeled HOH-$PO_4^{3-}$ exchange reaction [73], and this was 14 times more rapid than the calcium-dependent ATP hydrolysis.  The $^{18}O$-exchange reaction was in fact inhibited by $Ca^{2+}$, with a half-inhibition constant of 2 $\mu M$, similar to the measures of external calcium binding. Hill plots indicated that 2 $Ca^{2+}$ bind to effect inhibition of the $^{18}O$-exchange reaction [73] and the dephosphorylation of the phosphoprotein [67]. However, similar plotting of the $Ca^{2+}$ requirement for ATPase activity showed a Hill coefficient near 1, indicating a lack of cooperativity in that function [73]; and in similar experiments on the $Ca^{2+}$ requirement for phosphoprotein formation the Hill coefficient was appreciably greater than 2, which was interpreted as evidence for cooperation between different ATPase molecules, each of which bind two $Ca^{2+}$ ions [27]. Over 4 mM $PO_4^{3-}$ was required for half-saturation of the $^{32}PO_4^{3-}$ labeling of the protein and the steady-state level of phosphorylation here was only one-fiftieth of that activated with $^{32}P$-ATP and $Ca^{2+}$ [73]. Kinetic studies of the steps of ATP hydrolysis indicate that the dephosphorylation of the phosphoprotein is the rate-limiting step [27]. Proton-conducting uncouplers were without effect on the $^{18}O$ exchange or on $Ca^{2+}$ accumulation, indicating that the process is not proton driven in the sense that the bacterial accumulation of calcium is. The overall effect of a loop of Figure 2-4 is the movement of 2 $Ca^{2+}$ from outside to inside with only 1 $Mg^{2+}$ moving in the opposite direction. Electrical neutrality is considered likely to be balanced by the movement of 2 $K^+$ from inside to outside by an unspecified mechanism [90].

The sarcoplasmic reticulum calcium pump has been reconstituted starting with nonmembranous ATPase and phospholipids [92, 115, 152]. The process involved dissolving the membranes with deoxycholate and eventual replacement of the deoxycholate with phospholipids. Reconstitution of calcium transport was achieved with the ATPase protein containing less than one of the endogenous 90 lipid molecules per ATPase molecule [152]; the remaining lipid molecules could be replaced by synthetic dioleoyl lecithin with only partial loss of transport activity [152]. The reconstituted membranes were partially leaky, and calcium accumulation by these reconstituted membranes was much higher in the presence of oxalate as an internal trapping agent [152]. Reconstitution of membranes from purified ATPase and lipids derived from sarcoplasmic reticulum resulted in essentially full calcium transport function [95, 152]. The reconstitution experiments establish that the same polypeptide chain functions both as ATPase and as calcium transporter across the sarcoplasmic reticulum membrane.

It has even been possible to obtain the synthesis of a limited amount of ATP without a sealed membrane surface. Knowles and Racker [76] phosphorylated with $^{32}PO_4^{3-}$ leaky membranes made from purified ATPase to the extent of one $^{32}P$ per 2.5 to 5 ATPase molecules. After removal of excess phosphate and the rapid addition of ADP and $Ca^{2+}$, $^{32}P$-ATP was formed with about 80% efficiency and released from the membranes. These membranes were unable to establish a calcium gradient [76] and yet produced ATP from a calcium pulse. The nature of the energy for this ATP synthesis from a $Ca^{2+}$-induced change in protein conformation poses a difficulty in light of our standard ideas of energy coupling [76].

For those of us interested in calcium movement, the sacroplasmic reticulum system provides an ultimate model system where three primary objectives have been met: (1) apparently all of the components have been isolated and characterized; (2) the reaction sequence has been analyzed and subjected to experimental tests of partial reactions; and (3) the essential components have been reassembled to form a reconstituted membrane with transport activity. Although we anticipate that the microbial systems will be proton driven and not directly ATP coupled (cf. Sections II and III), the desired standard of resolution remains the same.

C.  Mitochondria

There are two reasons to be concerned with mitochondrial calcium transport here. First, mitochondria possess a chemiosmotically driven calcium transport system that is similar in many properties to that proposed for bacterial cells. Therefore, understanding of this system may shed light or offer models for bacterial studies. However, there is an important difference between the mitochondrial and bacterial coupling of the proton-motive force to calcium movement (Fig. 2-5). With mitochondria, the direction of calcium movement is opposite that of energy-dependent proton movement, and therefore the calcium flows into the mitochondria electrogenically uniport down the potential gradient [86,122]. With bacteria (Fig. 2-5), the energy-dependent calcium and proton movements are in the same direction, and therefore electroneutral movement of calcium antiport with protons might be driven by the pH gradient (Section III). Secondly, mitochondrial calcium accumulation dominates the intracellular calcium economy of most mammalian cells [11,26,137]. Red blood cells (Section VIII.A) are the exception since they lack mitochondria. Internal calcium levels with red blood cells are determined by

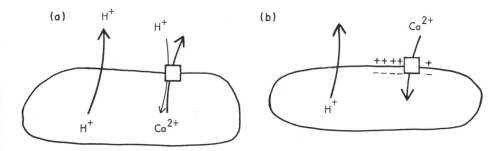

FIG. 2-5  Difference in orientation of (a) bacterial and (b) mitochondrial calcium transport systems.

the rate of inward leakage and the rate of functioning of the membrane calcium efflux system.  With other mammalian cells, however, the mitochondria provide an internal calcium "trap" [11] whose high rate of entry regulates the stable low calcium level in the cytoplasm.  The mitochondria must over long times, of course, empty their calcium store into the cytoplasm where it can be pumped back across the cytoplasmic membrane by the calcium transport ATPase [11,137].

Studies of mitochondrial calcium accumulation are not as extensive as those with sarcoplasmic reticulum.  A very entertaining and provocative lecture given in 1970 by Lehninger [85] provides the best introduction to mitochondrial calcium transport, and relatively little has changed conceptually since that time.  Calcium transport by mitochondria is unlike that of most mitochondrial transport systems that mediate thermodynamically passive exchanges driven solely by concentration gradients of metabolic intermediates.  Calcium transport is uphill against a concentration gradient and therefore requires energy that can be derived either from ATP hydrolysis or from oxidation of respiratory chain substrates.  The overall stoichiometry of calcium uptake is generally but not always 2 $Ca^{2+}$ per ATP molecule hydrolysed.  This ratio is identical to that of the sarcoplasmic reticulum system (although the mechanism is different).  Alternatively, 6 $Ca^{2+}$ are accumulated during the passage of a pair of electrons from NADH to oxygen, which according to chemiosmotic theory leads to the expulsion of 6 $H^+$ from the mitochondria.  The $Ca^{2+}/H^+$ stoichiometry is thus 1:1.  Continued calcium uptake requires the concomitant accumulation of a permeant anion.  During NADH-driven calcium uptake no ATP synthesis occurs, and calcium uptake is therefore a preferred alternative to oxidative phosphorylation in mitochondria.  This explains the early finding that calcium "uncouples" oxidative phosphorylation from respiration.

Binding and uptake experiments have defined the sites and steps of calcium transport in mitochondria.  The first stage appears to be binding to a small number (less than 1 nmol per milligram of protein) of membrane sites with a high affinity ($K_m$ less than 1 $\mu M$) [85].  This binding is specific for $Ca^{2+}$ or $Sr^{2+}$, and to a lesser degree $Mn^{2+}$; it is inhibited by $La^{3+}$ and ruthenium red [85-87].  Although this binding is independent of respiration, the high-affinity bound calcium can be released by uncouplers [87].  Upon the addition of a respiratory substrate, about 60 nmol of $Ca^{2+}$ per milligram of protein are accumulated by the mitochondria with high specificity and sensitivity again to the $La^{3+}$ and ruthenium red.  Although the calcium is retained by a membrane potential and can be released by uncouplers and ionophores that alter the potential [122], calcium is not totally free internally since it is released during the eversion of mitochondrial membranes by sonication and by alternative means [85,86].  The assumption that the internal calcium is osmotically free leads to an estimate of a 5,000:1 inside/outside ratio of calcium concentrations [122], which is much higher than the less than 100:1 ratio for $K^+$.  Both of these equilibrium ratios changed in parallel with nigericin, $K^+$-valinomycin, or uncouplers, which led to the conclusion that both $K^+$ were being retained simply in response to the membrane Nernst potential [25],

$$-\Delta\psi = \frac{2.3RT}{F} \ln \frac{[K^+] \text{ in}}{[K^+] \text{ out}} = \frac{2.3RT}{F} \ln \frac{[Ca^{2+}] \text{ in}}{[Ca^{2+}] \text{ out}}$$

($\Delta\psi$ is the transmembrane potential; R, the gas constant; T, absolute temperature; F, the Faraday constant)

and that the $Ca^{2+}$ ratio of 5,000:1 occurred with a transmembrane potential of -110 mV [122].

Only in the presence of a permeant anion such as $HPO_4^{2-}$ [61] does still further calcium uptake occur. With concomitant anion uptake, "massive loading" of mitochondria takes place so that up to 25% of the dry weight of the loaded mitochondria can be intramembrane precipitates of calcium phosphate that can be visualized in electron microscopy sections. This corresponds to 3 $\mu$mol $Ca^{2+}$ per milligram of protein. The calcium accumulation with a permeant anion showed the same specificity and inhibition patterns as did low-affinity and high-affinity binding. Some of the massively loaded calcium was osmotically free in solution, since swelling of the mitochondria was measured by light scattering. Calcium was not irreversibly precipitated, since uncouplers caused rapid efflux. When mitochondrial membranes were everted by sonication, this calcium was released into solution in the absence of uncouplers.

Most mitochondria have calcium uptake systems [85], including those of birds, amphibians, reptiles, mammals, and those from most invertebrates and plants. Exceptions are the mitochondria isolated from Neurospora, some yeasts, and those from the blowfly flight muscle [85,87]. Reasons for this difference are not known. Mitochondria from amoeba have a calcium transport system [85]. It would be interesting to try to correlate evolution of cellular function of the mitochondrial calcium system with these findings. In red muscles, which have many mitochondria but little sarcoplasmic reticulum, it has been proposed that the mitochondria play the primary role in calcium uptake (relaxation) and calcium release (contraction) in the muscle cycle. Roles for mitochondria in calcification processes such as bone formation have also been advanced [85], and indeed mitochondria from calcifying chondrocytes have an unusually high calcium-storage capacity [133].

Some progress has been made in characterization of a presumptive calcium-carrier protein from mitochondria. A high-affinity $Ca^{2+}$ binding activity was extracted from mitochondria by osmotic shock [85], a procedure analogous to that used to extract binding proteins involved in bacterial transport systems [9]. The binding activity belonged to a glycolipoprotein of about 100,000-dalton molecular weight that contains a 67,000-dalton polypeptide plus 27% of its weight as phospholipid and 12% carbohydrate [86]. A similar $Ca^{2+}$-binding protein was purified by alternative means by Prestipino and coworkers [112], and this preparation has been shown to increase conductivity of lipid bilayers in a $Ca^{2+}$-dependent association. The conductance increase was cation specific, like the transport system, and inhibited by ruthenium red [112]. These studies did not show a $Ca^{2+}$ transport function in the bilayers but only a calcium-dependent increase in conductivity. The calcium could be required for glycoprotein adsorption to the bilayer, and the increased conductivity might be due to disruption of the normal bilayer structure. There is still much room for doubt as to whether the mitochondrial $Ca^{2+}$-binding glycoprotein is the ligand-binding component of the calcium transport system. Another type of soluble calcium-binding protein was isolated and purified earlier from intestinal tissue [65,153], but its role in intestinal calcium transport is even less well established than that of the mitochondrial protein.

A recent source of difficulty in studies of mitochondrial calcium transport concerns polarity. Not only can intact mitochondria accumulate calcium, dependent on respiratory energy, but the everted submitochondrial membranes customarily prepared by sonication also accumulate calcium (e.g., see Refs. 86 and 109 for discussion of this). Perhaps the mitochondriacs share with microbiologists the problem of mixed populations of right-side-out and inside-out membranes. Alternatively, one might postulate a $Ca^{2+}/H^+$ antiport system which normally functions in calcium efflux in addition to the electrogenic uniport system for uptake [122]. In everted vesicles, the antiport system could bring about energy-dependent calcium accumulation by the same mechanism as that responsible for calcium efflux in bacterial cells (cf. Section II).

Two specific types of inhibitors have been introduced with mitochondrial calcium studies; both are supposed to bind with high affinity and to inactivate the calcium-binding sites. Lanthanides including $La^{3+}$, $Eu^{3+}$, $Tb^{3+}$, $Nd^{3+}$, and $Pr^{3+}$ constitute one class [87,117], and the hexavalent dye ruthenium red is the other [100,109,150]. $La^{3+}$ binds specifically with a $K_m$ below $10^{-7}$ M and reversibly to the high-affinity calcium-binding sites of mitochondria [87,117]. When added prior to $^{45}Ca$, $La^{3+}$ specifically inhibits calcium binding and accumulation. Ruthenium red also binds to the high affinity calcium-binding sites, this time noncompetitively with $Ca^{2+}$ [117,150]. Although the specificity and kinetic parameters of the interactions of lanthanides and ruthenium red with the mitochondrial calcium transport system have not been fully established, these materials have now been thoroughly studied and are widely used. Workers with bacterial [136] and fungal [20] calcium transport systems are just beginning to use them.

D.  Ionophores

Since we have no evidence for ionophores as components of microbial calcium transport systems, we will limit this section to studies with ionophores and calcium transport in nonmicrobial systems and a brief discussion of ionophores as hypothetical cation carriers. The calcium binding factor found by Bronner and his associates [14,15] may indeed be a calcium ionophore, but it has never been tested in this regard.

An ionophore is a lipid-soluble small molecule that functions to transport an ion across otherwise impermeable membranes. The two most thoroughly studied ionophores function with monovalent cations but in different ways: valinomycin acts as a mobile carrier bringing $K^+$ ions across membranes, whereas gramicidin forms a channel through which cations can pass across membranes. Gramicidin and valinomycin are antibiotics. With regard to calcium we need to consider two antibiotic ionophores that transfer calcium: X537A [ Lasol acid (Hoffman-La Roche)] and A23187 (Eli Lilly and Company). In addition, we should consider recent and tentative reports of the isolation of calcium ionophores from both mitochondria [10,10a] and from sarcoplasmic reticulum [132].

Many antibiotics are produced by species of <u>Streptomyces</u>, and A23187 is synthesized by <u>Streptomyces chartreusensis</u>. We do not know the organism that produces X537A. The structure of X537A is shown in Figure 2-6 along with that of another calcium-binding antibiotic, beauvericin [120]. Both X537A and A23187 are carboxylic acid antibiotics of molecular weight 500-600 daltons. The elemental analysis of A23187 gives a composition of $C_{29}H_{37}N_3O_6$ [118], but no structure has been published to our knowledge. X537A differs from A23187 in composition, as the former contains no nitrogen and somewhat more carbon and oxygen. X537A has been crystallized in a 2:1 neutral "sandwich" complex with $Ba^{2+}$ [111]. The other $Ca^{2+}$ ionophore, beauvericin, a neutral cyclic hexadepsipeptide containing alternating D-hydroxyisovaleric acid and N-methyl-L-phenylalanine residues has been crystallized and shown to complex and mobilize $Ca^{2+}$ and $Ba^{2+}$ as well as monovalent cations [120]. Thus a variety of structures can function as divalent cation ionophores. In experiments measuring the ability of the ionophores to extract divalent cations out of aqueous solutions into 30% butanol-70% toluene solutions, X537A shows an affinity series of $Ba^{2+} > Sr^{2+} > Ca^{2+} > Mg^{2+}$ [111] whereas A23187 shows a reverse affinity order of $Mg^{2+} > Ca^{2+}$ [118]. The order of cation selectivity ($Mg^{2+} > Ca^{2+} > Sr^{2+} > Ba^{2+}$) in altering lipid bilayer conductivity with X537A is different because there are two steps required for this process [111]. The first stage of cation binding to the membrane embedded ionophore shows the $Ca^{2+} > Mg^{2+}$ sequence; but unless the cation is readily released on the opposite membrane face the increase in conductivity will be less. The tighter binding cations are less effectively mobilized because of the tight binding. Such model studies with lipid bilayers and bulk phase organic solvents also have been of great use with monovalent cation ionophores.

In addition to providing a working model for one class of hypothetical calcium carrier, the calcium ionophores also have been widely utilized to alter flux rates across cellular and mitochondrial membranes in studies of calcium transport and function. Reed and Lardy [118] used A23187 in

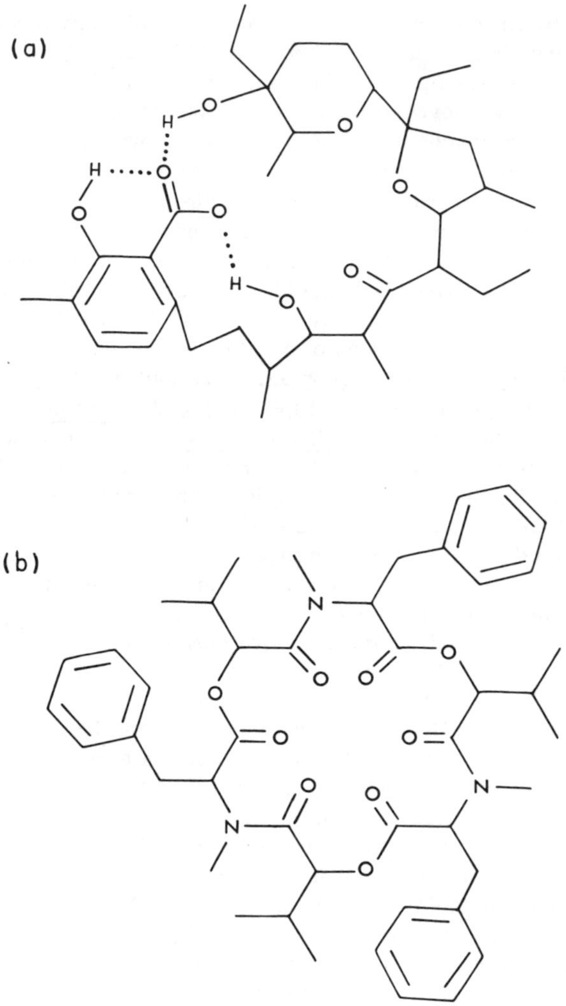

FIG. 2-6   Structures of calcium ionophores: (a) X537A; (b) beauvericin.

studies of mitochondrial respiration and oxidative phosphorylation. Caswell and Pressman [23] used X537A and A23187 in studies of calcium release with sarcoplasmic reticulum.   Release of $Ca^{2+}$ was first order with respect to ionophore concentration, thus showing that a 1:1 complex rather than a 2:1 sandwich was involved [23].   A23187 was approximately 60 times more potent on a molar basis than was X537A in stimulating calcium release from sarcoplasmic reticulum membrane.   However,

A23187 was less than 1/100 as effective as X537A in extracting $Ca^{2+}$ from an aqueous phase into butanol/toluene [23]. The extraction should show a 2:1 ionophore/$Ca^{2+}$ stoichiometry to obtain charge neutrality of the complex. Transport across the membrane appeared to be governed by complexation between cation and ionophore at the membrane surfaces rather than by the bulk solubility of the complexes in the membrane [23]. These ionophores have also come into use in a wide variety of physiological studies with animal cells and tissues, a few of which were noted by Pressman [111]. Closer to our microbial interests, A23187 inhibited cell division in yeast by altering the quantity of cellular magnesium and calcium [36]. Both X537A and A23187 induced cell fusion in red blood cells [4]. We are unaware of ionophores being used as yet in studies of bacterial calcium transport or function.

Although the antibiotic ionophores are useful both as models and as reagents in studies of calcium movement, low-molecular-weight ionophores may not function as part of "normal" calcium transport systems. In fact, it is difficult to imagine the tightly controlled regulation that is characteristic of cellular or organelle transport systems functioning with diffusion of mobile carriers through the membrane lipid bilayer regions. Much more likely, the calcium carriers are part of or are embedded in transmembrane protein structure and, in these proteins, the carriers are more apt to form channels than to be freely mobile. There have been two recent attempts to isolate calcium ionophores from materials with calcium transport function. Shamoo and MacLennan [132] isolated a material from sarcoplasmic reticulum membranes that showed a $Ba^{2+} > Ca^{2+} > Sr^{2+} > Mg^{2+}$ divalent cation specificity sequence in increasing conductivity of lipid bilayers. Succinylated (to make it water soluble) sarcoplasmic reticulum ATPase increased $Ca^{2+}$-dependent conductivity of bilayers. Trypsin digestion of the sarcoplasmic reticulum membranes released a small peptide that increased $Ca^{2+}$ conductance [132]. Blondin [10] partially purified a calcium ionophore released by trypsin from mitochondria (by a factor of 2,000 with respect to mitochondrial protein). This material facilitated calcium (and magnesium) movement across both mitochondrial membranes and chloroform layers [10]. In more recent studies [10a], the ionophorous material appeared not to be peptide in nature but rather lipid. It is still not known whether the natural cation channels across biological membranes are formed by the amino acids within membrane-embedded proteins or by the lipids of the membranes themselves. Although the sarcoplasmic reticulum activity is water soluble [132], the ionophore from mitochondria is both water insoluble and organic-solvent soluble [10], as are the antibiotic ionophores A23187 and X537A.

While the results just described with both antibiotic and with endogenous ionophores are encouraging in that such studies may lead to an understanding of the nature of the calcium carriers, much further work needs to be done. Most workers in the calcium transport field (except for the isolators of ionophores) remain quite skeptical as to whether the ionophorous materials released by trypsin digestion are parts of the natural calcium carriers rather than artifacts generated by the extraction procedures.

## IX. SUMMARY

This chapter is a first effort at reviewing calcium transport and, to a lesser extent, calcium function in microorganisms. As such it lacks any obvious and published precursor review. We have tried to be comprehensive and to place essentially all of what is known about microbial calcium metabolism into a single conceptual framework. To the degree that we have emphasized mechanism and function, we have made the models more explicit than the available data generally justifies and should add this caveat for the reader. We have tried to force all microbial cell calcium metabolism (and that of eucaryotic cells and membranes as well) into a minimal number of proposed mechanisms, and tomorrow's experiments may prove us wrong. However, Occam's tenet allows us to rationalize data that have not been previously organized.

We propose that all cells, including all microbial cells, minimize their internal calcium levels by means of membrane calcium transport systems oriented for energy-dependent calcium egress. Therefore, functions of calcium such as enzymatic activation and cell wall stabilization should be extracellular. The calcium efflux in bacteria is driven by the Mitchell chemiosmotic potential, presumably as a 2:1 $Ca^{2+}$/proton antiport; the mechanism is different in animal cells. Although the specificity and energy requirement of the bacterial transport system have been demonstrated, the nature of the carrier molecules remains for the future. In specialized cases such as bacterial sporulation (and mammalian sarcoplasmic reticulum), the orientation of calcium movement can be from outside to inside, or the opposite of the normal cellular situation. This calcium movement is again substrate specific and carrier mediated. The relationship of such reverse transport to normal cellular transport mechanisms remains to be determined.

## ACKNOWLEDGMENTS

The research in our laboratory and the writing of this review has been supported by National Science Foundation grant BMS71-01456 and National Institutes of Health grant AI08062. Many colleagues in these studies both within our laboratory (especially P. Bhattacharyya, D. Clark, E. Eisenstadt, and K. Toth) and in other laboratories (notably E. Barnes, F. Bronner, E. Golub, and B. P. Rosen) have contributed to the development of ideas by frequent and easy exchange of both data and hypotheses.

Drs. Bronner and Rosen provided detailed critiques of an earlier draft of this manuscript. Kathleen Farrelly drew the figures and watched over the bibliography.

# REFERENCES

1.  A. Abrams and J. B. Smith, Bacterial membrane ATPase. In The Enzymes, 3rd ed. (P. D. Boyer, ed.), Vol. 10, Academic Press, New York, 1974, pp. 395-429.
2.  M. H. Adams, Bacteriophages. Wiley (Interscience), New York, 1959.
3.  J. Adler, Chemotaxis in bacteria. Ann. Rev. Biochem., 44, 341-356 (1975).
4.  Q. F. Ahkong, W. Tampion, and J. A. Lucy, Promotion of cell fusion by divalent cation ionophores. Nature, 256, 208-209 (1975).
5.  K. H. Altendorf and L. A. Staehelin, Orientation of membrane vesicles from Escherichia coli as detected by freeze-cleave electron microscopy. J. Bacteriol., 117, 888-899 (1974).
6.  A. J. Andreoli, J. Saranto, P. A. Baecker, S. Suehiro, E. Escamilla, and A. Steiner, Biochemical properties of forespores isolated from Bacillus cereus. In Spores VI (P. Gerhardt, R. N. Costilow, and H. L. Sadoff, eds.), American Society for Microbiology, Washington, D.C., 1975, pp. 418-424.
7.  C. B. Anfinsen, P. Cuatrecasas, and H. Taniuchi, Staphylococcal nuclease, chemical properties and catalysis. In The Enzymes, 3rd ed. (P. D. Boyer, ed.), Vol. 4, Academic Press, New York, 1971, pp. 177-204.
8.  E. A. Berger and L. A. Heppel, Different mechanisms of energy coupling for the shock-sensitive and shock-resistant amino acid permeases of Escherichia coli. J. Biol. Chem., 249, 7747-7755 (1974).
9.  P. Bhattacharyya and E. M. Barnes, Jr., ATP-dependent calcium transport in isolated membrane vesicles from Azotobacter vinelandii. J. Biol. Chem., 251, 5614-5619 (1976).
10. G. A. Blondin, Isolation of a divalent cation ionophore from beef heart mitochondria. Biochem. Biophys. Res. Commun., 56, 97-105 (1974).
10a. G. A. Blondin, Isolation, properties, and structural features of divalent cation ionophores derived from beef heart mitochondria. Ann. N.Y. Acad. Sci., 264, 98-111 (1975).
11. A. B. Borle, Calcium metabolism at the cellular level. Fed. Proc., 32, 1944-1950 (1973).
12. P. D. Boyer (ed.), The Enzymes, 3rd ed., 13 vols., Academic Press, New York, 1970-1976.

13.  F. Bronner, F. Botnick, and T. S. Freund, Calcium transport in
     Bacillus megaterium. Israel J. Med. Sci., 7, 1224-1229 (1971).
14.  F. Bronner and T. S. Freund, Calcium accumulation during sporula-
     tion of Bacillus megaterium. In Spores V (H. O. Halvorson,
     R. Hanson, and L. L. Campbell, eds.), American Society for Micro-
     biology, Washington, D.C., 1972, pp. 187-190.
15.  F. Bronner, W. C. Nash, and E. E. Golub, Calcium transport in
     Bacillus megaterium. In Spores VI (P. Gerhardt, R. N. Costilow,
     and H. L. Sadoff, eds.), American Society for Microbiology,
     Washington, D.C., 1975, pp. 356-361.
16.  J. L. Browning, D. L. Nelson, and H. G. Hansma, $Ca^{2+}$ influx across
     the excitable membrane of behavioural mutants of Paramecium.
     Nature, 259, 491-494 (1976).
17.  J. L. Browning and D. L. Nelson, Biochemical studies of the excit-
     able membrane of Paramecium aurelia I. $^{45}Ca^{++}$ fluxes across
     resting and excited membrane. Biochim. Biophys. Acta, 448, 338-
     351 (1976).
18.  R. R. Brubaker and M. J. Surgalla, The effect of $Ca^{++}$ and $Mg^{++}$ on
     lysis, growth, and production of virulence antigens by Pasteurella
     pestis. J. Infec. Dis., 114, 13-25 (1964).
19.  J. T. Buckley, Calcium ion transport by pig erythrocyte membrane
     vesicles. Biochem. J., 142, 521-526 (1974).
20.  L. E. Cameron and H. B. LéJohn, $Ca^{2+}$ is a specific regulator of
     amino acid transport and protein synthesis in the water-mould
     Achyla. Biochem. Biophys. Res. Commun., 48, 181-189 (1972).
21.  L. E. Cameron and H. B. LéJohn, On the involvement of calcium in
     amino acid transport and growth of the fungus Achlya. J. Biol. Chem.,
     247, 4729-4739 (1972).
22.  E. Carafoli, The interaction of $Ca^{2+}$ with mitochondria, with special
     reference to the structural role of $Ca^{2+}$ in mitochondrial and other
     membranes. Mol. Cell. Biochem., 8, 133-140 (1975).
23.  A. H. Caswell and B. C. Pressman, Kinetics of transport of divalent
     cations across sarcoplasmic reticulum vesicles induced by ionophores.
     Biochem. Biophys. Res. Commun., 49, 292-298 (1972).
24.  J. R. Chipley and H. M. Edwards, Jr., Cationic uptake and exchange
     in Salmonella enteritidis. Can. J. Microbiol., 18, 509-513 (1972).
25.  V. P. Cirillo, Membrane potentials and permeability. Bacteriol.
     Rev., 30, 68-79 (1966).
26.  A. Cittadini, A. Scarpa, and B. Chance, Calcium transport in intact
     Ehrlich ascites tumor cells. Biochim. Biophys. Acta, 291,
     246-259 (1973).
27.  R. L. Coffey, E. Lagwinska, M. Oliver, and A. Martonosi, The
     mechanism of ATP hydrolysis by sarcoplasmic reticulum. Arch.
     Biochem. Biophys., 170, 37-48 (1975).

28. W. J. Cook and C. E. Bugg, Calcium-carbohydrate bridges composed of uncharged sugars. Structure of a hydrated calcium bromide complex of α-fucose. Biochim. Biophys. Acta, 389, 428-435 (1975).

29. S. D. Cosloy and M. Oishi, The nature of the transformation process in Escherichia coli K12. Mol. Gen. Genet., 124, 1-10 (1973).

30. F. A. Cotton and E. E. Hazen, Jr., Staphylococcal nuclease X-ray structure. In The Enzymes, 3rd ed. (P. D. Boyer, ed.), Vol. 4, Academic Press, New York, 1971, pp. 153-176.

31. K. Cremer and D. Schlessinger, Ca$^{2+}$ ions inhibit messenger ribonucleic acid degradation, but permit messenger ribonucleic acid transcription and translation in deoxyribonucleic acid-coupled systems from Escherichia coli. J. Biol. Chem., 249, 4730-4736 (1974).

32. W. H. Crosby, R. A. Greene, and R. A. Slepecky, The relationship of metal content to dormancy, germination and sporulation in Bacillus megaterium. In Spore Research 1971 (A. N. Barker, G. W. Gould, and J. Wolf, eds.), Academic Press, New York, 1971, pp. 143-160.

33. M. Dixon and E. C. Webb, Enzymes, 2nd ed., Academic Press, New York, 1964.

34. G. J. Dring and G. W. Gould, Sequence of events during rapid germination of spores of Bacillus cereus. J. Gen. Microbiol., 65, 101-104 (1971).

35. G. J. Dring and G. W. Gould, Reimposition of heat-resistance on germinated spores of Bacillus cereus by osmotic manipulation. Biochem. Biophys. Res. Commun., 66, 202-208 (1975).

36. J. H. Duffus and L. J. Patterson, Control of cell division in yeast using the ionophore, A23187 with calcium and magnesium. Nature, 251, 626-627 (1974).

37. R. Eckert, Bioelectric control of ciliary activity. Science, 176, 473-481 (1972).

38. M. Ehrström, L. E. G. Eriksson, J. Israelachvili, and A. Ehrenberg, The effects of some cations and anions on spin labeled cytoplasmic membranes of Bacillus subtilis. Biochem. Biophys. Res. Commun., 55, 396-402 (1973).

39. E. Eisenstadt and S. Silver, Calcium transport during sporulation in Bacillus subtilis. In Spores V (H. O. Halvorson, R. Hanson and L. L. Campbell, eds.), American Society for Microbiology, Washington, D.C., 1972, pp. 180-186.

40. E. Eisenstadt and S. Silver, Restoration of cation transport during germination. In Spores V (H. O. Halvorson, R. Hanson, and L. L. Campbell, eds.), American Society for Microbiology, Washington, D.C., 1972, pp. 443-448.

41. D. J. Ellar, M. W. Eaton, C. Hogarth, B. J. Wilkinson, J. Deans, and J. La Nauze, Comparative biochemistry and function of forespore and mother-cell compartments during sporulation of Bacillus megaterium cells. In Spores VI (P. Gerhardt, R. N. Costilow, and H. L. Sadoff, eds.), American Society for Microbiology, Washington, D.C., 1975, pp. 425-433.

42. J. Ennever, J. J. Vogel, and J. L. Streckfuss, Calcification by Escherichia coli. J. Bacteriol., 119, 1061-1062 (1974).
43. W. Epstein, Membrane transport. In Biochemistry of Cell Walls and Membranes (C. F. Fox, ed.), Butterworths, London, 1975, pp. 249-278.
44. P. C. Fitz-James, Formation of protoplasts from resting spores. J. Bacteriol., 105, 1119-1136 (1971).
45. E. Frieden, The chemical elements of life. Sci. Amer., 227, 52-60 (1972).
46. F. Fuchs, Striated muscle. Ann. Rev. Physiol., 36, 461-502 (1974).
47. M. Futai, Orientation of membrane vesicles from Escherichia coli prepared by different procedures. J. Membrane Biol., 15, 15-28 (1974).
48. H. Gest, Metabolic patterns in photosynthetic bacteria. Bacteriol. Rev., 15, 183-210 (1951).
49. E. E. Golub and F. Bronner, Bacterial calcium transport: energy-dependent calcium uptake by membrane vesicles from Bacillus megaterium. J. Bacteriol., 119, 840-843 (1974).
50. E. E. Golub, W. C. Nash, L. S. Cutler, and F. Bronner, Bacterial calcium transport: the membrane-bound calcium efflux system of Bacillus megaterium. Manuscript submitted (1976).
51. A. Gottschalk and A. S. Bhargava, Neuraminidases. In The Enzymes, 3rd ed. (P. D. Boyer, ed.), Vol. 5, Academic Press, New York, 1971, pp. 321-342.
52. G. W. Gould and G. J. Dring, Heat resistance of bacterial endospores and concept of an expanded osmoregulatory cortex. Nature, 258, 402-405 (1975).
53. N. Grecz, T. Tang, and K. S. Rajan, Relation of metal chelate stability to heat resistance of bacterial spores. In Spores V (H. O. Halvorson, R. Hanson, and L. L. Campbell, eds.), American Society for Microbiology, Washington, D.C., 1972, pp. 53-60.
54. P. J. Hall, G. C. H. Yang, R. V. Little, and R. R. Brubaker, Effect of $Ca^{2+}$ on morphology and division of Yersinia pestis. Infect. Immun., 9, 1105-1113 (1974).
55. D. J. Hanahan, Phospholipases. In The Enzymes, 3rd ed. (P. D. Boyer, ed.), Vol. 5, Academic Press, New York, 1971, pp. 71-85.
56. R. S. Hanson, M. V. Curry, J. V. Garner, and H. O. Halvorson, Mutants of Bacillus cereus strain T that produce thermoresistant spores lacking dipicolinate and have low levels of calcium. Can. J. Microbiol., 18, 1139-1143 (1972).
57. F. M. Harold, Conservation and transformation of energy by bacterial membranes. Bacteriol. Rev., 36, 172-230 (1972).
58. F. M. Harold and K. Altendorf, Cation transport in bacteria: $K^+$, $Na^+$, and $H^+$. In Current Topics in Membranes and Transport (F. Bronner and A. Kleinzeller, eds.), Vol. 5, Academic Press, New York, 1974, pp. 1-50.

59. F. M. Harold, J. R. Baarda, and E. Pavlasova, Extrusion of sodium and hydrogen ions as the primary process in potassium ion accumulation by Streptococcus faecalis. J. Bacteriol., 101, 152-159 (1970).
60. F. M. Harold and E. Levin, Lactic acid translocation: terminal step in glycolysis by Streptococcus faecalis. J. Bacteriol., 117, 1141-1148 (1974).
61. E. J. Harris, J. M. Wimhurst, and I. Landaeta, Ionic and non-ionic mitochondrial phosphate in relation to $Ca^{2+}$ and arsenate accumulation. Eur. J. Biochem., 45, 561-565 (1974).
62. D. G. Harrison and C. Long, The calcium content of human erythrocytes. J. Physiol., 199, 367-381 (1968).
63. A. Haug and B. Larsen, Biosynthesis of alginate, part 2. Polymannuronic acid C-5-epimerase from Azotobacter vinelandii (Lipman). Carbohyd. Res., 17, 297-308 (1971).
64. A. Haug, B. Larsen, and O. Smidsrød, Uronic acid sequence in alginate from different sources. Carbohyd. Res., 32, 217-225 (1974).
65. W.-Y. Huang, D. V. Cohn, J. W. Hamilton, C. Fullmer, and R. H. Wasserman, Calcium-binding protein of bovine intestine. The complete amino acid sequence. J. Biol. Chem., 250, 7647-7655 (1975).
66. B. Humphrey and J. M. Vincent, Calcium in cell walls of Rhizobium trifolii. J. Gen. Microbiol., 29, 557-561 (1962).
67. N. Ikemoto, Transport and inhibitory $Ca^{2+}$ binding sites on the ATPase enzyme isolated from the sarcoplasmic reticulum. J. Biol. Chem., 250, 7219-7224 (1975).
68. G. Inesi, S. Blanchet, and D. Williams, ATPase and ATP binding sites in the sarcoplasmic reticulum membrane. In Organization of Energy-Transducing Membranes (M. Nakao and L. Packer, eds.), University Park Press, Baltimore, Maryland, 1973, pp. 93-105.
69. P. L. P. Jasper, Cation transport systems of Rhodopseudomonas capsulata. Ph.D. Thesis, Washington University, St. Louis, Missouri, 1975.
69a. T. Ito and S.-I. Ohnishi, $Ca^{2+}$-induced lateral phase separations in phosphatidic acid-phosphatidylcholine membranes. Biochim. Biophys. Acta, 352, 29-37 (1974).
70. H. R. Kaback, Bacterial membranes. In Methods in Enzymology (W. B. Jakoby, ed.), Vol. 22, Academic Press, New York, 1971, pp. 99-120.
71. H. R. Kaback, Transport studies in bacterial membrane vesicles. Science, 186, 882-892 (1974).
72. H. R. Kaback and J.-s. Hong, Membranes and transport. CRC Critical Rev. Microbiol., 2, 333-376 (1973).
73. T. Kanazawa and P. D. Boyer, Occurrence and characteristics of a rapid exchange of phosphate oxygens catalyzed by sarcoplasmic reticulum vesicles. J. Biol. Chem., 248, 3163-3172 (1973).

74. R. F. Keeler, L. B. Carr, and J. E. Varner, Intracellular localization of iron, calcium, molybdenum and tungsten in Azotobacter vinelandii. Exp. Cell Res., 15, 80-84 (1958).

75. P. A. Knauf, F. Proverbio, and J. F. Hoffman, Electrophoretic separation of different phosphoproteins associated with Ca-ATPase and Na, K-ATPase in human red cell ghosts. J. Gen. Physiol., 63, 324-336 (1974).

76. A. F. Knowles and E. Racker, Formation of adenosine triphosphate from $P_i$ and adenosine diphosphate by purified $Ca^{2+}$-adenosine triphosphatase. J. Biol. Chem., 250, 1949-1951 (1975).

77. M. Kojima, S. Suda, S. Hotta, and K. Hamada, Induction of pleomorphy and calcium ion deficiency in Lactobacillus bifidus. J. Bacteriol., 102, 217-220 (1970).

78. W. N. Konings, A. Bisschop, M. Veenhuis, and C. A. Vermuelen, New procedure for the isolation of membrane vesicles of Bacillus subtilis and an electron microscopy study of their ultrastructure. J. Bacteriol., 116, 1456-1465 (1973).

79. C. Kung, S.-Y. Chang, Y. Satow, J. Van Houten, and H. Hansma, Genetic dissection of behavior in Paramecium. Science, 188, 898-904 (1975).

80. J. F. Lamb and R. Lindsay, Effect of Na, metabolic inhibitors and ATP on Ca movements in L cells. J. Physiol., 218, 691-708 (1971).

81. J. M. LaNauze, D. J. Ellar, G. Denton, and J. A. Posgate, Some properties of forespores isolated from Bacillus megaterium. In Spore Research 1973 (A. N. Barker, G. W. Gould, and J. Wolf, eds.), Academic Press, New York, 1974, pp. 41-46.

82. Y. T. Lanni, Invasion by bacteriophage T5. II. Dissociation of calcium-independent and calcium-dependent processes. Virology, 10, 514-529 (1960).

83. B. Larsen and A. Haug, Biosynthesis of alginate, part 1. Composition and structure of alginate produced by Azotobacter vinelandii (Lipman). Carbohyd. Res., 17, 287-296 (1971).

84. G. Leanz and C. Gilvarg, Dipicolinic acid location in intact spores of Bacillus megaterium. J. Bacteriol., 114, 455-456 (1973).

85. A. L. Lehninger, Mitochondria and calcium ion transport. Biochem. J., 119, 129-138 (1970).

86. A. L. Lehninger, The coupling of $Ca^{2+}$ transport to electron transport in mitochondria. In The Molecular Basis of Electron Transport (J. Schultz and B. F. Cameron, eds.), Academic Press, New York, 1972, pp. 133-151.

87. A. L. Lehninger and E. Carafoli, The interaction of $La^{3+}$ with mitochondria in relation to respiration-coupled $Ca^{2+}$ transport. Arch. Biochem. Biophys., 143, 506-515 (1971).

88. H. B. LéJohn, L. E. Cameron, R. M. Stevenson, and R. U. Meuser, Influence of cytokinins and sulfhydryl group-reacting agents on calcium transport in fungi. J. Biol. Chem., 249, 4016-4020 (1974).

89. J. C. Lewis, $Ca^{2+}$-peptidoglycan interactions. In Spores VI (P. Gerhardt, R. N. Costilow, and H. L. Sadoff, eds.), American Society for Microbiology, Washington, D. C., 1975, pp. 547-549.

90. D. H. MacLennan and P. C. Holland, Calcium transport in sarcoplasmic reticulum. Ann. Rev. Biophys. Bioeng., 4, 377-404 (1975).

91. Y. Maeda and M. Maeda, The calcium content of the cellular slime mold, Dictyostelium discoideum, during development and differentiation. Exp. Cell Res., 82, 125-130 (1973).

92. A. Martonosi, The structure and function of sarcoplasmic reticulum membranes. In Biomembranes (L. A. Manson, ed.), Vol. 1, Plenum Press, New York, 1971, pp. 191-256.

93. J. W. Mason, H. Rasmussen, and F. Dibella, 3'5' AMP and $Ca^{2+}$ in slime mold aggregation. Exp. Cell Res., 67, 156-160 (1971).

94. Ph. Matile, Vacuoles. In Plant Biochemistry, 3rd ed. (J. Bonner and J. E. Varner, eds.), Academic Press, New York, 1976, pp. 189-224.

95. G. Meissner and S. Fleischer, Dissociation and reconstitution of functional sarcoplasmic reticulum vesicles. J. Biol. Chem., 249, 302-309 (1974).

96. A. S. Mildvan, Metals in enzyme catalysis. In The Enzymes, 3rd ed. (P. D. Boyer, ed.), Vol. 2, Academic Press, New York, 1970, pp. 445-536.

97. L. Minkoff and R. Damadian, Actin-like proteins from Escherichia coli: concept of cytotonus as the missing link between cell metabolism and the biological ion-exchange resin. J. Bacteriol., 125, 353-365 (1976).

98. P. Mitchell, Chemiosmotic Coupling and Energy Transduction, Glynn Research Ltd., Bodmin, Cornwall, United Kingdom, 1968.

99. J. L. Mongar and A. Wassermann, Influence of ion exchange on optical properties, shape, and elasticity of fully-swollen alginate fibres. J. Chem. Soc., London, pp. 500-510 (1952).

100. C. L. Moore, Specific inhibition of mitochondrial $Ca^{++}$ transport by ruthenium red. Biochem. Biophys. Res. Commun., 42, 298-305 (1971).

101. J. F. Morrison, Arginine kinase and other invertebrate guanidino kinases. In The Enzymes, 3rd ed. (P. D. Boyer, ed.), Vol. 8, Academic Press, New York, 1973, pp. 457-486.

102. R. W. Moyer and J. M. Buchanan, Effect of calcium ions on synthesis of T5-specific proteins, J. Biol. Chem., 245, 5897-5903 (1970).

103. W. G. Murrell, The biochemistry of the bacterial endospore. Advan. Microbial Physiol., 1, 133-251 (1967).

104. J. R. Norris and H. L. Jensen, Calcium requirements of Azotobacter. Nature, 180, 1493-1494 (1957).

105. W. J. Page and H. L. Sadoff, Relationship between calcium and uronic acids in the encystment of Azotobacter vinelandii. J. Bacteriol., 122, 145-151 (1975).
106. B. A. Palevitz, J. F. Ash, and P. K. Hepler, Actin in the green alga, Nitella. Proc. Nat. Acad. Sci. U.S., 71, 363-366 (1974).
107. M. H. Park, B. B. Wong, and J. E. Lusk, Mutants in three genes affecting transport of magnesium in Escherichia coli: genetics and physiology. J. Bacteriol., 126, 1096-1103 (1976).
108. S. M. Pearce and P. C. Fitz-James, Spore refractility in variants of Bacillus cereus treated with actinomycin D. J. Bacteriol., 107, 337-344 (1971).
109. P. L. Pedersen and W. A. Coty, Energy-dependent accumulation of calcium and phosphate by purified inner membrane vesicles of rat liver mitochondria. J. Biol. Chem., 247, 3107-3113 (1972).
110. M. R. Pollock, Exoenzymes. In The Bacteria (I. C. Gunsalus and R. Y. Stanier, eds.), Vol. 4, Academic Press, New York, 1962, pp. 121-178.
111. B. C. Pressman, Properties of ionophores with broad range cation selectivity. Fed. Proc., 32, 1698-1703 (1973).
112. G. Prestipino, D. Ceccarelli, F. Conti, and E. Carafoli, Interactions of a mitochondrial $Ca^{2+}$-binding glycoprotein with lipid bilayer membranes. FEBS Letters, 45, 99-103 (1974).
113. E. E. Quist and B. D. Roufogalis, Determination of the stoichiometry of the calcium pump in human erythrocytes using lanthanum as a selective inhibitor. FEBS Letters, 50, 135-139 (1975).
114. A. S. Rae and K. P. Strickland, Uncoupler and anaerobic resistant transport of phosphate in Escherichia coli. Biochem. Biophys. Res. Commun., 62, 568-576 (1975).
115. E. Racker and E. Eytan, Reconstitution of an efficient calcium pump without detergents. Biochem. Biophys. Res. Commun., 55, 174-178 (1973).
116. H. Rasmussen, Cell communication, calcium ion, and cyclic adenosine monophosphate. Science, 170, 404-412 (1970).
117. K. C. Reed and F. L. Bygrave, The inhibition of mitochondrial calcium transport by lanthanides and ruthenium red. Biochem. J., 140, 143-155 (1974).
118. P. W. Reed and H. A. Lardy, A23187: a divalent cation ionophore. J. Biol. Chem., 247, 6970-6977 (1972).
119. K. R. Robinson and L. F. Jaffe, Polarizing fucoid eggs drive a calcium current through themselves. Science, 187, 70-72 (1975).
120. R. W. Roeske, S. Isaac, T. E. King, and L. K. Steinrauf, The binding of barium and calcium ions by the antibiotic beauvericin. Biochem. Biophys. Res. Commun., 57, 554-561 (1974).

121. B. P. Rosen and J. S. McClees, Active transport of calcium in in-
     verted membrane vesicles of Escherichia coli. Proc. Nat. Acad.
     Sci. U.S., 71, 5042-5046 (1974).

122. H. Rottenberg and A. Scarpa, Calcium uptake and membrane poten-
     tial in mitochondria. Biochemistry, 13, 4811-4817 (1974).

123. H. L. Sadoff, Encystment and germination in Azotobacter vinelandii.
     Bacteriol. Rev., 39, 516-539 (1975).

124. M. H. Saier, Jr., B. U. Feucht, and M. T. McCaman, Regulation
     of intracellular adenosine cyclic 3':5'-monophosphate levels in
     Escherichia coli and Salmonella typhimurium. J. Biol. Chem.,
     250, 7593-7601 (1975).

125. C. J. Scandella and A. Kornberg, A membrane-bound phospholipase
     A1 purified from Escherichia coli. Biochemistry, 10, 4447-4456
     (1971).

126. H. J. Schatzmann, ATP-dependent $Ca^{++}$-extrusion from human red
     cells. Experientia, 22, 364-365 (1966).

127. H. J. Schatzmann, Active calcium transport and $Ca^{2+}$-activated
     ATPase in human red cells. In Current Topics in Membranes and
     Transport (F. Bronner and A. Kleinzeller, eds.), Vol. 6, Academic
     Press, New York, 1975, pp. 125-168.

127a. S. J. Schein, Nonbehavioral selection for pawns, mutants of
     Paramecium aurelia with decreased excitability. Genetics, 84,
     453-468 (1976).

128. R. Scherrer and P. Gerhardt, Location of calcium within Bacillus
     spores by electron probe x-ray microanalysis. J. Bacteriol., 112,
     559-568 (1972).

129. P. Scholes, P. Mitchell, and J. Moyle, The polarity of proton
     translocation in some photosynthetic microorganisms. Eur. J.
     Biochem., 8, 450-454 (1969).

130. H. E. Scribner, J. Mogelson, E. Eisenstadt and S. Silver, Regula-
     tion of cation transport during bacterial sporulation. In Spores VI
     (P. Gerhardt, R. N. Costilow, and H. L. Sadoff, eds.), American
     Society for Microbiology, Washington, D.C., 1975, pp. 346-355.

131. H. Schraer (ed.), Biological Calcification: Cellular and Molecular
     Aspects, Appleton-Century-Crofts, New York, 1970.

132. A. E. Shamoo and D. H. MacLennan, A $Ca^{++}$-dependent and
     -selective ionophore as part of the $Ca^{++} + Mg^{++}$-dependent
     adenosinetriphosphatase of sarcoplasmic reticulum. Proc. Nat.
     Acad. Sci. U.S., 71, 3522-3526 (1974).

133. I. M. Shapiro and N. H. Lee, Effects of $Ca^{2+}$ on the respiratory
     activity of chondrocyte mitochondria. Arch. Biochem. Biophys.,
     170, 627-633 (1975).

134. S. Silver and M. L. Kralovic, Manganese accumulation by
     Escherichia coli: evidence for a specific transport system. Bio-
     chem. Biophys. Res. Commun., 34, 640-645 (1969).

135.  S. Silver, K. Toth, P. Bhattacharyya, E. Eisenstadt, and
      H. Scribner, Changes and regulation of cation transport during bac-
      terial sporulation.  In Comparative Biochemistry and Physiology of
      Transport (L. Bolis, K. Bloch, S. E. Luria, and F. Lynen, eds.),
      North-Holland Publ., Amsterdam, 1974, pp. 393-408.
136.  S. Silver, K. Toth, and H. Scribner, Facilitated transport of calcium
      by cells and subcellular membranes of Bacillus subtilis and
      Escherichia coli. J. Bacteriol., 122, 880-885 (1975).
137.  K. Simkiss, Calcium translocation by cells. Endeavour, 33, 119-
      123 (1974).
138.  C. W. Slayman, The genetic control of membrane transport.  In
      Current Topics in Membranes and Transport (F. Bronner and
      A. Kleinzeller, eds.), Vol. 4, Academic Press, New York, 1973,
      pp. 1-174.
139.  O. Smidsrød and A. Haug, Dependence upon the gel-sol state of the
      ion-exchange properties of alginates. Acta Chem. Scand., 26,
      2063-2074 (1972).
140.  J. Steinberg, E. J. Masoro, and B. P. Yu, Role of sarcoplasmic
      reticulum phospholipids in calcium ion binding activity. J. Lipid
      Res., 15, 537-543 (1974).
141.  J. L. Streckfuss, W. N. Smith, L. R. Brown, and M. M. Campbell,
      Calcification of selected strains of Streptococcus mutans and
      Streptococcus sanguis. J. Bacteriol., 120, 502-506 (1974).
142.  R. L. Switzer, Phosphoribosylpyrophosphate synthetase and related
      pyrophosphokinases. In The Enzymes, 3rd ed. (P. D. Boyer, ed.),
      Vol. 10, Academic Press, New York, 1974, pp. 607-629.
143.  T. Takagi, H. Toda, and T. Isemura, Bacterial and mold amylases.
      In The Enzymes, 3rd ed. (P. D. Boyer, ed.), Vol. 5, Academic
      Press, New York, 1971, pp. 235-271.
144.  D. L. Taylor, J. S. Condeelis, P. L. Moore, and R. D. Allen,
      The contractile basis of amoeboid movement. I. The chemical
      control of motility in isolated cytoplasm. J. Cell Biol., 59, 378-
      394 (1973).
145.  Y. Tonomura and S. Yamada, Molecular mechanism of $Ca^{2+}$ trans-
      port through the membrane of the sarcoplasmic reticulum.  In
      Organization of Energy Transducing Membranes (M. Nakao and
      L. Packer, eds.), University Park Press, Baltimore, Maryland,
      1973, pp. 107-115.
146.  T. Tsuchiya and B. P. Rosen, Characterization of an active trans-
      port system for calcium in inverted membrane vesicles of
      Escherichia coli. J. Biol. Chem., 250, 7687-7692 (1975).
147.  T. Tsuchiya and B. P. Rosen, Restoration of active calcium trans-
      port in vesicles of an $Mg^{2+}$-ATPase mutant of Escherichia coli by
      wild-type $Mg^{2+}$-ATPase. Biochem. Biophys. Res. Commun., 63,
      832-838 (1975).

148. T. Tsuchiya and B. P. Rosen, Energy transduction in Escherichia coli. The role of the $Mg^{2+}$ ATPase. J. Biol. Chem., 250, 8409-8415 (1975).

149. T. Tsuchiya and B. P. Rosen, Calcium transport driven by a proton gradient in inverted membrane vesicles of Escherichia coli. J. Biol. Chem., 251, 962-967 (1976).

150. F. D. Vasington, P. Gazzotti, R. Tiozzo, and E. Carafoli, The effect of ruthenium red on $Ca^{2+}$ transport and respiration in rat liver mitochondria. Biochim. Biophys. Acta, 256, 43-54 (1972).

151. V. Vinter, Physiology and biochemistry of sporulation. In The Bacterial Spore (G. W. Gould and A. Hurst, eds.), Academic Press, New York, 1969, pp. 73-123.

152. G. B. Warren, P. A. Toon, N. J. M. Birdsall, A. G. Lee, and J. C. Metcalfe, Reconstitution of a calcium pump using defined membrane components. Proc. Nat. Acad. Sci. U.S., 71, 622-626 (1974).

153. R. H. Wasserman, R. A. Corradino, and A. N. Taylor, Vitamin D-dependent calcium-binding protein: purification and some properties. J. Biol. Chem., 243, 3978-3986 (1968).

154. E. L. Watson, F. F. Vincenzi, and P. W. Davis, $Ca^{2+}$-activated membrane ATPase: selective inhibition by ruthenium red. Biochim. Biophys. Acta, 249, 606-610 (1971).

155. R. R. Weihing and E. D. Korn, Acanthamoeba actin. Isolation and properties. Biochemistry, 10, 590-600 (1971).

156. M. L. Weiner and K. S. Lee, Active calcium ion uptake by inside-out and right side-out vesicles of red blood cell membranes. J. Gen. Physiol., 59, 462-475 (1972).

157. I. C. West and P. Mitchell, Proton/sodium ion antiport in Escherichia coli. Biochem. J., 144, 87-90 (1974).

158. J. E. Zajic, Microbial Biogeochemistry, Academic Press, New York, 1969.

159. T. H. Zytkovicz and H. O. Halvorson, Some characteristics of dipicolinic acid-less mutant spores of Bacillus cereus, Bacillus megaterium, and Bacillus subtilis. In Spores V (H. O. Halvorson, R. Hanson, and L. L. Campbell, eds.), American Society for Microbiology, Washington, D.C., 1972, pp. 49-52.

Chapter 3

MANGANESE TRANSPORT IN MICROORGANISMS

Simon Silver and Paula Jasper[1]

Biology Department
Washington University
St. Louis, Missouri

---

[1] Present address: Department of Biochemistry, St. Louis University, St. Louis, Missouri.

I.   INTRODUCTION

Manganese and iron and as many as a dozen other cation-forming elements
are required in trace quantities for the growth of most or all cells [41].
In contrast to iron, which occurs primarily in heme proteins, there is no
unique and universal function for manganese.  For example, in chloro-
plast-containing higher photosynthetic organisms, $Mn^{2+}$ functions in the
process of oxygen evolution from photosystem II [19]; but photosynthetic
bacteria require $Mn^{2+}$ [130] although they lack photosystem II and do not
evolve $O_2$.
     In this chapter, we will review what is currently known about the spe-
cific intracellular functions of manganese that may account for the variety
of microbes that require manganese [e.g., 28,113].  Then we will describe the highly specific manganese transport systems that were discovered
in our laboratory and have been found in every cell type that we have examined.  Finally, we will try to draw comparisons with roles and functions of
manganese in higher forms (plants and animals).  In a sense, too little is
known about manganese to permit the kind of comprehensive coverage that
we attempted with microbial calcium metabolism (Chapter 2).  The purpose
of this chapter, then, is to place current knowledge of manganese metabolism in a rational framework and, by so doing, to stimulate research in
this area.


II.  MANGANESE REQUIREMENTS
     AND FUNCTIONS

A.   Effects on Growth

During the early "nutritional" era of microbial growth and physiology
bacterial growth requirements for $Mn^{2+}$ were frequently demonstrated
[e.g., 130].  Under conditions of $Mg^{2+}$ limitation of bacterial growth,
$Mn^{2+}$ is able to lower but not to eliminate the $Mg^{2+}$ requirement for
Bacillus subtilis and Bacillus megaterium [124].  Recently, a requirement for high $Mn^{2+}$ for growth of Pediococcus cerevisiae was found [28].
A tenfold increase in growth yield occurred upon addition of $Mn^{2+}$, and
$Zn^{2+}$ could not replace the needed $Mn^{2+}$ [28].  The physiological basis for
the manganese requirement is unknown, although involvement of citrate in
accentuating the $Mn^{2+}$ requirement was suggested [28] by parallels to
earlier work [72] with lactic acid bacteria.  Some strains of Escherichia
coli also have a high $Mn^{2+}$ requirement in citrate-containing medium.
E. coli strain W lacks a citrate-dependent iron transport system
(P. Wookey and F. Gibson, unpublished data); and in the presence of
citrate, trace $Fe^{3+}$ is tied up in a form unavailable to the cells.  When
blocked in the alternative high-affinity enterochelin $Fe^{3+}$ accumulation

system [94] by a mutation affecting aromatic $Fe^{3+}$ chelate synthesis, this strain shows a requirement for either high $(1,000\,\mu M)$ $Fe^{3+}$ concentrations or lower $Mn^{2+}$ concentrations [$1\,\mu M$ $Mn^{2+}$ in the absence of added $Fe^{3+}$ (F. Gibson and S. Silver, unpublished data)]. Although the basis of this $Mn^{2+}$ requirement is unknown, the need for citrate to express the $Mn^{2+}$ requirement suggests that either $Mn^{2+}$ is replacing $Fe^{3+}$ in a critical intracellular process or that $Mn^{2+}$ is inducing the synthesis of an alternative iron transport compound.

Brevibacterium ammoniagenes, which is used industrially to produce 5'-inosinic acid as a flavor additive, required $2\,\mu M$ $Mn^{2+}$ for full growth yield and inosinic acid formation [79]. In the absence of added $Mn^{2+}$, DNA synthesis was first curtailed; and after subsequent unbalanced RNA, protein, and cell wall synthesis, the cells swelled and ultimately lysed [79]. There was a rapid and complete recovery of DNA synthesis when $2\,\mu M$ $Mn^{2+}$ was added to cells 5 to 10 hr after the cessation of DNA synthesis [79]. Other coryneform bacteria appeared similar to B. ammoniagenes in $Mn^{2+}$ requirement [79], and this careful work is one of the most clear-cut associations of which we are aware of a $Mn^{2+}$ requirement for growth with a specific biochemical process (DNA synthesis).

Equally elegant is the assignment of the $Mn^{2+}$ requirement of Lactobacillus curvatus to transcription of RNA by the divalent cation requiring RNA polymerase (see Section II.D) [113]; at $0.5\,\mu M$ $Mn^{2+}$ L. curvatus showed less than 40% of the growth yield that it had with greater than $10\,\mu M$ $Mn^{2+}$. This requirement late in growth involved the induction of some new enzymes such as L(+)-D(-)-lactate racemase and enzymes for ribose fermentation, as indicated by the decreased activity of these enzymes in the absence of added $Mn^{2+}$. $Mn^{2+}$ was specifically required for this induced enzyme formation, which was actinomycin D sensitive; $Mg^{2+}$ in excess did not suffice. Early growth depleted the medium of needed $Mn^{2+}$ unless excess levels were available. In the absence of $Mn^{2+}$, the cellular RNA polymerase could not produce RNA for synthesis of proteins needed for additional late growth [113].

In this compilation of the various effects of $Mn^{2+}$, we must note that $Mn^{2+}$ cations were among the earliest studied chemical mutagens [93]. High concentrations of $Mn^{2+}$ are mutagenic both to E. coli [22,93] and to yeast [85,88]. In the latter case, both mitochondrial [88] and nuclear [85] mutations were produced. The predominance of mitochondrial mutations was ascribed to the higher intramitochondrial $Mn^{2+}$ concentration due to mitochondrial $Mn^{2+}$ transport (Section V.A). It was suggested that $Mn^{2+}$ could be mutagenic by increasing the error frequency of the divalent cation requiring DNA polymerase [88]. High levels of manganese are also toxic, but manganese-resistant mutants of yeast [85] and E. coli [109] have been isolated. Since these result from a change in the magnesium transport system and not in the manganese transport system [109], the manganese-resistant mutants are considered in Chapter 1.

Of the variety of manganese requirements and functions in microbial cells, the three best studied classes are (a) an absolute $Mn^{2+}$ requirement for the synthesis of many secondary metabolites; (b) an absolute $Mn^{2+}$ requirement for bacterial sporulation; and (c) a $Mn^{2+}$ requirement for the functioning of many enzymes. Other $Mn^{2+}$ requirements and $Mn^{2+}$ effects on other microbial activities include a $Mn^{2+}$ requirement for the production of specific antigens and toxins [126]. Manganese is also required for transformation by exogenous DNA [126].

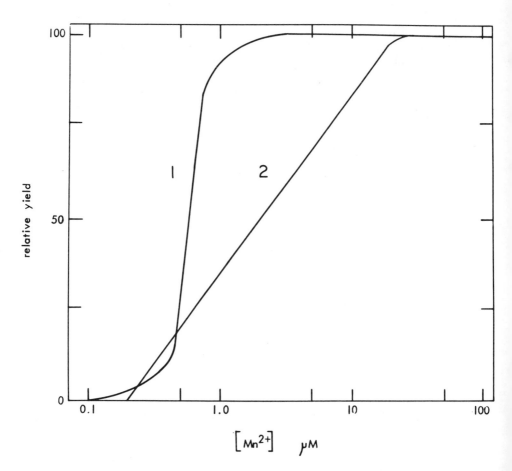

FIG. 3-1  $Mn^{2+}$ dependent secondary metabolite synthesis: (1) spore formation [31]; (2) synthesis of either mycobacillin [73] or bacitracin [128].

## B.  Manganese Requirements for
Secondary Metabolite Production

Secondary metabolites were defined by Weinberg [126], who has reviewed their synthesis and functions extensively, as those "natural products that have a restricted taxonomic distribution, possess no obvious function in cell growth, and are synthesized by cells that have stopped dividing." Often the production of secondary metabolites is controlled by trace levels of divalent cations, especially $Mn^{2+}$, $Fe^{3+}$, and $Zn^{2+}$. The antibiotics produced by bacterial cells and fungi are some of the best-known secondary metabolites. Although $Mn^{2+}$ is required for the production of other secondary metabolites, we will consider only the peptide antibiotics synthesized by species of Bacillus [97,126]. There is a requirement for $Mn^{2+}$ in the growth medium for the synthesis of many, if not all, of these antibiotics [125,126]. In the two cases where careful dose-response curves have been studied (Fig. 3-1), medium $Mn^{2+}$ had to be raised from about 0.1 μM to about 10 μM for antibiotic synthesis to occur. It is remarkable that the $Mn^{2+}$ requirement for the synthesis of two distinct antibiotics (mycobacillin and bacitracin) by different bacterial species (B. subtilis and Bacillus licheniformis, respectively) in experiments tested 6 years and 16,000 km apart can be represented by the same line in Figure 3-1. Weinberg [126] noted the log-linear relationship between $Mn^{2+}$ levels and antibiotic production; but there is no further clue as to the nature of the $Mn^{2+}$ requirement from this than from the log-log relationship between $Mn^{2+}$ level and spore production (Fig. 3-1). It is likely that divalent cations, including $Mn^{2+}$, also stabilize antibiotic polypeptides from inactivation after their production [125].

The polypeptide structures of five antibiotics produced by species of Bacillus are presented in detail in Figure 3-2 for three reasons. Firstly, they are the best defined of the $Mn^{2+}$ dependent secondary metabolites. The second reason is that these polypeptides specifically bind divalent cations including manganese [115]. This cation binding by the polypeptides has been hypothesized as serving a function in cation transport during bacterial sporulation [51-53,97]. Because antibiotic-negative mutants of Bacillus have been isolated which are still able to sporulate [54], however, it is clear that antibiotic production is not required for sporulation. The B. subtilis strains whose $Mn^{2+}$ transport we have studied (see Section III. C) produce at least five such polypeptides [98]. The third reason for describing these Bacillus polypeptides is that they provide us with a current model for the $Mn^{2+}$ transport stimulator described in Section III.D. However, this factor is produced during $Mn^{2+}$ starvation, whereas the polypeptide antibiotics require manganese for synthesis. Mycobacillin and bacitracin are not made in the absence of $Mn^{2+}$, as shown in Figure 3-1; and other polypeptide antibiotics are produced in $Mn^{2+}$-containing medium, although the $Mn^{2+}$ requirement has not been carefully determined.

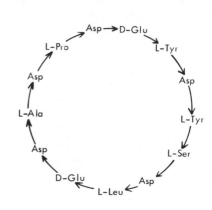

(a)    Bacitracin A

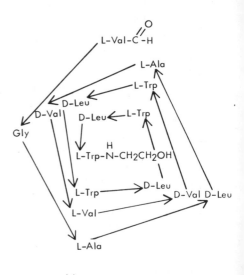

(d)    Mycobacillin

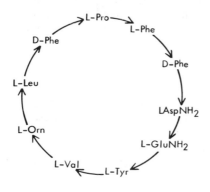

(b)    Tyrocidine A

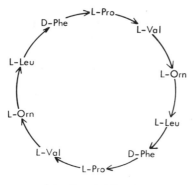

(c)    Gramicidin S

(e)    Gramicidin A

FIG. 3-2

Some of the polypeptide antibiotics form covalently closed circles (Fig. 3-2b,c,d), for example, the 10 amino acid rings of tyrocidine A (synthesized by Bacillus brevis during sporulation [90]) and gramicidin S (synthesized by another "Soviet" strain of B. brevis during sporulation [90]) and the 13 amino acid ring of mycobacillin (synthesized by a strain of B. subtilis during sporulation [5]). The other gramicidins are neutral antibiotics synthesized by the same B. brevis strain that produces tyrocidines (basic polypeptides containing free amino groups). One, gramicidin A, is a noncyclized polypeptide chain that can form a spiral structure in membranes (Fig. 3-2e) [121]. The decapeptides tyrocidine A and gramicidin share the pentapeptide sequence (L-Val)-(L-Orn)-(L-Leu)-(D-Phe)-(L-Pro), but the 13-member ring of mycobacillin [5] has a very different sequence. The last of the five structures in Figure 3-2 is that of bacitracin, which is synthesized by strains of B. licheniformis during $Mn^{2+}$-dependent sporulation. Bacitracin has a ring of six amino acids, with an additional asparagine residue attached to one position and a five amino acid chain to another position of the ring.

Both D and L amino acids are found in the polypeptide antibiotics, and therefore these polypeptides are not directly synthesized via messenger RNA [71]. The process is very different from that in protein synthesis on ribosomes because the specificity for each amino acid resides in the synthetase enzyme, not in an RNA sequence. The polypeptides are synthesized by protein aggregates with enzyme-thioester-linked intermediates by a process similar to that of fatty acid elongation [71,95,96,119]. Amino acids are activated by hydrolysis of ATP, forming aminoacyl-AMP; and the amino acids are attached to specific sulfhydryl groups on components of the synthetases. With gramicidin S, for example, the D-Phe is attached to a sulfhydryl residue on protein EII (molecular weight of 100,000 daltons) whereas the other four amino acids are attached to sulfhydryl groups on the other protein EI (molecular weight of 280,000 daltons); these two enzymes together constitute the gramicidin synthetase enzyme [71,119]. The first

---

FIG. 3-2   Structures of polypeptide antibiotics produced by the Bacilli: Asp, indicates an aspartate residue; $AspNH_2$, asparagine; Ala, alanine; Cys, cysteine; Glu, glutamic acid; $GluNH_2$, glutamine; Gly, glycine; His, histidine; Ile, isoleucine; Leu, leucine; Lys, lysine; Orn, ornithine; Phe, phenylalanine; Pro, proline; Ser, serine; Trp, tryptophan; Tyr, tyrosine; and Val, valine. Arrows indicate C $\longrightarrow$ N peptide linkages. The contributions of L-Ile and L-Cys to the thiazoline ring in bacitracin are drawn in detail, and the dashed line separates the L-Ile and L-Cys contributions to the thiazoline ring. The carboxyl terminal group of Trp and the amino terminal group of Val in gramicidin A are blocked by ethanolamine and formyl groups, respectively.

peptide bond formation involves the transfer of the EII-bound D-Phe to the
L-Pro site on EI forming the dipeptide (D-Phe)-(L-Pro). If the analogy to
fatty acid synthesis is appropriate to the steps of polypeptide synthesis,
then the growing peptide chain will be transferred to the single phosphopantetheine residue that is found on EI [71]. The subsequent steps will be
synthesis of the tripeptide, tetrapeptide, and pentapeptide (D-Phe)-(L-Pro)-
(L-Val)-(L-Orn)-(L-Leu) at each stage with a sequential transfer from the
amino acid-specific sulfhydryl to the phosphopantetheine sulfhydryl group
which acts as a carrier from position to position on the EII enzyme [71].
The relative positions of the amino acid-specific sulfhydryls (cysteines
probably) determines the sequence of amino acids in the growing polypeptide [71]. Finally two pentapeptide-containing enzymes interact to form
the cyclic decapeptide. The synthesis of tyrocidine [71,95,96], mycobacillin [5], and bacitracin [42,43] all appear to be basically similar to
that found with gramicidin S [71]. Of these, tyrocidine A synthesis has
been most thoroughly studied [71,95,96]. This decapeptide does not contain a repeated sequence (Fig. 3-2) and is synthesized sequentially as one
chain by a complex of three enzymes [69a,71,95,96]. The formation of
unusual linkages such as the C-S bond in bacitracin or the γ-ε peptide
linkage completing the bacitracin ring and the attachment of the C-terminal
formyl group and the N-terminal ethanolamine on gramicidin A are all accomplished enzymatically before (presumably) the polypeptide antibiotics
are freed from the biosynthetic complex. With mycobacillin four of the
five aspartyl residues are D and only one L, but which is the L in the sequence in Figure 3-2 is not known. The γ carboxyl group of the subterminal
L-Asp of bacitracin is involved in the γ-ε linkage with the ε amino group
of lysine.

The interaction of gramicidin A with cations and with membranes provides a useful model for ion-translocating channels. Gramicidin A forms
channels in lipid bilayers, as was demonstrated in conductance measurements showing discrete fluctuations up and down in conductance indicative
of the formation and loss of these channels [59]. Two molecules of
gramicidin A were needed to bridge the thickness of the bilayer [48,49].
Urry [120] proposed a helical structure with a lipophilic exterior and an
internal channel lined with oxygens that could associate with cations. A
head-to-head model with the formylated valines at the center and the
ethanolamine-substituted tryptophans (see Fig. 3-2e) on the aqueous surfaces was favored [120]. Urry [121] further postulated that voltage
regulation of the cation-conducting capacity of the gramicidin A channel
might occur because of interchangeable conformations of gramicidin A
having opposite net dipole moments. Gramicidin A, however, conducts
monovalent and not divalent cations [57]. A cyclic 12 amino acid polypeptide with a sequence

$$\overline{\underline{(Gly\text{-}Gly\text{-}Val\text{-}Pro)_3}}$$

has been synthesized and shown to bind $Ca^{2+}$ [122]. Physicochemical studies showed binding specificity for $Ca^{2+}$ and $Sr^{2+}$ but not for $Mg^{2+}$. Two structural models were proposed from the physical measurements and from analogies with the structure of the $K^+$-carrying ionophore antibiotic valinomycin [122]. Calcium was considered complexed with the neutral oxygens of the peptide backbone chain, and this polypeptide was considered a potentially useful model in the understanding of calcium-binding proteins [122] and perhaps also of $Ca^{2+}$ transport proteins in membranes. Ion conducting channels in natural membranes are not likely to form in lipid bilayer regions because channels in the lipid bilayer would not show a high degree of specificity for particular cations. Either dissociable polypeptide chains analogous to the antibiotic polypeptides or amino acid sequences forming channels within larger transmembrane proteins provide working models for the structure of membrane-bound cation transport systems.

C. Manganese Requirements
   for Bacterial Sporulation

Since the original determination by Charney et al. [18] that $Mn^{2+}$ is an essential micronutrient for sporulation in a variety of Bacillus species, this work has been extended and confirmed. It has become clear that the requirement for sporulation is completely specific for $Mn^{2+}$. No alternative cations can replace $Mn^{2+}$ in enabling sporulation to occur [76]. Although other cation requirements also occur (e.g., for $Fe^{3+}$ or $Mg^{2+}$), these are similar to the requirements for growth in general. "High" $Mn^{2+}$ is required only for sporulation; and B. subtilis W23, at least, will grow equally as rapidly and to an equal yield at the end of log-phase growth in $0.02\ \mu M\ Mn^{2+}$ as with the $1.0\ \mu M\ Mn^{2+}$ required for sporulation (Fig. 3-1). No demonstrable change in the amount or level of cellular $Mn^{2+}$ occurs during the critical time at the end of log-phase growth or into the early post-log-phase sporulation stages. Cellular $Mn^{2+}$ is maintained at a relatively constant level during this time [31] with exchange of intracellular for extracellular $Mn^{2+}$. If the total medium $Mn^{2+}$ is adequate, the cells sporulate. However, if the total medium $Mn^{2+}$ is inadequate, it will have been entirely accumulated by the cells before the end of log-phase growth [31]. As growth continues in the absence of added $Mn^{2+}$, the cellular $Mn^{2+}$ concentration will decrease by dilution with increasing cell mass. If growth causes too great a decrease in cellular $Mn^{2+}$ (less than about 100 nmol per milliliter of cell water) [31], then the cells do not sporulate.

Axial filament and prespore septum formation, the first discernible events in sporulation, do not occur in the absence of $Mn^{2+}$. The key function in sporulation for which $Mn^{2+}$ is absolutely and specifically required is unknown, although it frequently has been suggested that $Mn^{2+}$ acts as a cofactor in an essential enzymatic reaction (see also Section II. C); $Mn^{2+}$-activated enzymes have been studied, but until recently no highly

$Mn^{2+}$-specific enzyme was demonstrated. Oh and Freese [78] have shown that B. subtilis Marburg cells grown on nutrient sporulation medium plus glucose but without manganese stop growing at a reduced turbidity relative to sporulating cells and accumulate 3-phosphoglyceric acid. When manganese is added to these cells, the 3-phosphoglycerate disappears, and the cells sporulate. Phosphoglycerate phosphomutase isolated from stationary phase cells (in the absence of $Mn^{2+}$) had a specific activity at least 40 times higher when assayed with added $Mn^{2+}$ than with $Mg^{2+}$, $Co^{2+}$, $Ca^{2+}$, or $Zn^{2+}$ (no detectable activity with any of these four cations) [78]. This enzyme is also present in growing cells, but its cation-specificity has not been studied. The activation of phosphoglycerate phosphomutase by $Mn^{2+}$ appears to be an essential role of $Mn^{2+}$ in sporulation, although $Mn^{2+}$ may have other functions as well.

Another function of $Mn^{2+}$ in promoting sporulation is that in regulating the potassium content of the cells. At the onset of stationary phase, cells in the absence of $Mn^{2+}$ began to lose potassium; but cells in media containing $Mn^{2+}$ maintained a high concentration of $K^+$ throughout sporulation [30]. The addition of 10 μM $Mn^{2+}$ to growing or to stationary phase cells (in media previously lacking $Mn^{2+}$) resulted in the maintenance or reestablishment of the high intracellular $K^+$ concentration [30]. The addition of $Mn^{2+}$ to stationary phase cells rapidly stimulated the accumulation of $^{42}K^+$ by these cells [30]. Since the addition of chloramphenicol (an inhibitor of protein synthesis) did not affect the stimulation of $^{42}K^+$ accumulation [30], it may be concluded that $Mn^{2+}$ directly stimulated the preexisting potassium transport system.

Bacillus fastidiosus appears to be an exception to the rule that all bacilli need $Mn^{2+}$ for sporulation [3]. Not only did B. fastidiosus sporulate (although more slowly and less completely) without detectable $Mn^{2+}$ (probably less than 0.1 μM) in the medium, but the spores produced in the absence of $Mn^{2+}$ germinated more rapidly and without the usual heat shock required by spores made in the presence of detectable $Mn^{2+}$ [3]; B. fastidiosus spores produced in the presence of high $Mn^{2+}$ concentrations contained up to 0.6% of their dry weight as manganese. Many other species of bacilli also accumulate $Mn^{2+}$ to a comparable level in their spores in the form of a chelate with dipicolinic acid (DPA) [132]. During spore formation, the calcium and DPA levels in B. fastidiosus spores were constant, and the amount of $Mn^{2+}$ appeared to rise as the $Mg^{2+}$ level decreased [3]. The $Mn^{2+}$-containing spores were heat resistant in proportion to their $Mn^{2+}$ content. Thus B. fastidiosus shows an unusual variation on the requirement for $Mn^{2+}$; in this case the $Mn^{2+}$ requirement is fully met by the 0.1 μM or less $Mn^{2+}$ in the unsupplemented medium.

D.  Manganese as an Enzymatic Cofactor

Because of the large number of enzymes for which $Mn^{2+}$ can serve as an alternative cofactor to $Mg^{2+}$ in activating enzymatic activity [13, 21, 68] and the equally large number of enzymes whose activity is inhibited by high manganese, we will not attempt a comprehensive treatment of $Mn^{2+}$ in enzymes similar to the treatment of $Ca^{2+}$ in enzymes in Chapter 2. Sometimes an enzyme from one microbial source appears to have a specific requirement for $Mn^{2+}$, whereas the comparable enzyme from another bacterium utilizes $Mn^{2+}$ or $Mg^{2+}$ with equal efficacy [114] or even utilizes other divalent cations, notably $Zn^{2+}$ and $Co^{2+}$, as well [13]. With such a variable pattern, it is difficult to ascribe $Mn^{2+}$ growth or sporulation requirements to any specific enzyme(s). Our more limited goal is to describe the superoxide dismutase of E. coli, a manganese-containing metalloenzyme, and to discuss two enzymes that may play key roles in the regulation of sporulation and/or secondary metabolite synthesis and for which there is evidence for regulatory changes by varying $Mg^{2+}$ and $Mn^{2+}$ levels. These two enzymes are the bacterial DNA-dependent RNA polymerases, which appear to change in relative stimulation by $Mn^{2+}$ or $Mg^{2+}$ during sporulation [15], and glutamine synthetase, which regulates cellular nitrogen metabolism and plays a key role in the early stages of sporulation [34].

1.  Superoxide Dismutase

Superoxide dismutases are found in all oxygen-metabolizing cells (plant, animal, and bacterial) and catalyze the inactivation of the toxic partial-reduction product of $O_2$, the superoxide radical $O_2^-$:

$$2O_2^- + 2H^+ \rightarrow H_2O_2 + O_2$$

Catalase converts the relatively less toxic hydrogen peroxide to $O_2$ and water. The ubiquity, roles, and chemical properties of various superoxide dismutase enzymes have been recently reviewed by Fridovich, their principal investigator [40]. The superoxide dismutase first isolated from E. coli differs from the $Cu^{2+}$- or $Zn^{2+}$-containing enzymes from the cytosol of eucaryotic cells in that the former is a manganese enzyme [40]. The "resting" enzyme consists of a single $Mn^{3+}$ in association with a 40,000-dalton molecular weight protein. The manganese is tightly bound and is cyclically reduced and reoxidized during successive interactions with $O_2^-$ [40]. The $Mn^{3+}$ superoxide dismutase of E. coli is an intracellular enzyme, but E. coli also has a distinct $Fe^{3+}$-containing superoxide dismutase that is a periplasmic enzyme (removable from the cells by osmotic shock). Determinations of the sequences of the first 25 aminoterminal amino acids show that 9 of the 25 amino acids are identical, not only in the two E. coli enzymes but also in a mitochondrial $Mn^{3+}$ superoxide dismutase [40]. Thus there is strong evidence for a common evolutionary origin for the genes determining these proteins. The $Mn^{3+}$, but not the $Fe^{3+}$, superoxide dismutase level in E. coli was induced by exposure of the cells to high $O_2$; and enzyme-rich cells were far more resistant than were other cells to the killing effects of hyperbaric $O_2$ [40]. The activity

of the B. subtilis superoxide dismutase does not increase on exposure to
hyperbaric $O_2$, nor does $O_2$ resistance increase [40]. Thus it is possible
that B. subtilis contains only an $Fe^{3+}$ enzyme. To date only E. coli and
and Streptococcus faecalis have been examined for the $Mn^{3+}$-superoxide
dismutase, although all aerobically growing cells studied contain superoxide dismutase activity. The relative distribution of $Fe^{3+}$ and $Mn^{3+}$ enzymes amongst bacterial species remains to be determined.


2.  Glutamine Synthetase

Glutamine synthetase is the key enzyme in regulating cellular nitrogen
metabolism [47] and also plays a critical role in the commitment to sporulation at the end of bacterial growth [34]. Glutamine synthetase is involved in the prevention of or "repression" of sporulation in the presence
of adequate nitrogen for growth, as indicated by the failure of glucose and
ammonia to repress sporulation in a mutant lacking glutamine synthetase
[34]. Bypassing this block by addition of glutamine results in repression
of sporulation [34]. Glutamine synthetase from a variety of bacteria
including E. coli [47], B. subtilis [24-26] and B. licheniformis has been
purified and studied in the laboratory of E. R. Stadtman. The regulatory
properties of this key enzyme vary depending upon whether the enzyme is
activated with $Mg^{2+}$ or with $Mn^{2+}$ [24-26,47]. Table 3-1 gives a comparison of the properties of the enzyme isolated from E. coli with those of the
B. subtilis enzyme.
    The glutamine synthetase molecule of E. coli consists of twelve
50,000-dalton subunits arranged in a double hexagon shape that can be
seen by electron microscopy [47]. Each subunit has a tyrosine position
that can be adenylated by a highly specific transferase enzyme. The fully
(12 per enzyme) adenylated glutamine synthetase functions maximally with
$Mn^{2+}$ as activator and not at all with $Mg^{2+}$. The completely adenylate-free
enzyme functions only with $Mg^{2+}$, and it does so at a rate four times
greater than the maximum rate of the adenylated enzyme with $Mn^{2+}$ [47].
In the absence of either cation, the E. coli glutamine synthetase assumes
a "relaxed" form that is enzymatically inactive and relatively unstable
[47]. Either $Mn^{2+}$ or $Mg^{2+}$ can convert the enzyme to a "tightened"
enzymatically active form with the association of one divalent cation per
subunit [47].
    At least four proteins and a cascade type of enzyme regulation are in-
volved in the regulation of the adenylation-deadenylation cycle [47]. The
adenyltransferase functions with either $Mg^{2+}$ or $Mn^{2+}$ but is itself subject
to regulation by another highly specific modifying enzyme $P_{II}$, which is in
turn governed by uridylation-deuridylation by another specific enzyme.
The enzyme adds uridine monophosphate (from UTP) to $P_{II}$ when functioning with either $Mg^{2+}$ or $Mn^{2+}$. The deuridylating enzyme activity specifically requires $Mn^{2+}$ to remove the uridyl moiety from the $P_{II}$ [1a]. In

TABLE 3-1   Properties of Glutamine Synthetase from E. coli and B. subtilis

| E. coli | B. subtilis |
|---|---|
| 600,000 daltons molecular weight<br>12 subunits<br>50,000 daltons per subunit | 600,000 daltons molecular weight<br>12 subunits<br>50,000 daltons per subunit |
| Two superimposed hexagon rings:<br>  (a) adenylated<br>    one 5'-AMP/subunit<br>    (activated by $Mn^{2+}$)<br>  (b) Unadenylated<br>    (activated by $Mg^{2+}$)<br><br>Adenylation cycle regulators:<br>  (a) adenyltransferase (ATase)<br>    activated by $Mg^{2+}$ or $Mn^{2+}$<br>  (b) regulatory protein $P_{II}$<br>    (i) uridylated form ($P_{IID}$)<br>      interacts with ATase to<br>      stimulate deadenylation<br>    (ii) deuridylated form ($P_{IIA}$)<br>      interacts with ATase to<br>      stimulate adenylation<br>  (c) uridyltransferase<br>    attaches UMP to $P_{IIA}$ to form<br>    $P_{IID}$<br>    (activated by $Mg^{2+}$ or $Mn^{2+}$,<br>    ATP, and $\alpha$-ketoglutarate)<br>  (d) deuridylating enzyme<br>    removes UMP from $P_{IID}$<br>    (activated by $Mg^{2+}$) | Two superimposed hexagon rings:<br>Not adenylated<br>  (activated by $Mn^{2+}$ or $Mg^{2+}$)<br>  (a) $Mn^{2+}$ enzyme activity<br>    optimal at $Mn^{2+}$/ATP ratio<br>    of 1:1<br>    activity stable<br>    feedback inhibition by alanine<br>      and glycine, not by<br>      glutamine<br>  (b) $Mg^{2+}$ enzyme activity<br>    optimal at $Mg^{2+}$/ATP=4-5:1<br>    less stable than $Mn^{2+}$ enzyme<br>    feedback inhibition by gluta-<br>      mine, potentiated by AMP |

its nonuridylated form $P_{II}$ stimulates the adenylating reaction and therefore converts the glutamine synthetase to its fully adenylated ($Mn^{2+}$-requiring) form. Here then are little fleas upon the backs of bigger fleas [117] in order to provide E. coli cells with tight regulation of glutamine synthesis. Manganese is required for the functioning of the adenylated enzyme and also is required for the conversion of the regulatory protein to a form favoring the adenylation process [1a,47].

The B. subtilis glutamine synthetase [24], like the E. coli enzyme, consists of twelve 50,000-dalton subunits and can be activated by either $Mg^{2+}$ or $Mn^{2+}$. Only the $Mn^{2+}$-enzyme of E. coli was regulated by the cation $(Mn^{2+})$/ATP ratio; but with either $Mg^{2+}$ or $Mn^{2+}$, the synthetic activity of the B. subtilis enzyme depends on the cation/ATP ratio [25, 26, 47]. Unlike the E. coli enzyme, for which either $Mg^{2+}$ or $Mn^{2+}$ leads to the conformational change to a stable enzyme form, in B. subtilis the $Mn^{2+}$ enzyme is more stable in solution than is the $Mg^{2+}$ form and the addition of divalent cations does not lead to a measurable conformational change [24]. The B. subtilis $Mg^{2+}$ glutamine synthetase is completely inhibited by feedback inhibition by glutamine and/or adenosine monophosphate (AMP), but glutamine is essentially without effect on the same enzyme functioning with $Mn^{2+}$ [25, 26]. This differs from the E. coli glutamine synthetase, which is only indirectly inhibited by glutamine via the adenylation-deadenylation reactions [47], since high glutamine favors the $Mn^{2+}$-active fully adenylated form by inhibiting the uridyltransferase in E. coli. The B. subtilis enzyme is not adenylated, nor can it function as substrate for the E. coli adenylating enzyme [25, 26]. However, in both types of cells, regulation of glutamine synthetase activity in response to glutamine level involves changes in responses to $Mg^{2+}$ and $Mn^{2+}$. We can postulate that the sharp decrease in the ratio of B. subtilis $Mg^{2+}$/$Mn^{2+}$ transport rates at the end of log-phase growth [41] might be reflected in changes in free intracellular cations that could, in turn, help regulate key enzymes such as glutamine synthetase [25].

3.   RNA Polymerase

The very first published reports on the existence of a DNA-dependent enzyme that synthesizes RNA [17] indicated striking differences between $Mn^{2+}$ and $Mg^{2+}$ forms of the enzyme. The concentrations of $Mn^{2+}$ that are optimal are generally tenfold lower than the optimal $Mg^{2+}$ level, and higher in vitro RNA synthesis rates are obtained with $Mn^{2+}$ [17, 44]. Nevertheless, the assumption has been that $Mg^{2+}$ is the physiologically normal cofactor in the cell. The most striking exception to this is the RNA polymerase from L. curvatus, which functions with six times greater specific activity with $Mn^{2+}$ under assay conditions where the E. coli RNA polymerase functions with three times greater specific activity with $Mg^{2+}$ [114]. This coupled with the $Mn^{2+}$ requirement for late protein synthesis in L. curvatus (cf. Section II. A) has led Stetter and Kandler [113] and Stetter and Zillig [114] to propose that the L. curvatus RNA polymerase functions in vivo as a highly specific $Mn^{2+}$-requiring enzyme. Other than the demonstrated high specificity for $Mn^{2+}$, which we believe will be shown in other RNA polymerases as well, the basic structure and characteristics of the L. curvatus polymerase are similar to the B. subtilis polymerase described next.

   We will develop briefly here the hypothesis that the change in properties of the RNA polymerase early in sporulation [44], coupled with a
change in the intracellular free $Mg^{2+}/Mn^{2+}$ ratio [101], may lead to a difference in specificity and thus to new and different "sporulation-specific"
messenger RNA and protein synthesis [84]. An alternative key role for
$Mn^{2+}$ early in sporulation would thus involve regulation at the level of RNA
transcription. The RNA polymerase found in sporulating cells differs in
composition and in template specificity from that found in vegetatively
growing cells [44]. The nature of these differences and whether some
have been introduced during enzyme isolation by sporulation proteases has
been the subject of much excitement and controversy in recent years.
Rather than summarize the large literature on the subject, we will refer
the reader to pp. 202-264 in the recent symposium Spores VI [46].
   Competitive hybridization experiments have shown that sporulation
specific RNA molecules are synthesized: 10 to 25% of the RNA synthesized
2 to 4 hr into the sporulation cycle ($T_{2-4}$) consists of sequences not found
in vegetatively growing cells [84]. Pero et al. [84], however, were unable to find qualitative differences in the RNA sequences transcribed in
sporulating cells and those transcribed during postlogarithmic phase of an
asporogenous (stage 0) mutant. Clearly, the final resolution of the question of qualitative as well as quantitative differences and "sporulation-
specific" genes remains to be accomplished.
   The RNA polymerase isolated early in sporulation appeared identical
with that from vegetatively growing cells [44]. However, by $T_{3.5}$ (3.5 hr
into the sporulation cycle) or so, a change in subunit structure of the
polymerase was found. The subunit structure of the vegetative cell polymerase may be designated $\alpha_2:\beta:\beta'$, representing two copies of an $\alpha$ subunit (molecular weight 42,000 daltons), one copy of a $\beta$ subunit (molecular
weight about 150,000 daltons), and one of a $\beta'$ subunit (molecular weight
about 150,000 daltons). At $T_{3.5}$ an additional protein subunit of molecular
weight 27,000 daltons was found associated with the RNA polymerase, and
the subunit composition could be represented as $\alpha_2:\beta:\beta':\delta^1$[44]. Two hours
later (at $T_{5.5}$), in addition to the $\alpha_2:\beta:\beta':\delta^1$ found at $T_{3.5}$, a third form
of the polymerase was found with the 27,000-dalton protein replaced by one
of 20,000 daltons, i.e., $\alpha_2:\beta:\beta':\delta^2$[44]. Both of the altered RNA polymerases had lowered in vitro activity with the DNA from bacteriophage $\phi e$,
which can grow only in vegetative and not in sporulating cells [44]. Another subunit, $\sigma$ (molecular weight 55,000 daltons), has been ascribed
regulatory roles with the vegetative cell polymerase; $\sigma$ activity was inhibited during sporulation [105] as still another polypeptide of molecular
weight 70,000 daltons was found associated with the sporulating cell RNA
polymerase [105]. Chloramphenicol restored $\sigma$ activity to sporulating
B. subtilis, apparently by depleting the cells of an unstable protein inhibitor of $\sigma$ [105]. There is as yet no agreement among laboratories studying
the problem as to the nature of the changes in subunit composition of RNA

polymerase during sporulation [44,105]. In addition to the changes in the structure of the polymerase during the sporulation cycle in wild-type cells, the existence of cells with mutant polymerases that were unable to sporulate or did so poorly [44,105] points to critical roles for the RNA polymerase(s) in sporulation.

The first suggestion of an involvement of $Mg^{2+}/Mn^{2+}$ changes in regulation of sporulation-specific RNA synthesis came from the finding of differences in response to $Mg^{2+}$ and to $Mn^{2+}$ with RNA polymerases isolated during sporulation and from mutants blocked in different stages of sporulation [15]. Sporulating cell RNA polymerase could function in vitro with poly d(AT) as substrate only with $Mn^{2+}$ as divalent cation [15]. However, $Mn^{2+}$ was much less effective than $Mg^{2+}$ in supporting RNA synthesis in cells made permeable to substrates and cofactors by a cold-shock treatment [37]. In another study, the altered polymerases from cells 5.5 hr into sporulation ($T_{5.5}$) did not have responses to $Mg^{2+}$ and $Mn^{2+}$ differing from those of vegetative cell RNA polymerase [44]. These observations are clearly so variable and preliminary that more and careful experiments will be needed in order to support or discard the hypothesis of a role for $Mn^{2+}$ regulating the sporulating cell RNA polymerase. Another B. subtilis enzyme that appears to be regulated by the available $Mn^{2+}/Mg^{2+}$ ratio is anthranilate synthetase, which is the first enzyme of the tryptophan biosynthetic pathway and is regulated as well both by tryptophan and by histidine [66a].

## E.   Oxidation of Manganese

Bacterial oxidation of $Mn^{2+}$ occurs both in soil [2,11,123] and at the bottom of the sea, where microbial activity has been implicated in the formation of ferromanganese nodules (Fig. 3) [32,133]. These mineral deposits are of appreciable commercial interest at the moment and may contain up to 63% of their weight as $MnO_2$ [32,133]. Bacterial enzymes function in both directions in the oxidation-reduction cycle between $Mn^{2+}$ and $Mn^{4+}$ [32,33]. However, various bacterial strains are probably involved. For example, Ehrlich [33] isolated an Arthrobacter strain from a Mn nodule that oxidized $Mn^{2+}$ under laboratory conditions. Other entirely different bacteria were also isolated from manganese nodules as oxidizers. Among the $Mn^{4+}$ reducers were a Bacillus strain and a Gram-negative rod [33]. The conclusion is that a wide variety of marine bacteria are either $Mn^{2+}$ oxidizers or $Mn^{4+}$ reducers. Because of the long times involved and difficult conditions of low temperature and high pressure, relatively little is known about the mechanisms of nodule formation, bacterial roles in the process, or the possible utilization by the bacteria of the energy derived from $Mn^{2+}$ oxidation (see Ref. 33 for access to this literature).

Somewhat more is known about the numerous manganese-oxidizing bacteria that abound in the soil [e.g., 75]. Such bacteria have been

FIG. 3-3  Manganese nodules on the floor of the Pacific Ocean.  Nodules
range from 5 to 10 cm in size.  (Courtesy, E. Dowell, Kennecott Copper
Corporation, New York.)

known since the studies of S. Winogradsky and M. W. Beijerinck; but
more current literature also concerns characterization of microbial
manganese oxidation [2,123], in addition to the isolation of bacteria [99].
In addition to the numerous soil bacteria and fungi [75,123] which oxidize
manganous salts to insoluble $MnO_2$, there are other bacteria in the soil
which are able to convert $MnO_2$ to soluble manganous salts.  It is thought
that oxidation of $Mn^{2+}$ to insoluble unavailable $MnO_2$ by soil microbes may
result in the manganese starvation of plants that do not harbor $Mn^{4+}$-
reducing bacteria on their roots [123].  Some bacteria require the presence of hydroxycarboxylic acids for $Mn^{2+}$ oxidation, whereas other species
do not [123].  However, Mulder [75] has concluded that the organisms
which require organic acids in order to oxidize manganese do not play an
important role in the oxidation of manganese in the soil.

The sheathed filamentous iron bacterium Leptothrix discophorus
(sometimes referred to as Sphaerotilus discophorus) oxidizes $Mn^{2+}$ as well
as $Fe^{2+}$ and accumulates microscopically visible deposits of ferric and
manganic oxides within its sheath [2].  The oxidation of $Mn^{2+}$ by
L. discophorus is due to an inducible enzyme system formed by rough
strains only; smooth variants do not oxidize $Mn^{2+}$.  It has not been conclusively shown whether the energy liberated in the oxidation process

$(Mn^{2+} \rightarrow Mn^{4+})$ can be utilized by the bacteria for growth and other metabolic processes. With L. discophorus, it was suggested that $Mn^{2+}$ oxidation
can provide the sole energy source for chemoautotrophic growth [2].
Under $Mn^{2+}$-limiting conditions some 70% of the available $Mn^{2+}$ was
oxidized to $MnO_2$, and the yield of cell mass was proportional to the $Mn^{2+}$
added [2]. However, these results have been discounted, and other arguments against bacterial utilization of energy from $Mn^{2+}$ oxidation have
been made [75]. One such argument is that the oxidation occurs extracel-
lularly, since it is thought that the oxidation is effected by an extracellular
protein [75].

III.  MANGANESE TRANSPORT SYSTEMS

We have found highly specific manganese active transport systems in each
and every type of cell that we have examined. The current list includes
E. coli [108,110], B. subtilis [31], B. cereus (unpublished data from our
laboratory), Rhodopseudomonas capsulata [64], Staphylococcus aureus
[129], Euglena gracilis (D. Kohl and S. Silver, unpublished data) and human
KB cells growing in culture [6 (also R. S. Beauchamp and S. Silver, unpublished data)]. Since plants also have manganese transport systems [89],
we can predict that energy-dependent transport systems specific for manganese will be found in all or essentially all living cells. These transport
systems will be in addition to the active transport of manganese as an
alternative low-affinity substrate for the cellular magnesium transport
systems [109] (see Chapter 1). The high specificity $Mn^{2+}$ transport systems are of particular interest in that they enable organisms to concentrate
$Mn^{2+}$, even in the presence of much higher levels of calcium and magnesium. Thus, for example, $^{54}Mn$ in radioactive fallout [107] would be concentrated by living organisms.

A.  Escherichia coli

The E. coli manganese transport system has a high affinity ($K_m$ of 0.2 $\mu$M)
and relatively low rate of accumulation ($V_{max}$ of 4 to 16 nmol/min per
gram dry weight)[1] [108]. We found this system only by chance [110],
while looking for an alternative substrate for the magnesium transport system. Because it saturates at submicromolar levels, we might have missed
the manganese transport system if tryptone broth had had more than the
low (less than 0.02 $\mu$M) level of $Mn^{2+}$ that it does. It is likely that the lack

---

[1] $K_m$ is the concentration for half-saturation of rate and $V_{max}$ the
maximum rate of uptake.

of reports of specific transport systems for other essential micronutrient cations is primarily due to the presence in normal growth media of contaminants at transport system saturating levels. The E. coli system is highly specific for manganese and is unaffected by 100,000-fold molar excess of the related "macronutrients," magnesium and calcium [108]; $Ni^{2+}$, $Cu^+$, $Cu^{2+}$, and $Zn^{2+}$ are not alternative high-affinity substrates for the $Mn^{2+}$ transport system of E. coli. Although iron ($Fe^{2+}$ or $Fe^{3+}$?) and cobalt ($Co^{2+}$) appear to be competitive inhibitors of $Mn^{2+}$ uptake and perhaps alternative substrates for the system, the $K_i$ values for this inhibition are some 100 times greater than the $K_m$ for $Mn^{2+}$ [108]. The system is thus essentially specific for manganese. In addition to saturation kinetics and substrate specificity, the manganese transport system of E. coli showed the other characteristics standard for bacterial active transport systems, namely, temperature dependence [110] and inhibition by poisons of energy-dependent processes, such as cyanide and the uncouplers dinitrophenol and m-chlorophenyl carbonylcyanide hydrazone (CCCP) [108]. The manganese transport system is unusual among the several active transport systems that we have studied in that inhibition by the respiratory chain inhibitor cyanide was only partial (about 50%) [110], and this raises questions about the energy coupling analogous to those we have discussed with regard to calcium transport (Chapter 2). The critical experiments concerning the coupling of $Mn^{2+}$ transport to proton movement and membrane potential [55, 74] have not been done.

The manganese which is accumulated by E. coli to a level of about 15 nmol per milliliter of cell water is probably not osmotically free, since $Mn^{2+}$ will bind to intracellular polyanions [106] in a way similar to intracellular magnesium (cf. Chapter 1). However, at least 85% of this accumulated $Mn^{2+}$ was exchangeable for added extracellular manganese [108]; and more than 80% of it was released by toluene in the absence of visible cell lysis [108]. The exchange of intracellular for extracellular manganese was also inhibited by uncouplers, suggesting that manganese cannot freely leak from the cells. Formaldehyde, an agent thought to "seal" the cell surface and inactivate membrane transport systems, seals in more than 80% of the accumulated manganese so that it can no longer exchange with extracellular manganese [108]. These characteristics of the E. coli manganese transport system are just those expected for any micronutrient transport system; and the only surprise was that E. coli, an organism without a demonstrable $Mn^{2+}$ requirement, has such a system.

In addition to $Mn^{2+}$ uptake by the high-affinity $Mn^{2+}$-specific transport system, there are alternative mechanisms for acquiring cellular $Mn^{2+}$. One is as a low-affinity substrate for the $Mg^{2+}$ transport system [100, 109]. In some bacteria (B. subtilis but not E. coli), the citrate-inducible citrate-$Mg^{2+}$ or citrate-$Mn^{2+}$ cotransport system [131] is still another transport system for $Mn^{2+}$. This system functions at higher than trace levels and probably is usually a $Mg^{2+}$-citrate system, since uptake of $Mn^{2+}$-citrate is inhibitory of further functioning [131]. Under the usual conditions for growth of E. coli and for our studies of the system (low manganese concentrations), the high-affinity, high-specificity $Mn^{2+}$ system provides most of the cellular $Mn^{2+}$. At high external $Mn^{2+}$ concentrations, however, the uptake of $Mn^{2+}$ by the $Mg^{2+}$ system ($K_m$ of about $2,000 \mu M$ $Mn^{2+}$, i.e., the $K_i$ for inhibition of $Mg^{2+}$ transport [109]) will be greater than that by the micronutrient system. Assuming equal $V_{max}$ values for $Mn^{2+}$ and for $Mg^{2+}$ transport by the $Mg^{2+}$ system, we calculated that below $40 \mu M$ $Mn^{2+}$ the high-affinity $Mn^{2+}$ system would predominate and that above this level the $Mg^{2+}$ system would be the principal source of cellular $Mn^{2+}$ [109]. The extracellular level below which the $Mn^{2+}$-specific system predominates varies with growth medium and with bacterial species, but the principle that the highly specific $Mn^{2+}$ system provides cellular $Mn^{2+}$ required at low external concentrations remains the same.

Cotransport of $Mn^{2+}$ chelated with tetracycline antibiotics is a possible route for bacterial accumulation of manganese that deserves a direct experimental test. Tetracyclines are prominent among the classes of antibiotics that are accumulated by bacterial cells by active transport systems [39] analogous to those systems for biosynthetic precursors and cations. Although it is unclear why bacterial cells should have such transport systems, it is apparently their ability to concentrate tetracycline by an energy-dependent process that makes this antibiotic toxic for bacteria [39]. The resistance of animal cells to tetracycline, on the other hand, is due to their inability to accumulate the antibiotic, rather than to an insensitivity of their protein synthesis machinery to this antibiotic. Early work showed that tetracycline uptake required a divalent cation and that $Mn^{2+}$ could replace $Mg^{2+}$ in that role [63]. Although it was once considered likely that the tetracycline was cotransported with a divalent cation [63], as it now stands the available information does not distinguish between an actual cotransport mechanism (analogous to the citrate-$Mn^{2+}$ transport system in B. subtilis [131]) or a more limited "delivery" mechanism (analogous to the citrate-dependent $Fe^{3+}$ uptake system in E. coli [94]).

## B.   Bacillus subtilis

In investigating manganese transport systems, we turned from E. coli, which has no known growth requirement for $Mn^{2+}$ to B. subtilis, an organism with a well-known requirement for $Mn^{2+}$ for sporulation [18,69]. The manganese transport system in B. subtilis shares the basic properties of specificity and energy dependence with those of other organisms. Uptake of $Mn^{2+}$ by B. subtilis was unaffected by 100,000-fold molar excess of $Mg^{2+}$ or $Ca^{2+}$ [31] but was a saturable function of the $Mn^{2+}$ concentration with a $K_m$ of about $1 \mu M$ (five times the E. coli value) and a $V_{max}$ that varied with the growth conditions from less than $0.1\,\mu$mol/min per gram dry weight of cells to more than $5.0\,\mu$mol/min per gram [36,101] (see Section III.A). Manganese uptake was temperature dependent and inhibited by both cyanide and uncouplers [31]. The manganese accumulated by log phase cells or nonsporulating stationary phase cells could be exchanged for extracellular $Mn^{2+}$ and was released by treatment with toluene or lysis with lysozyme [31]. During sporulation, however, cellular $Mn^{2+}$ was converted to a toluene- and lysozyme-insensitive form synchronously with the packaging of $Ca^{2+}$ into the developing spore. By $T_9$ (9 hr into the sporulation cycle) 100% of the $Mn^{2+}$ from media containing low levels of $Mn^{2+}$ was found in the mature spores, and none was left in the culture medium or mother-cell cytoplasm [29]. The two positions of cell $Mn^{2+}$ were associated with two phases of $Mn^{2+}$ accumulation during sporulation. First, $Mn^{2+}$ was accumulated by B. subtilis cells during log-phase growth. This was followed by a period of equilibrium during which relatively little $Mn^{2+}$ was accumulated by the cells, although exchange of extracellular with intracellular $Mn^{2+}$ occurred. At $T_3$ (3 hr into the sporulation cycle), when massive amounts of $Ca^{2+}$ were accumulated, the remaining available $Mn^{2+}$ was also taken up by the sporulating cells; and all of the $Mn^{2+}$ eventually transferred into the developing spore [29,31].

Because $1\,\mu$M or more $Mn^{2+}$ is specifically required for sporulation in species of Bacillus (see Section II.B), we measured the level of cell $Mn^{2+}$ to determine if a movement of $Mn^{2+}$ is associated with early stages of commitment to sporulation. No such movement was found. At very low $Mn^{2+}$ concentrations (below $0.5\,\mu$M), cellular $Mn^{2+}$ was limited by the available $Mn^{2+}$, and the cells did not sporulate. With inadequate medium $Mn^{2+}$, the intracellular $Mn^{2+}$ is lowered by post-log-phase cell growth to levels below that needed for sporulation. However, for cells grown in concentrations of $Mn^{2+}$ sufficient for sporulation (1 to $10\,\mu$M $Mn^{2+}$), the cellular level of $Mn^{2+}$ was essentially constant during log-phase growth and early sporulation stages at 200 nmol per milliliter of cell water [31]. This does not imply that all of the intracellular $Mn^{2+}$ was free in solution, but simply that a high and constant $Mn^{2+}$ level was maintained. The $Mn^{2+}$ requirement for spore formation appeared to involve the maintenance of this high level, rather than an extra increase in $Mn^{2+}$ content during early sporulation.

C. Regulation of $Mn^{2+}$ Transport
   in Bacillus subtilis

The apparent $V_{max}$ for $Mn^{2+}$ accumulation by B. subtilis cells is regulated as a function of extracellular $Mn^{2+}$ by both "derepression" to high potential transport rates during $Mn^{2+}$ starvation and by inactivation of existing transport function upon the addition of high ($1\,\mu$M or more) $Mn^{2+}$ to derepressed cells [36,101,111]. No changes in affinity ($K_m$) were

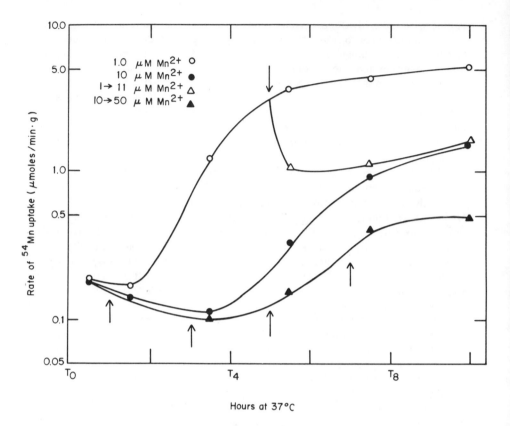

FIG. 3-4  Repression and derepression of manganese transport during
sporulation in B. subtilis.  Three sporulating cultures were grown in the
presence of 1 or 10 $\mu$M $Mn^{2+}$.  Additional $Mn^{2+}$ was added at the times
shown to raise the concentration of $Mn^{2+}$ in the medium by 10 $\mu$M per step.
Samples of the cultures were centrifuged and resuspended at a constant
turbidity in medium "spent" by nonsporulating overnight growth.  The initial rates of accumulation of 1.0 $\mu$M $^{54}$Mn were determined.  (From Ref.
101 with permission.)

observed during regulation of $Mn^{2+}$ transport [101].  Such regulation occurred in both sporulating and nonsporulating cells, and we have studied
these processes in some detail.  The derepression of manganese transport
function upon $Mn^{2+}$ starvation is shown in Figure 3-4 [101].  Early post-
log-phase cells [$T_0$ marks the end of log-phase growth] grown in 1 or
10 $\mu$M $Mn^{2+}$ had a low $Mn^{2+}$ transport rate (Fig. 3-4).  In parallel cultures
(to those of Fig. 3-4) containing radioactive $Mn^{2+}$, we determined the time

course of accumulation of the available $Mn^{2+}$ and found that it was only after $Mn^{2+}$ depletion that the $V_{max}$ for $Mn^{2+}$ transport increased. The amount of the rate increase was dependent on the concentration of available $Mn^{2+}$. Thus the increase in rates in Figure 3-4 ranges from a fivefold increase (less than $0.1 \mu mol/min$ per gram cells to $0.5 \mu mol/min$ per gram) in the culture to which $50 \mu M Mn^{2+}$ was added (in five steps: at the beginning of log-phase growth and 1, 3, 5, and 7 hr after log phase) to an increase of 25-fold (from 0.2 to more than $5 \mu mol/min$ per gram cells) in the cells sporulating with $1.0 \mu M Mn^{2+}$. It was possible partially to prevent the increase in the rate of $Mn^{2+}$ accumulation by addition of $10 \mu M$ aliquots of $Mn^{2+}$ to cells sporulating in $Mn^{2+}$. It was not possible, however, by additions of $Mn^{2+}$ every 2 hr, to maintain a high concentration of manganese in the medium, since the cells accumulated manganese so rapidly. Thus at $T_5$, 20% of the $30 \mu M Mn^{2+}$ already had been accumulated; at $T_7$, over 80% of the $40 \mu M Mn^{2+}$ was in the cells; and at our final measurement of spore $Mn^{2+}$ at $T_{20}$, fully 99.4% of the $50 \mu M Mn^{2+}$ was in the spores. Only 0.3 $\mu M$ of the added $50 \mu M$ $^{54}Mn^{2+}$ was in solution [101]. It appears that when the extracellular concentration of $Mn^{2+}$ is high, the rate of $Mn^{2+}$ uptake is low; but as the extracellular $Mn^{2+}$ level decreases, the rate of $Mn^{2+}$ accumulation increases. The process of derepression of $Mn^{2+}$ transport function was reversible; and 15 min after the addition of $10 \mu M Mn^{2+}$ to the cells sporulating in $1.0 \mu M Mn^{2+}$, the rate of $Mn^{2+}$ transport had dropped some 70% (Fig. 3-4). Protein synthesis was required for the derepression process, since inhibitors such as chloramphenicol or rifamycin prevented the rise in rate. However, since the rise in rate occurred slowly (over several hours) in cells depleted of $Mn^{2+}$ by accumulation during growth and since such cells have a rather low rate of protein synthesis, we have not pursued the mechanism of the derepression process.

We were able to study the decrease in $Mn^{2+}$ transport function in more detail since it occurred over a brief time (10 to 60 min, depending on conditions) and also required RNA and protein synthesis. Figure 3-5 depicts our current working model for the $Mn^{2+}$ transport cycle. The $V_{max}$ for $Mn^{2+}$ transport increases during $Mn^{2+}$ starvation in the absence of a change in $K_m$; we interpret this as being due to the synthesis of additional $Mn^{2+}$ transport carriers in the cell membrane (Fig. 3-5). The equilibrium $Mn^{2+}$ level in the cells follows an absorption isotherm:

$$[Mn^{2+}]_{in} = \frac{[Mn^{2+}]_{in}^{max} [Mn^{2+}]_{out}}{K_m + [Mn^{2+}]_{out}}$$

The $K_m$ in this equation is identical to that describing the Michaelis-Menten kinetics of $Mn^{2+}$ uptake:

$$\frac{d[Mn^{2+}]_{in}}{dt} = \frac{V_{max} [Mn^{2+}]_{out}}{K_m + [Mn^{2+}]_{out}}$$

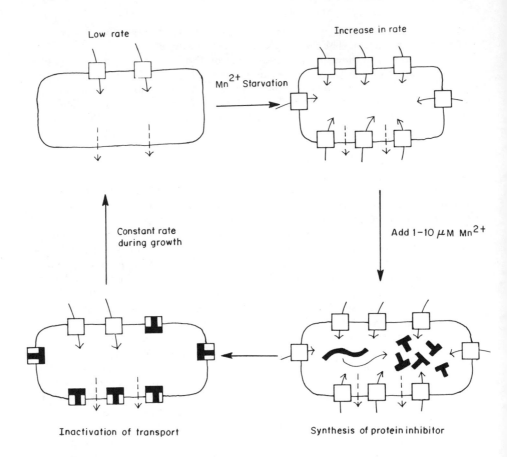

FIG. 3-5  Model for the regulation of $Mn^{2+}$ transport function during the cycle of manganese starvation followed by addition of exogenous manganese.

Since we have never observed saturation of the egress of $Mn^{2+}$, this process can be represented as $[Mn]_t/[Mn]_0 = \exp(-k_{exit}t)$; and by setting the rate of uptake equal to the rate of egress, one calculates the equilibrium maximum cellular $Mn^{2+}$ to be $[Mn^{2+}]_{in}^{max} = V_{max}$ for uptake/$k_{exit}$.  In our experiments, all of these numbers were constants ($K_m = 1\,\mu M$; $k_{exit} = 0.1/min$; and $[Mn^{2+}]_{in}^{max} = 200$ nmol per milliliter cell water), except for the regulated variable, the $V_{max}$ for uptake.  The first-order kinetics for $Mn^{2+}$ efflux do not mean that the egress process is by passive diffusion (Fig. 3-5), since efflux was temperature dependent and coupled to uptake in an energy-dependent manner.  It means only that system-saturating levels of internal $Mn^{2+}$ were not established.  Figure 3-6a shows that

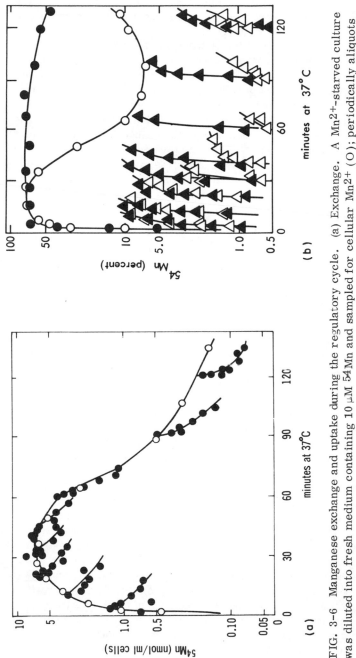

FIG. 3-6  Manganese exchange and uptake during the regulatory cycle.  (a) Exchange.  A Mn²⁺-starved culture was diluted into fresh medium containing 10 μM ⁵⁴Mn and sampled for cellular Mn²⁺ (O); periodically aliquots were diluted a further 10-fold into nonradioactive 10 μM Mn²⁺-containing medium and the exchange of radioactive for nonradioactive Mn²⁺ followed (●).  (b) RNA synthesis requirement for inhibition of Mn²⁺ transport function. A Mn²⁺-starved culture was diluted into fresh medium containing 1.0 μM ⁵⁴Mn with (●) or without (O) 5 μg/ml rifamycin (rif).  Alternatively, the Mn²⁺-starved culture was diluted into 1.0 μM nonradioactive Mn²⁺-containing medium with (▲) or without (△) rifamycin and periodically samples were diluted into the same medium with 1.0 μM ⁵⁴Mn in order to determine the rate of Mn²⁺ uptake.  (From Ref. 36 with permission.)

the exit rate remained the same throughout a regulatory cycle during which the internal $Mn^{2+}$ varied by about 25-fold. This result has been repeatedly observed in similar experiments [36].

Since we observed a 90% lowering of the $V_{max}$ for uptake with no concomitant change in $K_m$ and no change in the exit $t/2$, there are several possibilities for the relationship between the $Mn^{2+}$-uptake system and the $Mn^{2+}$-egress system. The first possibility is that $Mn^{2+}$ uptake and $Mn^{2+}$ egress occur by physically different systems (as diagrammed in Fig. 3-5). Alternatively, if the same system is functioning in both directions and if the $V_{max}$ for exit is reduced in proportion to the change in entrance rate, then the $K_m$ for exit must be reduced so as to give constant $t/2$ values over a regulatory cycle. A third alternative, of course, is that a single system is operative in both directions but that the regulatory protein shown in Figure 3-5 only affects entrance rate. Available data do not allow us to distinguish among the possibilities, but they clearly indicate that a protein regulator is involved in lowering the $V_{max}$ for uptake during a regulatory cycle. Since inhibitors of RNA and protein synthesis prevented the reduction in $V_{max}$ for $Mn^{2+}$ transport (Fig. 3-6b and Ref. 36), the synthesis of a new protein inhibitor of $Mn^{2+}$ transport was suggested (Fig. 3-5). This protein inhibitor is specific for the $Mn^{2+}$ transport system and is without effect on the functioning of the $Mg^{2+}$ transport system of the same cell [36]. Two alternatives come to mind as to the functional nature of the protein inhibitor: it might bind on a one-to-one basis with part of the $Mn^{2+}$-transport system, inactivating the transport function in the process; or possibly the inhibitor may be a highly specific protease destroying part of the $Mn^{2+}$ transport system. Regulation by highly specific protease action is well known both with bacterial cells [117a, 117b] and in yeast [60, 61, 117a], and intracellular proteases are thought to serve other regulatory functions in bacteria and especially in sporulating Bacillus [61, 117b]. The results of one type of experiment argue in favor of an inhibitor rather than a protease in reducing $Mn^{2+}$ transport rate. Inhibitors of RNA and protein synthesis added during the cycle of accumulation and release immediately froze the intracellular $Mn^{2+}$ content at the level of the time of addition [36], demonstrating that continuous protein synthesis was required for inactivation of $Mn^{2+}$ transport function. It is surprising that RNA synthesis inhibitors acted so rapidly, showing no period of messenger RNA function followed by decay. If a protease is involved, then it must have a self-destructive or self-inactivating function, in that it must be destroyed as rapidly as it functions. An example of such a highly specific one to one inhibitor has been found in post-log-phase B. subtilis cells. B. subtilis excretes proteases, amylase, and ribonuclease from the cells along with secondary metabolites. A highly specific inhibitor of the B. subtilis ribonuclease is synthesized at about the same time as the ribonuclease [112]. This protein is retained by the cells but can be quantitatively released by cold shock, during which process only 4% of the cellular protein is released. The inhibitor appears to inactivate the ribonuclease in a one-to-one complex that is essentially irreversible [112]; it is thought to be a protective device of the cells against reentry of ribonuclease that has been excreted [112]. The ribonuclease inhibitor, like the hypothetical $Mn^{2+}$ transport inhibitor, is an intracellular protein. Whether one could develop an in vitro assay for this function is problematical.

The isolation of a mutant that is temperature sensitive in regulation of $Mn^{2+}$-transport function promises another route to studying the inhibitory protein [101]. This mutant was selected as being inhibited in growth in the presence of high $Mn^{2+}$ at $42°$ C but was normal in its response at $30°$ C [101]. The mutant showed wild-type derepression of $Mn^{2+}$ transport function on $Mn^{2+}$ starvation and a normal release of accumulated $Mn^{2+}$ (i.e., lowering of the uptake rate) at $30°$ C. At $42°$ C, however, the accumulated $Mn^{2+}$ was maintained, indicative of a continued high rate of $Mn^{2+}$ uptake [101]. Upon shifting the mutant from $42°$ C to $30°$ C, the net loss of accumulated $Mn^{2+}$ started within minutes [101]. Whether this was due to de novo synthesis of the temperature-sensitive inhibitor or reversible inactivation of the inhibitor at $42°$ C remains to be determined.

The timing of the cycle of massive accumulation followed by release of $Mn^{2+}$ seen in Figure 3-6 for wild-type B. subtilis depends upon the concentration of added $Mn^{2+}$ used to start the cycle. With $1 \mu M$ or less $Mn^{2+}$ at $37°$ C, the peak of uptake occurred within 5 min and the trough following net loss was complete by 15 min. With $10 \mu M$ or more $Mn^{2+}$, however, the cycle took 90 to 120 min; the exact timing depended somewhat upon the bacterial concentration as well as upon that of $Mn^{2+}$, as expected since half or more (Fig. 3-6) of $10 \mu M$ $Mn^{2+}$ may be accumulated by the cells. The slowing of the cycle with increasing concentrations of $Mn^{2+}$ was due to an inhibition of protein synthesis by the accumulated $Mn^{2+}$. The level of $Mn^{2+}$ in derepressed cells can be more than 10 times the intracellular concentration found during growth at the same extracellular concentration [31,36]. RNA synthesis was inhibited more than 80% in derepressed cells exposed to $10 \mu M$ $Mn^{2+}$ as compared with the same cells diluted in low $Mn^{2+}$ broth [36]. Protein synthesis was also inhibited, as expected since RNA synthesis is reduced. The rate of RNA synthesis slowly increased over a 2-hr period so that growth began only after the establishment of a low equilibrium $Mn^{2+}$ level [36].

D.  The $Mn^{2+}$ Transport Stimulator

Early experiments with derepressed $Mn^{2+}$ transport were often carried out in fresh growth media. More recently, careful measurements have shown a difference in the rate of $Mn^{2+}$ uptake by derepressed cells in fresh or in spent growth medium (medium in which cells had previously been grown to stationary phase) that has led to the characterization of a low molecular weight $Mn^{2+}$ transport stimulator which is excreted by

derepressed cells and can function to facilitate $Mn^{2+}$ uptake by derepressed
— not by repressed — B. subtilis cells or to bind $Mn^{2+}$ in vitro. Only a
preliminary report of this work has appeared [101]. The stimulator is
synthesized during $Mn^{2+}$ starvation under both sporulating and nonsporulating conditions. It can be assayed in vitro by its ability to carry radioactive
$^{54}Mn$ through a gel filtration (G15 or G25) column. The rate of movement
through the column was indicative of a molecular weight of the order of
1,500 daltons [101]. The $Mn^{2+}$-binding activity was unaffected by excess
$Mg^{2+}$ or $Ca^{2+}$ and therefore is highly specific for $Mn^{2+}$, just as is the
cellular transport system. The material in the culture medium or that
partially purified by gel filtration stimulated the uptake of $Mn^{2+}$ by derepressed cells by about a factor of 3; it was without effect on $Mn^{2+}$-repressed
log-phase cells [101]. The chemical nature of this material has not been
determined. The activity was lost upon dialysis, but it was heat stable.
The factor was not adsorbed to activated charcoal, and it was water soluble and not extractable into organic solvents at physiological pH [101].
The material partially purified by gel filtration was ninhydrin positive;
but since a broth growth medium was used, further purification is required
before concluding that the stimulator is a polypeptide. The B. subtilis
$Mn^{2+}$ transport stimulator is not the polypeptide antibiotic bacitracin that
is produced under postgrowth conditions by B. licheniformis [53,125-128];
$Mn^{2+}$-derepressed B. subtilis cells were not stimulated in $^{54}Mn$ accumulation by added commercially prepared bacitracin (H. E. Scribner and
D. Clark, unpublished data). Note that bacitracin, mycobacillin, and the
other antibiotics are made only when $Mn^{2+}$ has been added to the growth
medium, but the $Mn^{2+}$ stimulator studied by Scribner and Clark is made
when $Mn^{2+}$ is absent. Therefore, although the polypeptide antibiotics
provide our current working model for the stimulator, none of the currently
known antibiotics is identical with the stimulator.


E.  Subcellular Membrane Vesicles

The $Mn^{2+}$ transport system was the first cation transport system demonstrated to function in the subcellular membrane vesicles isolated and characterized by Kaback and coworkers [65,66]. Bhattacharyya studied
energy-dependent $Mn^{2+}$ accumulation by vesicles prepared from both
E. coli [7] and B. subtilis [8]. Basically, the properties of energy dependence and substrate specificity in the vesicles resembled the properties
of the $Mn^{2+}$ transport systems of the intact cells and the general properties
were very similar to those of respiratory energy-dependent amino acid
transport in the same vesicles [65,66]. The $K_m$ values for the $Mn^{2+}$ systems were, however, 13 (B. subtilis [8]) or 40 (E. coli [7]) times higher
than the affinity constants for the intact cells. The most striking difference
between the intact cells and the subcellular membranes was a requirement
for high $Ca^{2+}$ levels for $Mn^{2+}$ uptake with E. coli membranes [7]. Levels

near 10 mM were optimal, and $Sr^{2+}$ but not $Mg^{2+}$ could substitute for $Ca^{2+}$. This calcium requirement was not seen with B. subtilis membranes [8].


F. Other Bacteria

Studies with bacteria other than E. coli and B. subtilis have been initiated both to establish the range of organisms with $Mn^{2+}$ transport systems (all that we have tested have high-affinity $Mn^{2+}$ transport systems) and to address specific problems unique to other organisms.

Manganese is required for the growth of $O_2$-evolving photosynthetic organisms. Although Rhodopseudomonas and other photosynthetic bacteria do not have a photosystem II (see Section IV) and hence do not evolve $O_2$, photosynthetic bacteria do accumulate $Mn^{2+}$ during growth [67]. About 25% of the intracellular $Mn^{2+}$ is found in the chromatophores [67], which contain the photosynthetic reaction centers [35]. When Rhodopseudomonas spheroides was grown in high (0.5 mM) $Mn^{2+}$, $Mn^{2+}$ was a functional substitute for about 30% of the reaction center $Fe^{2+}$ [35] (a net level of about 1 $Mn^{2+}$ per 20 chlorophyll molecules [67]). This $Mn^{2+}$ does not appear to be porphyrin embedded, nor does it function in a redox cycle. It is likely that this $Mn^{2+}$ facilitates electron transport through a ubiquinone complex [35] rather than participating as a primary electron acceptor itself. Because $Mn^{2+}$ is not the primary acceptor, it is not surprising that a requirement for $Mn^{2+}$ for growth is difficult to demonstrate. Wiessner [130] did show retarded growth of R. spheroides under some conditions with $Mn^{2+}$ starvation, but the level of $Mn^{2+}$ required for normal growth was less than 1/100 of that required by algae. Chromatophores isolated from R. spheroides grown in low $Mn^{2+}$ had only 5% of the $Mn^{2+}$ content of chromatophores from cells grown on excess $Mn^{2+}$. Rhodopseudomonas spheroides surprisingly had more than 10 times as much cellular $Mn^{2+}$ as Rhodospirillum rubrum or Chromatium, whether grown photosynthetically in the light or aerobically in the dark [67].

We have studied $Mn^{2+}$ transport in Rhodopseudomonas capsulata [64] and S. aureus [129], but not as extensively as in E. coli and B. subtilis. The R. capsulata $Mn^{2+}$ transport system functions both in aerobic heterotrophically growing cells and in anaerobic photosynthetically growing cells [64]. Photosynthetically growing R. capsulata showed a $K_m$ of 0.5 μM $Mn^{2+}$ and a $V_{max}$ of 20 nmol/min per gram dry weight of cells. The kinetic parameters are very similar to those found for $Mn^{2+}$ transport in E. coli. The specificity of this highly specific $Mn^{2+}$ transport system in R. capsulata [64] was similar to that in E. coli. Toluene caused the release of $Mn^{2+}$; uncouplers prevented uptake, and cyanide inhibited $Mn^{2+}$ uptake by aerobic but not by anaerobically growing R. capsulata [64]. It would be interesting to correlate the differences in $Mn^{2+}$ content of Rhodospirillum rubrum, Chromatium, and Rhodopseudomonas spheroides with differences in transport rates.

The characteristics of the $Mn^{2+}$ transport system in S. aureus [129] were similar to those already described for highly specific energy-dependent $Mn^{2+}$ uptake. However, plasmid-bearing strains have yielded additional information about $Mn^{2+}$ transport in S. aureus. Certain "resistance" plasmids (small DNA molecules which are separate from the chromosome and carry genes determining resistance to antibiotics and heavy metals) confer resistance to the heavy metal cadmium. Plasmidless strains are killed by about 10 μM $Cd^{2+}$, whereas resistant strains are killed by 1,000 μM $Cd^{2+}$ concentrations. Cadmium has been found to be a highly specific inhibitor of $Mn^{2+}$ accumulation in $Cd^{2+}$-sensitive (plasmidless) S. aureus, and $Cd^{2+}$ also stimulates the loss (egress) of $^{54}Mn$ from these cells [129]. The effects of $Cd^{2+}$ on cellular $Mn^{2+}$ appear rather specific in that $Cd^{2+}$ is without effect or has quantitatively diminished effects on cellular $^{86}Rb^+$, $^{65}Zn^{2+}$, and $^{28}Mg^{2+}$ [129]. Those S. aureus cells which are $Cd^{2+}$ resistant as a result of harboring "resistance plasmids" accumulate $Cd^{2+}$ to much lower levels than do plasmidless cells, and therefore $Cd^{2+}$-resistant S. aureus cells do not show either $Cd^{2+}$ inhibition of $Mn^{2+}$ uptake or $Cd^{2+}$-stimulated $Mn^{2+}$ loss [129].

## IV.  MICROBES OTHER THAN BACTERIA

Cation transport system(s) in yeast have been studied by A. Rothstein and his collaborators, starting in the late 1950s. These studies have been summarized by Armstrong [4], and the results relevant to this review showed that various divalent cations were accumulated by Saccharomyces cerevisiae with low affinities and low specificities. Manganese was reported to be an alternative substrate for a primary $Mg^{2+}$-$PO_4^{3-}$ transport system [4]. This system had broad specificity for $Mg^{2+}$ and $Co^{2+} > Zn^{2+} > Mn^{2+} > Ni^{2+} > Ca^{2+} > Sr^{2+}$ [4] and was studied mainly with $Mn^{2+}$, $Co^{2+}$, $Ni^{2+}$, and $Zn^{2+}$, rather than with $Mg^{2+}$, because of the availability of radioisotopes of these four cations. Accumulation of $Mn^{2+}$ by respiring yeast, unlike that by bacteria, required $PO_4^{3-}$ [4]. The accumulation of one $Mn^{2+}$ ion was associated with the egress of two $K^+$ or $Na^+$ ions; and the $Mn^{2+}$, once accumulated, was not exchangeable [4]. The experiments which led to all these results were carried out using previously starved cells and high substrate concentrations (300 μM or greater). It is clear that under such conditions, high-affinity micronutrient systems are not revealed. The more recent findings with bacterial manganese (see Section III) and magnesium (Chapter 1) allow a reinterpretation of Rothstein's results with the conclusion that he was observing the primary $Mg^{2+}$ system in yeast, analogous to those discussed in Chapter 1: $Mn^{2+}$ is a low-affinity alternative substrate for primary $Mg^{2+}$ systems at near millimolar concentrations. Hence Rothstein and coworkers produced data which indicate that $Mn^{2+}$ and also $Co^{2+}$, $Ni^{2+}$, and $Zn^{2+}$ are alternative substrates of the

$Mg^{2+}$ transport system. At submicromolar levels of $Mn^{2+}$, the micronutrient $Mn^{2+}$ transport system should be revealed.

There are no additional published reports of $Mn^{2+}$ transport systems
in algae, fungi, or protozoans. Euglena and other algae require $Mn^{2+}$ for
maximum growth [20,83], but the standard Euglena medium contains
nearly 100 $\mu M$ $Mn^{2+}$, far more than is growth limiting. Since preliminary
experiments by D. Kohn and S. Silver (unpublished) show energy-
dependent $Mg^{2+}$ accumulation that is inhibitable by high $Mn^{2+}$ and is similar
in many other properties to the bacterial systems, it would be surprising
if Euglena did not also have a trace level $Mn^{2+}$-specific system.

Normal chloroplasts of algae — or spinach for that matter — contain
one $Mn^{2+}$ ion per 50 chlorophyll molecules. The most critical role for
$Mn^{2+}$ in algae, as in higher plants, is in the photochemical complex in
photosystem II, which normally contains from 2.5 to 3 $Mn^{2+}$ ions per system II complex [19]. Chloroplasts under extreme deprivation of $Mn^{2+}$ (a
greater than 95% reduction is possible by growing the alga Scenedesmus
without $Mn^{2+}$) have lower quantum yields for photosynthesis and do not
generate $O_2$ [19]. Photosystem I, however, is not affected by $Mn^{2+}$
starvation [19]. Manganese can also be removed from chloroplasts by
washing with Tris-HCl or hydroxylamine. Normal fresh chloroplasts contain three Mn atoms per 200 chlorophylls of system II, indicating that each
system II trapping and $O_2$-evolving center contains three Mn atoms. Loss
of two-thirds of the original $Mn^{2+}$ results in total loss of $O_2$ evolution [19].
This finding allowed the assignment of two of the three manganese ions
closer to the site of $O_2$ evolution [19]; the third system II manganese was
placed closer chemically to the chlorophyll and probably is involved in
stabilizing the initial light-induced charge separation [19]. The two $Mn^{2+}$
involved in $O_2$ evolution may possibly function simultaneously to donate
two electrons each to two water molecules, yielding the four electrons required for $O_2$ evolution.

Classical nutritional studies with a variety of fungi, like classical
studies with bacteria, frequently demonstrated a growth requirement for
$Mn^{2+}$ without an explanation of the function of $Mn^{2+}$ in the cell [70].
Omission of $Mn^{2+}$ from growth medium led to a decreased growth yield
and the absence of secondary processes such as sporulation [118] and antibiotic and pigment [91] formation. As recently as 1968, the growth response of Aspergillus niger to $Mn^{2+}$ [116] was proposed as a convenient
"bioassay" method for determination of manganese in the range from
0.01 to 0.1 $\mu M$. Such a method is certainly feasible and is a versatile and
sensitive alternative to atomic absorption spectroscopy and radioisotope
analyses.

Developmental changes in fungi are regulated by manganese in the
range from 1 to 10 $\mu M$. In the absence of added $Mn^{2+}$ both Phialophora
verrucosa [91] and Aspergillus parasiticus [23,45] fail to form both normal mycelial growth and normal spores. Instead, these fungi undergo

a "dimorphic" change and produce shortened yeast-like forms [23,45,91]. Ultrastructural studies of both fungi showed that the shortened forms were abnormal and did not divide by budding, thus suggesting a role for $Mn^{2+}$ in affecting cell wall structure. As with bacterial (B. subtilis) sporulation, essentially all of the available 10 μM $Mn^{2+}$ was accumulated by P. verrucosa during a 7-day sporulation cycle [91]. However, only 23% of the accumulated $Mn^{2+}$ was found in the spores; and more than 50% was in the mycelial cytoplasm, essentially free or chelated with low molecular weight materials in a dialyzable form [91]. The characteristics of a $Mn^{2+}$ transport system in these fungi were not determined.


## V.  NONMICROBIAL SYSTEMS

The extensive literature on effects of manganese on plant [e.g., 58] and animal [e.g., 68] materials almost exclusively concerns nutrition and effects of manganese deficiency on complex multicellular organisms. We shall briefly consider cellular, subcellular, and biochemical aspects.


### A.  Mitochondria

Manganese appears to be an alternative substrate for the mitochondrial divalent cation transport system that we discussed in Chapter 2 on calcium transport. But neither dietary nor genetic manganese deficiency seems to have a significant effect on mitochondrial structure or function [62]. Strontium ($Sr^{2+}$) but not $Mg^{2+}$ is a substrate for this mitochondrial transport system [92]. Competitive binding and uptake experiments indicated a single transport system with an affinity for $Mn^{2+}$ somewhat lower than that for $Ca^{2+}$ but higher than that for $Sr^{2+}$ [92]. The paramagnetic properties of the $Mn^{2+}$ ion have provided an extra vehicle for studies of mitochondrial divalent cation transport [16,50,86,87]. The electron spin resonance spectra of $Mn^{2+}$ bound to mitochondrial membranes and/or accumulated by mitochondria have been compared with the complex spectra characteristic of "free" hexahydrated $Mn^{2+}(6H_2O)$. Under "limited loading" conditions with 100 nmol $Mn^{2+}$ accumulated per milligram of mitochondrial protein, less than 10% of the internal $Mn^{2+}$ was free; and the remainder appeared to be bound to mitochondrial protein with a mean separation distance between $Mn^{2+}$ ions of only 4.0 Å [50]. Even the 10% "free" $Mn^{2+}(6H_2O)$ level represents an appreciable concentration gradient of 20 to 500 times greater internal free $Mn^{2+}$ than external $Mn^{2+}$ [86,87]. The 500:1 gradient requires the presence of the permeant anion acetate [87] that is also accumulated by the mitochondria. Mitochondria in media containing phosphate buffer accumulate large amounts of $Mn^{2+}$. Under such "massive loading" conditions the spin resonance spectra disappeared, indicating the precipitation of $Mn^{2+}$ [50]. Water proton spin relaxation by neutron magnetic resonance was used to measure the initial tight binding of $Mn^{2+}$ to the outer mitochondrial surface [16]. These measurements led to the conclusion that $Mn^{2+}$ cannot be transported by the movement of a whole protein but may be moved via an "arm" of the protein carrier that moves independently of the main body of the protein [16]. (Bragadin et al. [14] introduced corrections to the method of Gunter and Puskin [50] and calculated matrix concentrations of $Mn^{2+}$ four to five times lower than reported previously. Even at high cation/protein ratios, the matrix $Mn^{2+}$ concentration as calculated by Bragadin et al. [14] never exceeded 10 mM.) The structure of this "carrier" and even the equivalence of the $Mn^{2+}$ and the $Ca^{2+}$ carriers are still in question.

Intracellular $Mn^{2+}$ in eukaryotic cells is accumulated by the mitochondria and therefore is largely compartmentalized [21]. This may be related to the presence of manganese-containing enzymes such as pyruvate carboxylase [104] and $Mn^{2+}$-superoxide dismutase [40] exclusively in the mitochondria. Manganoenzymes are those which contain tightly bound manganese that remains with the enzyme during purification. The pyruvate carboxylase isolated from mitochondria was the first manganoenzyme isolated and characterized as such [104]; and the mitochondrial superoxide dismutase is, as far as we know, only the second such manganoenzyme of animal origin. The bound manganese of pyruvate carboxylase is in addition to the divalent cation ($Mn^{2+}$ or $Mg^{2+}$) required for the association of substrate ATP to the enzyme [104]. Although $Mg^{2+}$ or $Mn^{2+}$ are equally effective in this role with the mitochondrial enzyme, the pyruvate carboxylase from Bacillus stearothermophilus specifically requires $Mn^{2+}$ [104].

The bound $Mn^{2+}$ of pyruvate carboxylase occurs in equimolar amounts with bound biotin (four each per enzyme complex with four active sites). This manganese functions in the second stage of the reaction sequence of pyruvate carboxylase in transferring biotin-bound $CO_2$ to enzyme-bound pyruvate to form oxaloacetate [104]. Although bound manganese was found associated with the pyruvate carboxylase from chicken liver mitochondria, other cations can replace $Mn^{2+}$ in this enzyme from other sources [104]. The enzyme from yeast appears to be a $Zn^{2+}$ metalloenzyme; but those from various bacteria have not been analyzed for cation content. Even with the chicken enzyme, the $Mn^{2+}$ was replaced by $Mg^{2+}$ when the pyruvate carboxylase was isolated from animals fed a $Mn^{2+}$-deficient diet [103,104]. The enzymes containing $Mg^{2+}$ or $Zn^{2+}$ are enzymatically active, although with some minor differences in kinetic properties. The critical requirement for the functioning of pyruvate carboxylase is for equal (4:4) levels of bound divalent cations and biotin in the enzyme [103].


B.  Plant and Animal Cells

Most studies of manganese in plants and animals have emphasized nutritional aspects and the effects of nutritional deficiencies [e.g., 68].

Published reports of $^{54}$Mn uptake by animal cells are rather preliminary [9,77,82] and essentially no additional work appears to have been done in the last 15 years. The first report was of $^{54}$Mn uptake by human erythrocytes [9] in vivo over a period of days following the rapid (within hours) disappearance of injected $^{54}$Mn from the blood plasma. This in vivo incorporation was contrasted with a failure of red blood cells in vitro to accumulate $^{54}$Mn during an 18-hr exposure. However, in a later study with duck erythrocytes, in vitro uptake of $^{54}$Mn over a 2-hr period was readily measured and studied [77]. About half of the $^{54}$Mn accumulated by duck erythrocytes was exchangeable [77], but the existence of a nonexchangeable pool of $Mn^{2+}$ that was tightly hemoglobin bound suggested that the question of $Mn^{2+}$-porphyrin complexes might merit further study [9,77]. Although $Mn^{3+}$-porphyrin complexes can be synthesized in the laboratory and have been extensively studied by chemists [10], there is no indication that we are aware of that a manganese-porphyrin complex is present or functions in any living cells. Models of $Mn^{3+}$-porphyrins as intermediates in photosystem II oxidation of $H_2O$ and in $NO_2^-$ oxidation to $NO_3^-$ have been imaginatively developed [80]. It seems likely at present, however, that redox-functional manganese in enzymes is not embedded in porphyrins [10].

Preliminary studies of $^{54}$Mn uptake by Ehrlich ascites cells [82] were aimed primarily at elucidating the presumptive role of $Mn^{2+}$ in lysine and diaminobutyrate uptake by these mammalian cells in vitro. However, the ascites tumor cells incubated in balanced salts without amino acids also rapidly accumulated trace levels of $^{54}$Mn, achieving a ratio of intracellular/ extracellular $Mn^{2+}$ of as much as 100:1 during an hour's incubation [82]. The basic amino acids stimulated the accumulation of radioactive $^{54}$Mn, but not of $^{65}$Zn, $^{59}$Fe or $^{64}$Cu. Equally preliminary experiments in our laboratory (R. S. Beauchamp and S. Silver, unpublished data) demonstrated a highly specific $Mn^{2+}$ micronutrient transport system in tissue culture cells.

Manganese uptake measurements have also been reported with plant tissues [12,81,89]. The uptake of $^{54}$Mn by rice root or leaf tissue was inhibited by standard energy poisons (e.g., azide and dinitrophenol) [89] and showed complex kinetics as a function of concentration and time of exposure, which was indicative of more than a single simple system. The $Mn^{2+}$ accumulated by roots was transported upwards into shoot tissue with kinetics similar to the "highly mobile" cation $K^+$ [89]. Manganese accumulation by sugarcane leaf tissue was also a function of external $Mn^{2+}$ concentration and was inhibited by uncouplers of oxidative phosphorylation [12]. We anticipate that transport systems analogous to those studied in bacterial cells will be demonstrated in both plant and animal cells.

The existence of truly obligatory manganoenzymes is still open to question, although the mitochondrial pyruvate carboxylase discussed in Section V.A is probably an excellent first example from animal sources, and enzymes of photosystem II may be such for plant sources. One critical plant enzyme that has an absolute requirement for $Mn^{2+}$ ($Mg^{2+}$ cannot replace $Mn^{2+}$) is the NAD-dependent malic enzyme found in the mitochondria of aspartate-type $C_4$ plants [56]. This enzyme participates in the $CO_2$ fixation cycle that enables $C_4$ plants to have a higher photosynthetic efficiency than do plants having only the Calvin cycle.

Some $Mn^{2+}$-containing proteins have been isolated and purified but lack known functions. It is sometimes suggested that these may be $Mn^{2+}$ transport or $Mn^{2+}$ storage forms. "Manganin" is the trivial name for a peanut seed globulin that contains 1 $Mn^{2+}$ ion per protein molecule [27]. Jack bean concanavalin A is another plant $Mn^{2+}$-containing globulin [1] of unknown function. "Transmanganin" is the name given [21] to a human serum protein that specifically binds $Mn^{2+}$ in a form that is exchangeable with added $Mn^{2+}$ but not with $Mg^{2+}$ [38]. This $\beta$-globulin is responsible for the essentially completely bound state of serum $Mn^{2+}$ [38] and is thought to function as a transport protein determining the differences in the fate and turnover between serum $Mg^{2+}$ and serum $Mn^{2+}$ [21,38]. Another animal manganoprotein of unknown function was isolated from chicken liver mitochondria and named "avimanganin" [102] to distinguish it from the other manganins. Avimanganin contains one $Mn^{3+}$ bound per protein molecule [102].

## VI.  SUMMARY

We have attempted to organize and rationalize the limited information available about $Mn^{2+}$ function and transport in microbial cells. Although a casual reading of the material we have discussed may lead to the conclusion that such a treatment is premature, we hope that this chapter will stimulate further studies on manganese metabolism. The existence of micronutrient $Mn^{2+}$ transport systems by all bacterial cells seems well established, if a bit too recently to be comfortably called dogma. The role and characteristics of extracellular chelates or ionophores in such systems is in the realm of preliminary hypotheses. That $Mn^{2+}$ functions in sporulation and secondary metabolism is well documented but not yet understood. At the level of enzyme activation, $Mn^{2+}$ can function with a wide variety of enzymes, but usually alternative divalent cations will also work. The number of manganoenzymes and obligatory $Mn^{2+}$-requiring enzymes is still small enough to raise the question as to whether $Mn^{2+}$ actually functions in this role in vivo. We invite the reader to let us know what should have been included in this chapter but was neglected from ignorance. More than that, we encourage you to fill the appreciable gaps in our understanding of microbial $Mn^{2+}$ metabolism by new experimentation.

ACKNOWLEDGMENTS

The research in this laboratory and the writing of this chapter were supported by National Science Foundation grant BMS71-01456 and National Institutes of Health grant AIO8062. Colleagues in development of the research and ideas were P. Bhattacharyya, D. Clark, E. Eisenstadt, S. Fisher, H. Scribner, and K. Toth. H. Scribner helped in writing the first draft, and we are grateful to Thomas Kinscherf for his thorough critical reading of this chapter. Kathleen Farrelly drew the figures and watched over the bibliography.

REFERENCES

1.  B. B. L. Agrawal and I. J. Goldstein, Protein-carbohydrate interaction. VII. Physical and chemical studies on concanavalin A, the hemagglutinin of the jack bean. Arch. Biochem. Biophys., 124, 218-229 (1968).

1a. S. P. Adler, D. Purich, and E. R. Stadtman, Cascade control of Escherichia coli glutamine synthetase: properties of the $P_{II}$ regulatory protein and the uridylyltransferase uridylyl-removing enzyme. J. Biol. Chem., 250, 6264-6272 (1975).

2.  S. H. Ali and J. L. Stokes, Stimulation of heterotrophic and autotrophic growth of Sphaerotilus discophorus by manganous ions. Antonie van Leeuwenhoek, 37, 519-528 (1971).

3.  H. Aoki and R. A. Slepecky, Inducement of a heat-shock requirement for germination and production of increased heat resistance in Bacillus fastidiosus spores by manganous ions. J. Bacteriol., 114, 137-143 (1973).

4.  W. McD. Armstrong, Ion transport and related phenomena in yeast and other micro-organisms. In Transport and Accumulation in Biological Systems, 3rd ed. (E. J. Harris, ed.), Butterworths, London, 1972, pp. 407-445.

5.  A. B. Banerjee, S. K. Majumdar, and S. K. Bose, Mycobacillin. In Antibiotics: Biosynthesis (D. Gottlieb and P. D. Shaw, eds.), Vol. 2, Springer-Verlag, New York, 1967, pp. 271-275.

6.  R. S. Beauchamp, S. Silver, and J. W. Hopkins, Uptake of $Mg^{2+}$ by KB cells. Biochim. Biophys. Acta, 225, 71-76 (1971).

7.  P. Bhattacharyya, Active transport of manganese in isolated membranes of Escherichia coli. J. Bacteriol., 104, 1307-1311 (1970).

8.  P. Bhattacharyya, Active transport of manganese in isolated membrane vesicles of Bacillus subtilis. J. Bacteriol., 123, 123-127 (1975).

9.  D. C. Borg and G. C. Cotzias, Incorporation of manganese into erythrocytes as evidence for a manganese porphyrin in man. Nature, 182, 1677-1678 (1958).

10. L. J. Boucher, Manganese porphyrin complexes. V. Axial interactions in manganese(III) porphyrins. Ann. N.Y. Acad. Sci., 206,
409-419 (1973).

11. H. J. M. Bowen, Trace Elements in Biochemistry, Academic Press,
New York, 1966.

12. J. E. Bowen, Absorption of copper, zinc, and manganese by sugarcane leaf tissue. Plant Physiol., 44, 255-261 (1969).

13. P. D. Boyer (ed.), The Enzymes, 3rd ed., 13 vols. to date,
Academic Press, New York, 1970-1976.

14. M. Bragadin, P. Dell'Antone, T. Pozzan, O. Volpato, and G. F.
Azzone, ESR determination of $Mn^{++}$ uptake and binding in mitochondria.
FEBS Letters, 60, 354-358 (1975).

15. J. Brevet and A. L. Sonenshein, Template specificity of ribonucleic
acid polymerase in asporogenous mutants of Bacillus subtilis.
J. Bacteriol., 112, 1270-1274 (1972).

16. G. D. Case, Magnetic resonance studies of the mitochondrial divalent
cation carrier. Biochim. Biophys. Acta, 375, 69-86 (1975).

17. M. Chamberlin and P. Berg, Deoxyribonucleic acid-directed synthesis of ribonucleic acid by an enzyme from Escherichia coli.
Proc. Nat. Acad. Sci. U.S., 48, 81-94 (1962).

18. J. Charney, W. P. Fisher, and C. P. Hegarty, Manganese as an
essential element for sporulation in the genus Bacillus. J. Bacteriol.,
62, 145-148 (1951).

19. G. M. Cheniae and I. F. Martin, Sites of function of manganese within
photosystem II. Roles in $O_2$ evolution and system II. Biochim. Biophys. Acta, 197, 219-239 (1970).

20. G. Constantopoulos, Lipid metabolism of manganese-deficient algae.
1. Effect of manganese deficiency on the greening and the lipid composition of Euglena gracilis Z. Plant Physiol., 45, 76-80 (1970).

21. G. C. Cotzias, Manganese. In Mineral Metabolism (C. L. Comar and
F. Bronner, eds.), Vol. 2, Pt. B, pp. 403-442, Academic Press,
New York, 1962.

22. M. Demerec and J. Hanson, Mutagenic action of manganous chloride.
Cold Spring Harbor Symp. Quant. Biol., 16, 215-228 (1951).

23. R. W. Detroy and A. Ciegler, Induction of yeastlike development in
Aspergillus parasiticus. J. Gen. Microbiobiol., 65, 259-264 (1971).

24. T. F. Deuel, A. Ginsburg, J. Yeh, E. Shelton, and E. R. Stadtman,
Bacillus subtilis glutamine synthetase: Purification and physical
characterization. J. Biol. Chem., 245, 5195-5205 (1970).

25. T. F. Deuel and S. Prusiner, Regulation of glutamine synthetase
from Bacillus subtilis by divalent cations, feedback inhibitors, and
L-glutamine. J. Biol. Chem., 249, 257-264 (1974).

26. T. F. Deuel and E. R. Stadtman, Some kinetic properties of
Bacillus subtilis glutamine synthetase. J. Biol. Chem., 245, 5206-
5213 (1970).

27. J. W. Dieckert and E. Rozacky, Isolation and partial characterization of manganin, a new manganoprotein from peanut seeds. Arch. Biochem. Biophys., 134, 473-477 (1969).

28. C. J. Efthymiou and S. W. Joseph, Difference between manganese ion requirements of pediococci and enterococci. J. Bacteriol., 112, 627-628 (1972).

29. H. L. Ehrlich, Bacteriology of manganese nodules. I. Bacterial action on manganese in nodule enrichments. Appl. Microbiol., 11, 15-19 (1963).

30. H. L. Ehrlich, Response of some activities of ferromanganese nodule bacteria to hydrostatic pressure. In Effect of the Ocean Environment on Microbial Activities (R. R. Colwell and R. Y. Morita, eds.), University Park Press, Baltimore, Maryland, 1974, pp. 208-221.

31. E. Eisenstadt, Cation transport during sporulation and germination in Bacillus subtilis. Ph.D. thesis, Washington University, St. Louis, Missouri, 1971.

32. E. Eisenstadt, Potassium content during growth and sporulation in Bacillus subtilis. J. Bacteriol., 112, 264-267 (1972).

33. E. Eisenstadt, S. Fisher, C.-L. Der, and S. Silver, Manganese transport in Bacillus subtilis W23 during growth and sporulation. J. Bacteriol., 113, 1363-1372 (1973).

34. C. Elmerich and J.-P. Aubert, Involvement of glutamine synthetase and the purine nucleotide pathway in repression of bacterial sporulation. In Spores VI (P. Gerhardt, R. N. Costilow, and H. L. Sadoff, eds.), American Society for Microbiology, Washington, D.C., 1975, pp. 385-390.

35. G. Feher, R. A. Isaacson, J. D. McElroy, L. C. Ackerson, and M. Y. Okamura, On the question of the primary acceptor in bacterial photosynthesis: manganese substituting for iron in reaction centers of Rhodopseudomonas spheroides R-26. Biochim. Biophys. Acta, 368, 135-139 (1974).

36. S. Fisher, L. Buxbaum, K. Toth, E. Eisenstadt, and S. Silver, Regulation of manganese accumulation and exchange in Bacillus subtilis W23. J. Bacteriol., 113, 1373-1380 (1973).

37. S. Fisher, D. Rothstein, and A. L. Sonenshein, Ribonucleic acid synthesis in permeabilized mutant and wild-type cells of Bacillus subtilis. In Spores VI (P. Gerhardt, R. N. Costilow, and H. L. Sadoff, eds.), American Society for Microbiology, Washington, D.C., 1975, pp. 226-230.

38. A. C. Foradori, A. Bertinchamps, J. M. Gulibon, and G. C. Cotzias, The discrimination between magnesium and manganese by serum proteins. J. Gen. Physiol., 50, 2255-2266 (1967).

39. T. J. Franklin, Antibiotic transport in bacteria. CRC Critical Rev. Microbiol., 2, 253-272 (1973).

40. I. Fridovich, Superoxide dismutase. Advan. Enzymol., 41, 35-97 (1974).

41. E. Frieden, The chemical elements of life. Sci. Amer., 227, 52-60 (1972).

42. Ø. Frøyshov, Bacitracin biosynthesis by three complementary fractions from Bacillus licheniformis. FEBS Letters, 44, 75-78 (1974).

43. Ø. Frøyshov and S. G. Laland, On the biosynthesis of bacitracin by a soluble enzyme complex from Bacillus licheniformis. Eur. J. Biochem., 46, 235-242 (1974).

44. R. Fukuda, G. Keilman, E. McVey, and R. H. Doi, Ribonucleic acid polymerase pattern of sporulating Bacillus subtilis cells. In Spores VI (P. Gerhardt, R. N. Costilow, and H. L. Sadoff, eds.), American Society for Microbiology, Washington, D.C., 1975, pp. 213-220.

45. R. G. Garrison and K. S. Boyd, Ultrastructural studies of induced morphogenesis by Aspergillus parasiticus. Sabouraudia, 12, 179-187 (1974).

46. P. Gerhardt, R. N. Costilow, and H. L. Sadoff (eds.), Spores VI, American Society for Microbiology, Washington, D.C., 1975.

47. A. Ginsburg and E. R. Stadtman, Regulation of glutamine synthetase in Escherichia coli. In The Enzymes of Glutamine Metabolism (S. Prusiner and E. R. Stadtman, eds.), Academic Press, New York, 1973, pp. 9-43.

48. M. C. Goodall, Structural effects in the action of antibiotics on the ion permeability of lipid bilayers. II. Kinetics of tyrocidine B. Biochim. Biophys. Acta, 219, 28-36 (1970).

49. M. C. Goodall, Structural effects in the action of antibiotics on the ion permeability of lipid bilayers. III. Gramicidins "A" and "S", and lipid specificity. Biochim. Biophys. Acta, 219, 471-478 (1970).

50. T. E. Gunter and J. S. Puskin, Manganous ion as a spin label in studies of mitochondrial uptake of manganese. Biophys. J., 12, 625-635 (1972).

51. H. I. Haavik, On the function of the polypeptide antibiotic bacitracin in the producer strain Bacillus licheniformis. Acta Pathol. Microbial Scand. [Sec. B], 83, 519-524 (1975).

52. H. I. Haavik, On the role of bacitracin in trace metal transport by Bacillus licheniformis. J. Gen. Microbiol., 96, 393-399 (1976).

53. H. I. Haavik and Ø. Frøyshov, Function of peptide antibiotics in producer organisms. Nature, 254, 79-82 (1975).

54. H. I. Haavik and S. Thomassen, A bacitracin-negative mutant of Bacillus licheniformis which is able to sporulate. J. Gen. Microbiol., 76, 451-454 (1973).

55. F. M. Harold, Conservation and transformation of energy by bacterial membranes. Bacteriol. Rev., 36, 172-230 (1972).

56. M. D. Hatch and T. Kagawa, NAD malic enzyme in leaves with $C_4$-pathway photosynthesis and its role in $C_4$ acid decarboxylation. Arch. Biochem. Biophys., 160, 346-349 (1974).

57. P. J. F. Henderson, J. D. McGivan, and J. B. Chappell, The action of certain antibiotics on mitochondrial, erythrocyte and artificial phospholipid membranes: the role of induced proton permeability. Biochem. J., 111, 521–535 (1969).

58. E. J. Hewitt and T. A. Smith, Plant Mineral Nutrition, The English Universities Press, London, 1975.

59. S. B. Hladky and D. A. Haydon, Discreteness of conductance change in bimolecular lipid membranes in the presence of certain antibiotics. Nature, 225, 451–453 (1970).

60. H. Holzer, Chemistry and biology of macromolecular inhibitors from yeast acting on proteinases A and B, and carboxypeptidase Y. Adv. Enzyme Regulation, 13, 125–134 (1975).

61. H. Holzer, H. Betz, and E. Ebner, Intracellular proteinases in microorganisms. Curr. Topics Cell. Regulation, 9, 103–156 (1975).

62. L. S. Hurley, L. L. Theriault, and I. E. Dreosti, Liver mitochondria from manganese-deficient and pallid mice: function and ultrastructure. Science, 170, 1316–1318 (1970).

63. K. Izaki and K. Arima, Effect of various conditions on accumulation of oxytetracycline in Escherichia coli. J. Bacteriol., 89, 1335–1339 (1965).

64. P. L. P. Jasper, Cation transport systems of Rhodopseudomonas capsulata. Ph.D. Thesis, Washington University, St. Louis, Missouri, 1975.

65. H. R. Kaback, Transport studies in bacterial membrane vesicles. Science, 186, 882–892 (1974).

66. H. R. Kaback and J.-s. Hong, Membranes and transport. CRC Critical Rev. Microbiol., 2, 333–376 (1973).

66a. J. F. Kane, Metabolic interlock: mediation of interpathway regulation by divalent cations. Arch. Biochem. Biophys., 170, 452–460 (1975).

67. R. J. Kassner and M. D. Kamen, Trace metal composition of photosynthetic bacteria. Biochim. Biophys. Acta, 153, 270–278 (1968).

68. R. M. Leach, Jr., Biochemical role of manganese. In Trace Element Metabolism in Animals-2 (W. G. Hoekstra, J. W. Suttie, H. E. Ganther, and W. Mertz, eds.), University Park Press, Baltimore, Maryland, 1974, pp. 51–59.

69. K.-Y. Lee and E. D. Weinberg, Sporulation of Bacillus megaterium: roles of metal ions. Microbios, 3, 215–224 (1971).

69a. S. G. Lee, V. Littau, and F. Lipmann, The relation between sporulation and the induction of antibiotic synthesis and of amino acid uptake in Bacillus brevis. J. Cell Biol., 66, 233–242 (1975).

70. V. G. Lilly and H. L. Barnett, Physiology of the Fungi. McGraw-Hill, New York, 1951.

71. F. Lipmann, Attempts to map a process evolution of peptide biosynthesis. Science, 173, 875–884 (1971).

72. R. A. MacLeod and E. E. Snell, Some mineral requirements of the lactic acid bacteria. J. Biol. Chem., 170, 351-365 (1947).

73. S. K. Majumdar and S. K. Bose, Trace element requirements of Bacillus subtilis for mycobacillin formation. J. Bacteriol., 79, 564-565 (1960).

74. P. Mitchell, Chemiosmotic Coupling and Energy Transduction, Glynn Research Ltd., Bodmin, Cornwall, United Kingdom, 1968.

75. E. G. Mulder, Le cycle biologique tellurique et aquatique du fer et dü manganèse. Rev. Écol. Biol. Sol, I, IX, 3, 321-348 (1972).

76. W. G. Murrell, The biochemistry of the bacterial endospore. Advan. Microbial Physiol., 1, 133-251 (1967).

77. G. R. Norris and J. R. Klein, Incorporation of manganese into duck erythrocytes. Proc. Soc. Exp. Biol. Med., 106, 288-291 (1961).

78. Y. K. Oh and E. Freese, Manganese requirement of phosphoglycerate phosphomutase and its consequences for growth and sporulation of Bacillus subtilis. J. Bacteriol., 127, 739-746 (1976).

79. T. Oka, K. Udagawa, and S. Kinoshita, Unbalanced growth death due to depletion of $Mn^{2+}$ in Brevibacterium ammoniagenes. J. Bacteriol., 96, 1760-1767 (1968).

80. J. M. Olson, The evolution of photosynthesis. Science, 168, 438-446 (1970).

81. E. R. Page and J. Dainty, Manganese uptake by excised oat roots. J. Exp. Botany, 15, 428-443 (1964).

82. P. R. Pal and H. N. Christensen, Interrelationships in the cellular uptake of amino acids and metals. J. Biol. Chem., 234, 613-617 (1959).

83. H.-D. Payer and U. Trültzsch, Ein Beitrag zur Versorgung dichter Kulturen von Grünalgen mit Mangan, Vanadium und anderen Spurenelementen. Arch. Mikrobiol., 84, 43-53 (1972).

84. J. Pero, J. Nelson, and R. Losick, In vitro and in vivo transcription by vegetative and sporulating Bacillus subtilis. In Spores VI (P. Gerhardt, R. N. Costilow, and H. L. Sadoff, eds.), American Society for Microbiology, Washington, D.C., 1975, pp. 202-212.

85. P. P. Puglisi, G. Lucchini, and A. Vecli, Resistenza allo ione manganoso ($Mn^{2+}$) in Saccharomyces cerevisiae. Atti Assoc. Genet. Ital., 15, 159-170 (1970).

86. J. S. Puskin and T. E. Gunter, Evidence for the transport of manganous ion against an activity gradient by mitochondria. Biochim. Biophys. Acta, 275, 302-307 (1972).

87. J. S. Puskin and T. E. Gunter, Ion and pH gradients across the transport membrane of mitochondria following $Mn^{++}$ uptake in the presence of acetate. Biochem. Biophys. Res. Commun., 51, 797-803 (1973).

88. A. Putrament, H. Baranowska, and W. Prazmo, Induction by manganese of mitochondrial antibiotic resistance mutations in yeast. Mol. Gen. Genet., 126, 357-366 (1973).

89. S. Ramani and S. Kannan, Manganese absorption and transport in rice. Physiol. Plant, 33, 133-137 (1975).

90. P. P. Regna, The chemistry of antibiotics. In Antibiotics: Their Chemistry and Non-medical Uses (H. S. Goldberg, ed.), Van Nostrand, Princeton, New Jersey, 1959, pp. 58-173.

91. E. Reiss and W. J. Nickerson, Control of dimorphism in Phialophora verrucosa. Sabouraudia, 12, 202-213 (1974).

92. B. Reynafarje and A. L. Lehninger, High affinity and low affinity binding of $Ca^{++}$ by rat liver mitochondria. J. Biol. Chem., 224, 584-593 (1969).

93. R. B. Roberts and E. Aldous, Manganese metabolism of Escherichia coli as related to its mutagenic action. Cold Spring Harbor Symp. Quant. Biol., 16, 229-231 (1951).

94. H. Rosenberg and I. G. Young, Iron transport in the enteric bacteria. In Microbial Iron Metabolism (J. B. Neilands, ed.), Academic Press, New York, 1974, pp. 67-82.

95. R. Roskoski, Jr., W. Gevers, H. Kleinkauf, and F. Lipmann, Tyrocidine biosynthesis by three complementary fractions from Bacillus brevis (ATCC 8185). Biochemistry, 9, 4839-4845 (1970).

96. R. Roskoski, Jr., H. Kleinkauf, W. Gevers, and F. Lipmann, Isolation of enzyme-bound peptide intermediates in tyrocidine biosynthesis. Biochemistry, 9, 4846-4851 (1970).

97. H. L. Sadoff, Sporulation antibiotics of Bacillus species. In Spores V (H. O. Halvorson, R. Hanson, and L. L. Campbell, eds.), American Society for Microbiology, Washington, D.C., 1972, pp. 157-166.

98. R. Schmitt and E. Freese, Curing of a sporulation mutant and antibiotic activity of Bacillus subtilis. J. Bacteriol., 96, 1255-1265 (1968).

99. R. Schweisfurth, Manganoxydierende Bakterien. I. Isolierung und Bestimmung einiger Stämme von Manganbakterien. Z. Allg. Mikrobiol., 13, 341-347 (1973).

100. H. Scribner, E. Eisenstadt, and S. Silver, Magnesium transport in Bacillus subtilis W23 during growth and sporulation. J. Bacteriol., 117, 1224-1230 (1974).

101. H. E. Scribner, J. Mogelson, E. Eisenstadt, and S. Silver, Regulation of cation transport during bacterial sporulation. In Spores VI (P. Gerhardt, R. N. Costilow, and H. L. Sadoff, eds.), American Society for Microbiology, Washington, D.C. 1975, pp. 346-355.

102. M. C. Scrutton, Purification and some properties of a protein containing bound manganese (Avimanganin). Biochemistry, 10, 3897-3905 (1971).

103. M. C. Scrutton, P. Griminger, and J. C. Wallace, Pyruvate carboxylase: bound metal content of the vertebrate liver enzyme as a function of diet and species. J. Biol. Chem., 247, 3305-3313 (1972).

104.  M. C. Scrutton and M. R. Young, Pyruvate carboxylase. In The Enzymes, 3rd ed. (P. D. Boyer, ed.), Vol. 6, Academic Press, New York, 1972, pp. 1-35.

105.  J. Segall and R. Losick, Effect of asporogenous mutation on rate of bacteriophage φe transcription in stationary-phase Bacillus subtilis cells. In Spores VI (P. Gerhardt, R. N. Costilow, and H. L. Sadoff, eds.), American Society for Microbiology, Washington, D.C., 1975, pp. 221-225.

106.  B. Sheard, S. H. Miall, A. R. Peacocke, I. O. Walker, and R. E. Richards, Proton magnetic relaxation studies of the binding of manganese ions to Escherichia coli ribosomes. J. Mol. Biol., 28, 389-402 (1967).

107.  W. H. Shipman, P. Simone, and H. V. Weiss, Detection of manganese-54 in radioactive fallout. Science, 126, 971-972 (1957).

108.  S. Silver, P. Johnseine, and K. King, Manganese active transport in Escherichia coli. J. Bacteriol., 104, 1299-1306 (1970).

109.  S. Silver, P. Johnseine, E. Whitney, and D. Clark, Manganese-resistant mutants of Escherichia coli: physiological and genetic studies. J. Bacteriol., 110, 186-195 (1972).

110.  S. Silver and M. L. Kralovic, Manganese accumulation by Escherichia coli: evidence for a specific transport system. Biochem. Biophys. Res. Commun., 34, 640-645 (1969).

111.  S. Silver, K. Toth, P. Bhattacharyya, E. Eisenstadt, and H. Scribner, Changes and regulation of cation transport during bacterial sporulation. In Comparative Biochemistry and Physiology of Transport (L. Bolis, K. Bloch, S. E. Luria, and F. Lynen, eds.), North-Holland Publ., Amsterdam, 1974, pp. 393-408.

112.  J. R. Smeaton and W. H. Elliott, Isolation and properties of a specific bacterial ribonuclease inhibitor. Biochim. Biophys. Acta, 145, 547-560 (1967).

113.  K. O. Stetter and O. Kandler, Manganese requirement of the transcription processes in Lactobacillus curvatus. FEBS Letters, 36, 5-8 (1973).

114.  K. O. Stetter and W. Zillig, Transcription in Lactobacillaceae: DNA-dependent RNA polymerase from Lactobacillus curvatus. Eur. J. Biochem., 48, 527-540 (1974).

115.  K. J. Stone and J. L. Strominger, Mechanism of action of bacitracin: complexation with metal ion and $C_{55}$-isoprenyl pyrophosphate. Proc. Nat. Acad. Sci. U.S., 68, 3223-3227 (1971).

116.  C. B. Sulochana and M. Lakshmanan, Aspergillus niger technique for the bioassay of manganese. J. Gen. Microbiol., 50, 285-293 (1968).

117.  J. Swift, On Poetry: A Rhapsody (1733):

>   So, Nat'ralists observe, a Flea
>   Hath smaller Fleas that on him prey,
>   And these have smaller still to bite 'em,
>   And so proceed ad infinitum . . . .

117a. R. L. Switzer, The inactivation of microbial enzymes in vivo.  Ann. Rev. Microbiol., 31, in press (1977).

117b. R. L. Switzer, C. L. Turnbough, Jr., J. S. Brabson, and L. M. Waindle, Enzyme inactivation of de novo nucleotide biosynthesis in sporulating Bacillus subtilis cells.  In Spores VI (P. Gerhardt, R. N. Costilow, and H. L. Sadoff, eds.), American Society for Microbiology, Washington, D. C., 1975, pp. 327-334.

118.  W. H. Tinnell, B. L. Jefferson, and R. E. Benoit, The organic nitrogen exigency of and effects of manganese on coremia production in Penicillium clavigerum and Penicillium claviforme.  Can. J. Microbiol., 20, 91-96 (1974).

119.  S. Tomino, M. Yamada, H. Itoh, and K. Kurahashi, Cell-free synthesis of gramicidin S.  Biochemistry, 6, 2552-2559 (1967).

120.  D. W. Urry, The gramicidin A transmembrane channel: a proposed $\pi_{(L,D)}$ helix.  Proc. Nat. Acad. Sci. U.S., 68, 672-676 (1971).

121.  D. W. Urry, A molecular theory of ion-conducting channels: a field-dependent transition between conducting and nonconducting conformations.  Proc. Nat. Acad. Sci. U.S., 69, 1610-1614 (1972).

122.  D. W. Urry, W. D. Cunningham, and T. Ohnishi, A neutral polypeptide-calcium ion complex.  Biochim. Biophys. Acta, 292, 853-857 (1973).

123.  W. L. Van Veen, Biological oxidation of manganese in soils. Antonie van Leeuwenhoek, J. Microbiol. Serol., 39, 657-662 (1973).

124.  M. Webb, The influence of certain trace metals on bacterial growth and magnesium utilization.  J. Gen. Microbiol., 51, 325-335 (1968).

125.  E. D. Weinberg, Bacitracin.  In Antibiotics: Mechanism of Action (D. Gottlieb and P. D. Shaw, eds.), Vol. 1, Springer-Verlag, New York, 1967, pp. 90-101.

126.  E. D. Weinberg, Biosynthesis of secondary metabolites: Roles of trace metals.  Advan. Microbial Physiol., 4, 1-44 (1970).

127.  E. D. Weinberg and S. M. Tonnis, Action of chloramphenicol and its isomers on secondary biosynthetic processes of Bacillus.  Appl. Microbiol., 14, 850-856 (1966).

128.  E. D. Weinberg and S. M. Tonnis, Role of manganese in biosynthesis of bacitracin.  Can. J. Microbiol., 13, 614-615 (1967).

129.  A. Weiss and S. Silver, Plasmid-determined cadmium resistance in Staphylococcus aureus: Cadmium, manganese and zinc transport systems.  J. Bacteriol., in preparation (1977).

130. W. Wiessner, Wachstum und Stoffwechsel von Rhodopseudomonas spheroides in Abhängigkeit von der Versorgung mit Mangan und Eisen. Flora, 149, 1-42 (1960).

131. K. Willecke, E.-M. Gries, and P. Oehr, Coupled transport of citrate and magnesium in Bacillus subtilis. J. Biol. Chem., 248, 807-814 (1973).

132. J. J. Windle and L. E. Sacks, Electron paramagnetic resonance of manganese(II) and copper(II) in spores. Biochim. Biophys. Acta, 66, 173-179 (1963).

133. J. E. Zajic, Microbial Biogeochemistry, Academic Press, New York, 1969.

Chapter 4

ZINC: FUNCTIONS AND TRANSPORT IN MICROORGANISMS

Mark L. Failla[1]

Department of Microbiology
Indiana University
Bloomington, Indiana

---

[1] Present address: Department of Nutrition, Cook College, Rutgers University, New Brunswick, New Jersey 08903.

As with man, so some man-made materials seem destined for
humdrum existences far removed from the glittering halls of fame
of their peers.  Silver, mercury, gold take their places with the
gods, war, and royalty; tin, lead, and zinc are for plating, cheap
useful baths, and trays — unless as with lead they have a more
dramatic if sinister function reserved for them by poisoners and
toxicologists.  Now zinc edges uncertainly into the spotlight.
Antiquity decreed a modest place for it as zinc oxide in calamine,
and this is mentioned in the Ebers Papyrus of 1550 B.C.  Over
3000 years later dermatologists prescribe the same remedy,
securely believing in its bland inertness.  This belief could be less
resistant to erosion than the metal itself, for zinc is fast becoming
a focus of diverse and powerful interests.  [From Ref. 1.]

## I.  INTRODUCTION[2]

All living systems require small quantities of zinc for normal metabolism
and growth.  While its essentiality was first recognized more than 100
years ago, it is only within the past two decades that zinc research has
generated much interest.  As a result of the ubiquity of zinc, the trace
quantities required by organisms and the apparent lack of zinc deficiency
in animals and humans, the nutritionists concentrated on the more pressing
problems, such as iron metabolism and deficiency.  Unlike the other transition metals of biological interest, zinc has a full complement of d-elec-
trons, does not participate in single electron transfers, and is spectroscopically "silent."  Obviously, the biophysical and bioinorganic chemists
found manganese, iron, cobalt, and copper more exciting elements for
study.  The environmentalists and toxicologists were concerned with such
harmful metals as cadmium, mercury, silver, and lead.  Consequently,
much of the early knowledge concerning zinc resulted mainly from plant
and fungal nutritional studies.  Anthropomorphically, zinc had a rather
"humdrum" existence.

The discoveries in the 1950s of B. L. Vallee and his associates of
the importance of zinc as an intrinsic component of many enzymes and the
realization that zinc deficiency in animals and humans is not as uncommon

---

[2] Abbreviations used in this chapter: AAS, atomic absorption spectroscopy; cyt, cytochromes; DNP, 2,4-dinitrophenol; DTNB, 5,5-dithio-
bis(2-nitrobenzoic acid); DTT, dithriothreitol; EDTA, ethylenediamine-
tetraacetic acid; NADPH, reduced nicotinamide adenine dinucleotide
phosphate; NEM, N-ethylmaleimide; pCMB, p-chloromercuribenzoate;
PHA, phytohemagglutinin.

as previously believed provided the impetus for detailed studies into the roles of this metal at the molecular and cellular levels. It is now known that zinc is essential for: (a) the activity of more than 25 enzymes, including DNA and RNA polymerases; (b) nucleic acid metabolism and cell division; (c) the storage and/or function of certain hormones, including insulin and gonadotrophin; (d) vision; (e) wound healing; and (f) the synthesis of many industrially and medically significant microbial secondary metabolites. Thus, the biocatalytic function of zinc is now well recognized. Recently, Chvapil [2] has suggested that because of its unique chemical nature, viz., a transition metal that has only one oxidation state, zinc may also be essential in the stabilization of macromolecules and cell components.

In the first part of this review I will present data concerning the molecular and cellular functions of zinc. In particular, emphasis will be placed on supporting and extending the hypothesis that the stabilization of various cell components is a basic function of zinc. After the importance of this metal is established, I will discuss the mechanisms employed by microorganisms to accumulate this essential element. References to animal cell and whole animal studies will be included when salient. Specific problems requiring further investigation will be discussed throughout the chapter. I expect that it will soon become obvious to the reader that, while our knowledge of certain aspects of zinc metabolism and function has rapidly increased, we are only beginning to recognize other, and perhaps more important, biological roles of this metal. Extensive reviews are available on zinc chemistry [3] and quantitative analysis [4], zinc nutrition in plants [5] and animals and humans [6-9], and the importance of zinc in animal development [10,11]. Additionally, the microbial biogeochemistry of zinc has been reviewed [12] and the toxicity of zinc for various microorganisms has been the subject of two recent communications [13,14]. These topics will not be discussed here.

## II. MOLECULAR AND CELLULAR FUNCTIONS OF ZINC

### A. History

The earliest demonstration of a role for zinc in living systems occurred in 1869 when one of Louis Pasteur's students, Raulin, found the metal to be essential for the growth of <u>Aspergillus niger</u> [15]. Details of subsequent research concerning the essentiality and physiological responses of organisms to various levels of zinc have been reviewed [6-8,16]. The nature of the experimental studies that have lead to our present understanding of the functions of zinc have been characterized by three overlapping periods of awareness. During the first period, the nutritional requirement of all organisms for this metal was elucidated. The period began with Raulin's observations and culminated more than 90 years later with the demonstration that zinc is essential for human growth and development [6,7]. Interestingly, the zinc requirement for bacterial growth was not demonstrated until 1947 [17], well after its essentiality for fungal (1869), algal (1900), plant (1914), and mammalian (1934) growth had been shown.

The second period of awareness was initiated in 1940 when Keilin and Mann [18] reported that bovine carbonic anhydrase is a zinc metalloenzyme containing 0.33% zinc by weight and that the metal is specifically required for catalytic activity. This was the first demonstration of a biological function for this metal. The list of zinc metalloenzymes has continued to increase, and recent technical advances assure that zinc-dependent enzymes will continue to be identified. Thus, research during this period has demonstrated that one of the principle functions of zinc in biological systems is catalysis.

The third period began with two observations which suggested that zinc is required for cell processes other than biocatalysis. First, in 1959, Wacker and Vallee [19] reported that ribonucleic acid from phylogenetically diverse organisms each contained significant quantities of tightly bound metals; magnesium, calcium, iron, and zinc are present in the highest amounts. Do these metals have a role (or roles) in nucleic acid structure, function, or metabolism? Also, since many of the zinc metalloenzymes isolated and characterized during the 1950s were pyrimidine nucleotide-dependent dehydrogenases, it was generally assumed that zinc deficiency limits growth by decreasing the activity of such enzymes. In 1962, Price and Millar [20] showed that this assumption was only partially correct since the oxygen consumption of zinc-deficient Euglena, while lower than that of zinc-sufficient cells, was great enough to support more growth than was observed. Together, these observations served as the impetus for the elucidation of functions of zinc other than as a cofactor in many enzymatic reactions.

Identification of such roles has been the focus for many of the recent studies concerning zinc. During the past decade, knowledge of specific interactions of zinc with molecular components and organelles of cells has resulted in the realization that this metal often acts as a "stabilizer" of biological systems. The term is used loosely and does not merely imply a static, passive, or inhibitory role for zinc. Rather, the data suggest that the alteration of the zinc levels within localized regions of the cell may lead to marked changes in biochemical and physiological activities. This concept will be discussed next (Section II.B to II.F).

B.  Interactions with Proteins

One of the most active areas of investigation on the biological function of zinc during the past 20 years has been the isolation and characterization of

zinc metalloenzymes. Although the zinc requirement of carbonic anhydrase was demonstrated in 1940 [18], intense work in this area did not
begin until 15 years later when Vallee and Neurath [21] observed that
bovine pancreatic carboxypeptidase is a zinc metalloenzyme. Since that
time, Vallee and his associates have provided continued leadership in the
development of techniques for the purification and physicochemical study
of metalloenzymes, in particular zinc metalloenzymes. The properties of
these zinc-dependent enzymes have been extensively reviewed elsewhere
[8,22-31].

A comprehensive list of microbial and viral zinc metalloenzymes, as
defined by the criteria of Vallee and his coworkers [25,29,30], is given in
Table 4-1 [32-93]. Classification as a metalloprotein requires that: (a)
the stability constant for the binding of the cation to the protein be relatively
high, thereby assuring retention of the metal during purification of the
macromolecule; (b) the metal content of the protein must remain at a fixed
stoichiometric ratio during purification, whereas contaminating metals
should be present in very low, nonstoichiometric quantities; and (c) if the
protein is an enzyme, activity may be inhibited by either metal-binding
agents or removal of the intrinsic cation, and such inhibition can be reversed by the addition of the intrinsic metal.

Analysis of Table 4-1 reveals that zinc metalloenzymes are widely
distributed throughout the microbial world and that certain classes of enzymes, viz., aldolases, dehydrogenases, polymerases, and proteases
from various sources, include both intracellular, wall-associated (e.g.,
alkaline phosphatase) and extracellular (e.g., amylase and neutral proteases) proteins. Moreover, many of the same enzymes in higher organisms also require zinc, for example, alcohol dehydrogenase, DNA polymerase, and superoxide dismutase (see Ref. 30 for an extensive listing of
such proteins in higher animals). Certainly, as zinc-dependent enzymes
continue to be isolated from procaryotes and eucaryotes, the highly conservative nature of the metal requirement for homologous macromolecules
in evolutionary diverse groups will become more apparent.

In Table 4-2 [94-105] are listed other enzymes isolated from microbial or viral sources that may be zinc metalloenzymes but cannot be
placed in Table 4-1 because all of the given criteria have not so far been
met. Also, various authors have reported decreased activity of specific
enzymes in zinc-deficient organisms. Examples include alcohol [106] and
glutamate [107] dehydrogenases and also tryptophan synthetase [107,108]
and the other enzymes of the tryptophan biosynthetic pathway [108] from
Neurospora crassa, glucose-6-phosphate and 6-phosphogluconate dehydrogenases from Aspergillus niger [109], and glycerol and lactate dehydrogenases from Mycobacterium smegmatis [110]. However, decreased
activity alone fails to differentiate between a specific zinc requirement for
enzyme synthesis, structure, or function. For example, Rhizopus
nigricans requires zinc for the synthesis of pyruvate carboxylase, not for

156   M. L. FAILLA

TABLE 4-1  Microbial and Viral Zinc Metalloenzymes

| Enzyme | Source | Molecular weight (daltons) | Gram-atom zinc per mole | Cofactors[a] | References |
|---|---|---|---|---|---|
| Aldolases | Aspergillus niger | 50,000 | 1 | | 32 |
| | Bacillus stearothermophilus | 61,000 | 2 | | 33 |
| | Candida utilis | 67,500 | 1 | K | 34 |
| | Escherichia coli | 140,000 | 2 | | 35,36 |
| | Saccharomyces carlsbergensis | 80,000 | 2 | K | 37,38 |
| Alkaline phosphatase | B. subtilis | 100,000 | 2-3 | | 39 |
| | E. coli | 89,000 | 4 | Mg(1.3) | 40-42 |
| Amylase | B. subtilis | 96,000 | 1 | Ca(2) | 43,44 |
| Aspartate trans-carbamylase | E. coli | 310,000 | 6 | | 45 |
| Dehydrogenases | | | | | |
| Alcohol dehydrogenase | Saccharomyces cerevisiae | 150,000 | 4 | NAD(4) | 46-48 |
| D-Glyceraldehyde 3-phosphate dehydrogenase | S. cerevisiae | 140,000 | 2.7 | | 49 |

| Enzyme | Source | Mol wt | Zn atoms | Other | Ref. |
|---|---|---|---|---|---|
| D-Lactate cyt c reductase | S. cerevisiae | 50,000 | 4–6 | FAD(2) | 50,51 |
| NADH-dependent lactate de-hydrogenase | S. cerevisiae | | | FAD | 52,53 |
| Dihydroorotase | Clostridium oroticum | 110,000 | 4 | | 53a |
| β–Lactamase II | B. cereus | 22,000 | 2 | | 54 |
| Nuclease P$_1$ | Penicillium citrinum | 44,000 | 3 | | 55,56 |
| Phospholipase C | B. cereus | 23,000 | 2 | | 57 |
| Phosphomannose isomerase | Saccharomyces carlsbergensis | 45,000 | 1 | | 58,59 |
| Polymerases | | | | | |
| DNA polymerase | E. coli | 109,000 | 1 | Mg, Mn, Fe | 60–62 |
| | Phage T$_4$ | 112,000 | 1 | | 62 |
| RNA polymerase | E. coli | 370,000 | 2 | | 63 |
| | (II) Euglena gracilis | | | | 64 |
| | (B) S. cerevisiae | 460,000 | 1 | | 64a |
| | (I) Yeast | 650,000 | 2.4 | | 64b |
| | T$_7$ Phage | 107,000 | 1 | | 65,66 |
| Reverse trans-criptase | Avian myeloblastosis virus | 180,000 | 1.8 | | 67–69 |

TABLE 4-1 (continued)

| Enzyme | Source | Molecular weight (daltons) | Grams-atoms zinc per mole | Cofactors[a] | References |
|---|---|---|---|---|---|
| "Proteolytic" enzymes[b] | | | | | |
| alkaline protease | Escherichia freundii | 45,000 | 1 | Ca | 69a |
| Aminopeptidase | S. cerevisiae | 600,000 | 1.8 | | 69a |
| Carboxypeptidase | Streptomyces griseus | 30,300 | 0.86 | | 70 |
| Carboxypeptidase $G_1$ | Pseudomonas stutzeri | 92,000 | 4 | | 70a |
| Collagenase | Clostridium histolytica | 105,000 | 1 | Ca | 71 |
| Dipeptidase | E. coli | 100,000 | 2 | Co or Mn? | 72 |
| Elastase | Pseudomonas aeruginosa | 39,500 | 1 | | 73 |
| Megateriopeptidase | B. megaterium | 40,000 | 1 | | 74 |
| Metalloprotease | B. polymyxa | ND | 1 | | 75,76 |
| Neutral protease | B. cereus | 63,000 | 1 | | 77 |
| | B. subtilis | 40,000 | 1 | Ca(4) | 78–80 |
| Neutral protein-ase I | Aspergillus sojae | 41,700 | 1 | Ca(2) | 81 |
| Neutral protein-ase II | | 19,800 | 1 | Ca(2) | 81 |
| | Streptomyces naraensis | 37,000 | 1 | | 82 |
| Protease | Serratia species | 60,000 | 1 | | 83 |

| | | | | | |
|---|---|---|---|---|---|
| Proteinase | Aeromonas proteolytica | 29,500 | 1 | | 84,85 |
| Thermolysin | Bacillus thermoproteolyticus | 37,500 | 1 | Ca(4) | 86,87 |
| Pyruvate carboxylase | S. cerevisiae | 600,000 | 3 | Biotin(3) Fe(1) | 88 |
| Superoxide dismutase[c] | Fusarium oxysporum | ND[d] | 2 | Cu(2) | 89 |
| | Neurospora crassa | 33,000 | 2 | Cu(2) | 89,90 |
| | S. cerevisiae | 32,700 | 2 | Cu(2) | 91 |
| Transcarboxylase | Propionobacterium shermanii | 670,000 | 4 | Biotin(6) Co(2) | 92 |

[a] The number in parenthesis indicates the number of gram-atoms per mole enzyme; NAD, nicotinamide adenine dinucleotide; FAD, flavin adenine dinucleotide.

[b] All zinc-metalloenzymes that cleave peptide bonds have been placed into this category.

[c] Cytoplasmic superoxide dismutase from eucaryotes is a Zn-Cu enzyme. Mitochondrial superoxide dismutase is a Mn-dependent enzyme. Procaryotic superoxide dismutases are also manganoenzymes. This has provided support in favor of the procaryote origin of mitochondria. For a detailed review of the superoxide dismutases and the evolutionary implications, see Ref. 93.

[d] ND, not determined.

TABLE 4-2   Probable Microbial and Viral Zinc Metalloenzymes

| Enzyme | Organism | References |
|--------|----------|-----------|
| ATPase | E. coli | 94 |
| DNA polymerase | Ustilago maydis | 95 |
| Glutamate dehydrogenase | Bacillus cereus | 95a |
| Hexokinase | Neurospora Crassa | 96 |
| Lactate dehydrogenase | Euglena gracilis | 97 |
| Mercaptopyruvate sulfurtransferase | E. coli | 97a |
| Neutral protease | Bacillus pumilus | 98 |
| 5'-Nucleotidase | E. coli | 99 |
| Phosphatase | Penicillium chrysogenum | 100 |
| Cyclic phosphodiesterase | E. coli | 99 |
| Phosphodiesterase | Pellicularia species | 101 |
| Phosphoglucomutase | Yeast | 102 |
| Phospholipase C | Clostridium perfrigens | 103 |
| Reverse transcriptase | Rous sarcoma virus | 104 |
| | Rauscher murine leukemic virus | 105 |
| | Rickard feline leukemic virus | 105 |
| | Wooly monkey Type C virus | 105 |
| | RD-114 virus | 105 |

enzyme activity [111]. Furthermore, unaltered activity of specific enzymes during zinc deficiency need not imply that the species are not zinc
metalloenzymes. This point is salient in that although the DNA polymerases from various sources are zinc metalloenzymes (Table 4-1 [cf. 104]),
the activity of DNA polymerase from M. smegmatis does not decrease
during zinc deficiency [112]. Perhaps the binding affinity of the polymerase for zinc is greater than that of other cell components [112]. That the
polymerase may be a zinc-dependent enzyme was actually suggested, since
the activity of a partially purified preparation was inhibited 50% in the
presence of 0.05 mM o-phenanthroline.

The catalytic function of zinc interacting with proteins has been emphasized to this point. However, one of the primary purposes of this
review is to consider the stabilizing effect of the metal on certain cell

components, organelles, and processes. Are there examples wherein
zinc has structural roles in proteins? The function of metals in the maintenance of the native conformation of proteins has recently been reviewed
[113]. Among examples from Table 4-1, the alkaline phosphatase of
Escherichia coli is of interest in that zinc is required for both structure
and catalytic activity. The native enzyme is a dimer with two atoms of
zinc that stabilize monomer-monomer binding. Two additional zinc atoms
are present in the active site [26,30,39,40]. When E. coli is cultured in
"low" zinc ($0.2 \mu M$) medium, alkaline phosphatase activity is decreased,
but the cell yield is the same as in zinc-sufficient cultures [114]. Following addition of $Zn^{2+}$ to "low zinc" cells in the stationary phase, alkaline
phosphatase activity increased without concomitant cell growth. The additional zinc is required for the aggregation of the polypeptide monomers
that are synthesized and accumulate during the exponential phase.

Escherichia coli aspartate transcarbamylase [44] and Bacillus subtilis
amylase [42,43] are examples of enzymes that require zinc solely for the
stabilization of the native quaternary structure. The presence of a zinc-
insulin complex in the pancreas has been known for some time [8]. It has
been suggested that the metal is necessary for storage of the hormone.

Zinc may also have a role in the stabilization of micro- and neurotubules [2]. These organelles are essential in intracellular movements,
cell division processes, and cell motility [115]. Tubules are produced by
the orderly aggregation of the protein monomer tubulin. One-tenth millimolar concentrations of $Zn^{2+}$, $Cd^{2+}$, $Hg^{2+}$, and $Ag^{+}$ each inhibit colchicine
binding to rat brain neurotubules at pH 7.2; moreover, 1.0 mM $Zn^{2+}$ and
$Cd^{2+}$ stabilize intact neurotubules during homogenization [116]. However,
only neurotubules stabilized by $Zn^{2+}$ are morphologically undistinguish-
able from neurotubules found in intact cells or stabilized by 1.0 M hexylene
glycol in phosphate buffer. Isolated sea urchin spermatozoa tails and tail
microtubules contained significant levels of zinc and iron that are removed
by dialysis against 0.1 mM EDTA [117]. Likewise significant quantities
of zinc and iron are present in the mitotic apparatus of sea urchin eggs.
The addition of $Zn^{2+}$ and adenosine triphosphate (ATP) to suspensions of
intact flagella or microtubules from sea urchin spermatozoa causes fluctuations in the turbidity of the suspension; ATP alone does not induce such
a response [118]. These macroscopic fluctuations might represent aggregation and disaggregation of flagella microtubules that result from
conformational changes in flagella proteins [118]. Possibly, the ATP-
induced changes lead to "disaggregation," whereas zinc holds the proteins
in close proximity, thereby assuring the proper intermolecular interactions during the reaggregation process.

Interestingly, reverse transcriptases of viruses isolated from mammalian tumors appear to be zinc metalloenzymes (Table 4-2). Metal
analysis of the minute quantities of enzyme available could not be carried
out by standard atomic absorption spectrographic procedures since at

least 0.1 mg samples are required.  Kawaguchi and Vallee [119] have
employed a low-pressure, microwave-induced emission spectrometer that
is capable of accurately determining picogram levels of metals from
0.0001 to 0.0002 mg of purified protein.  Continued development of this
technique will facilitate identification of intrinsic metals in homogeneous
preparations of other proteins, as well as the microanalysis of the metal
content and localization of metals within single cells.  In addition, replacement of the spectroscopically "silent" zinc in metalloenzymes with spectroscopically "active" cobalt has increased our knowledge of the specific
roles of metals in enzyme reactions [26,31].

## C.    Interactions with Nucleic Acids

Analysis of the metal content of RNA from many procaryotes and eucaryotes by Wacker and Vallee [19] revealed that various cations, including
magnesium, calcium, iron, zinc, nickel, copper, chromium, strontium,
barium, and aluminum, are tightly bound to the polynucleotides.  These
same metals were also found in beef liver DNA.  The ratio of metal to
nucleic acid phosphate was 1:150 and 1:50 for DNA and RNA, respectively.
Also, at concentrations equivalent to that associated with native nucleic
acids, zinc, as well as other transition metals, directly stabilized the
tertiary structure of RNA [120].  These early findings implied that metals
may have specific roles in nucleic acid structure, function, and metabolism and therefore that metal metabolism might be essential to cell processes regulated by the expression of the information stored in such macromolecules.  Let us begin this section by discussing the interaction of zinc
with nucleic acids to learn if the data support the hypothesis that this
metal affects the structure of DNA and RNA.

Pyrimidines, purines, nucleosides, nucleotides, and polynucleotides
offer multiple oxygen and/or nitrogen ligands capable of complexing $Zn^{2+}$
For example, a metal can bind to the phosphate backbone or the bases of
polynucleotides; additionally, binding to the 2' hydroxyl group of ribose
and the 3' phosphate moiety is possible in RNA.  The specific locus of
attachment has different effects upon the structure of the molecule.  In
double-stranded nucleic acids, coordination of the metal by the nitrogen
bases decreases the stability of the polymer by reducing hydrogen bonding
between complementary strands.  On the other hand, binding of the cation
to negatively charged phosphate groups stabilizes the polymer by reducing
electrostatic repulsive forces between adjacent deprotonated phosphate
moieties [121-123].  Experimental studies indicate that at low concentrations the transition metals stabilize DNA by neutralizing the negatively
charged backbone.  As the metal-to-phosphate ratio increases, especially
for zinc and copper ions, the polymer is destabilized by proton displacement between complementary base pairs.  It has been suggested that zinc

and copper may have some role in the unwinding and subsequent rewinding of double-stranded polymers required for replication and transcriptional processes [122].

More recently, the effects of $Zn^{2+}$ on the conformation of DNA from sources with G+C ratios ranging from 29 to 72% were analyzed by ultraviolet and $CD^3$ spectroscopy and polymer sedimentation behavior [124]. The results show that zinc preferentially binds to the N-7 atom of guanine residues. Also, zinc binds to phosphate sites along regions rich in adenine thymine base pairs. This interaction alters the secondary conformation of the polymer by increasing the winding angles between base pairs and changing base tilting. Furthermore, the interaction between guanine N-7 of a DNA template and the zinc cofactor of DNA polymerase may be essential for the binding of the templace to the enzyme [124].

That zinc can stabilize the structure of RNA has already been mentioned. Conversely, when the zinc/phosphate ratio exceeds 1, both synthetic and natural RNA depolymerize [122]. Bond cleavage is mediated through chelation of the metal by the 3' phosphate and the 2' hydroxyl groups. Specificity is exhibited since the 2' hydroxyl group is essential; thus, a similar degradation of DNA does not occur. Also, cleavage at uracil sites is greater than that at guanine sites. This is because guanine residues compete with phosphate and hydroxyl groups for $Zn^{2+}$, thereby reducing the metal/phosphate ratio. Finally, $Zn^{2+}$ is inhibitory to RNase [30] and the binding of this metal to nucleic acids may prevent enzyme cleavage at localized regions.

Zinc has also been shown to be essential for nucleic acid biosynthesis. Such information has been obtained from studies using the following systems: (a) microorganisms cultured in zinc-deficient media; (b) animal cells cultured in the presence of chelating agents to induce zinc deficiency; and (c) animal or microbial cells grown in zinc-supplemented media. Specific results are given in Table 4-3 [125-147].

Various aspects of the data merit discussion. First, zinc deficiency adversely affects numerous cell characteristics including DNA and/or RNA levels, protein synthesis, carbohydrate and phosphate metabolism, and cell metabolism. Second, nonspecific metal sequestering agents, such as EDTA and o-phenanthroline, inhibit the growth of cultured animal cells, presumably by preventing cellular accumulation of one or more essential trace metals. Normally, addition of most transition metals, singly, in concentrations equivalent to that of the chelating agent reverses the inhibitory activity of the metal-binding agent. However, $Zn^{2+}$ is capable of reversing this effect at one-tenth the concentration of the chelator. This suggests that cell division may cease because the chelators withhold zinc from the cells. Finally, increased RNA and protein synthesis precedes DNA synthesis in zinc-deficient cells when $Zn^{2+}$ is added, suggesting that RNA, rather than DNA, metabolism is more sensitive to zinc deficiency [133].

---

$^3$ CD = circular dichroism.

TABLE 4-3  Effects of Zinc Deficiency and Zinc Supplementation on Cellular Nucleic Acid Levels [a]

| Zinc status | Organism or cell | DNA | RNA | Comments | References |
|---|---|---|---|---|---|
| Deficiency | Candida utilis | 0 | → | Protein level unaltered; increased level of insoluble polyphosphate. | 125 |
| | Euglena gracilis | ← | → | Decreased level of protein; increased amino acid pools and insoluble polyphosphate. | 126, 127 |
| | Mycobacterium smegmatis | → | 0 | Decreased level of protein; increased polyphosphate pool. | 128, 129 |
| | Nocardia opaca | → | 0 | Decreased carbohydrate utilization; filamentous growth with bulbous swellings; manganese deficiency also decreases DNA level. | 130 |
| | Rhodotorula gracilis | 0 | → | Decreased protein synthesis; degradation of mitochondria, ribosomes and cytoplasmic components. | 131 |

| | | | | | |
|---|---|---|---|---|---|
| Zinc supplementation of zinc-deficient cells[b] | Chick embryo cells[c] | ↑ | ↑ | Actinomycin D prevents stimulation by $Zn^{2+}$. | 132,133 |
| | Human lymphocytes[d] | ↑ | 0 | $Ni^{2+}$ also reverses chelator-induced growth inhibition. | 134 |
| | Pig lymphocytes[c] | ↑ | ↑ | $Cd^{2+}$ and $Fe^{2+}$ partially reverse growth inhibition. | 135,136 |
| | Rabbit kidney cells[c] | ↑ | ↑ | | 137,138 |
| | Rat liver cells | ↑ | NR | | 139 |
| | Rat liver nuclei | ↑ | NR | Isolated from zinc-deficient animals. | 140 |
| Zinc supplementation of zinc-sufficient cells | Candida utilis | 0 | ↑ | ≥3.8 μM $Zn^{2+}$ | 141 |
| | Rhizopus nigricans | 0 | ↑ | 17 μM $Zn^{2+}$ | 142 |
| | Chick embryo cells | ↑ | ↑ | 60 μM $Zn^{2+}$; increased 2-deoxyglucose uptake. | 143-145 |
| | Human lymphocytes | ↑ | ↑ | 100 μM $Zn^{2+}$; actinomycin D inhibits stimulation by $Zn^{2+}$; blast transformation. | 146 |
| | Human lymphocytes | ↑ | NR | 250 μM $Zn^{2+}$; blast transformation. | 147 |

a ↑, increased; ↓, decreased; 0, same as control cells; NR, not reported.
b Zinc-deficiency was induced by the addition of chelators to the culture medium.
c EDTA added to induce deficiency.
d o-Phenanthroline added to induce deficiency.

Lymphocytes normally do not divide in vivo; in vitro, the addition of either mitogens (such as PHA) or certain antigens triggers macromolecular synthesis that ultimately results in transformation of the lymphocyte to an actively dividing lymphoblast. Likewise, the addition of subacute toxic quantities of $Zn^{2+}$ to lymphocytes induces this same series of events in the absence of other stimuli [143-147]. However, subacute toxic levels of $Cd^{2+}$, $Hg^{2+}$, $Mn^{2+}$, or the carcinogenic hydrocarbon 9,10-dimethyl-1,2-benzanthracene [143,145], as well as elevated pH [144], are each capable of stimulating 2-deoxyglucose uptake and DNA and RNA synthesis for a single generation in chick embryo cells. Thus, this nonspecific stimulation is different than that observed in zinc-deficient cells requiring low levels of the cation for nucleic acid biosynthesis. Rubin [145] has proposed that nonspecific stimulation can be induced by numerous physical and chemical agents that serve to accelerate ongoing reactions in a cell that has "slowed down" for various reasons.

Although the above studies support the proposal that zinc is required for nucleic acid metabolism, they fail to elucidate the nature of the roles of this metal in cell multiplication and division. Recently, Falchuk et al. [148,149] have initiated a series of experiments to answer this question. Euglena gracilis was chosen for study since the effects of zinc deficiency in this organism have been intensely examined [19,97,126,127,150-152]. In the recent work, the final number of cells in zinc-deficient medium was found to be less than 8% that of a control culture, and the intracellular zinc content of zinc-deficient cells was only 15% that of zinc-sufficient cells [148]. However, zinc-deficient cells are much larger and weigh 13 times as much as control cells. Also, zinc deficiency causes accumulation of large paramylon granules, decreases in the level of RNA, and accumulation of significantly greater intracellular quantities of magnesium, calcium, manganese, nickel, and chromium. Addition of $Zn^{2+}$ to such cells results in complete recovery [152]. The investigators concluded that zinc has essential roles in biochemical and morphological events throughout the cell cycle [148].

In a series of elegant experiments Falchuk et al. [149] have studied the effects of zinc deficiency on DNA synthesis and cell division in situ, utilizing laser-induced cytofluorometry. Cytofluorometry is a technique in which naturally fluorescent molecules, such as chlorophyll, or intracellular compounds to which a fluorescent dye can be attached in situ are placed in a beam of light at the excitation wavelength and the resultant fluorescent light pulses are viewed by a photomultiplier. In these experiments cells were fixed and treated with RNase prior to incubation with propidium diiodide, a fluorescent dye that specifically binds to nucleic acids. The cells are then passed singly through an argon laser beam, and the dye-DNA complex is excited. The intensity of the signal is directly proportional to the quantity of DNA in the cell, and the signals from 10,000 cells are subsequently analyzed by a computer. Obviously, cells containing a replicated genome, i.e., either in the $G_2$ or the M (mitotic) phase of the cell cycle, contain twice as much DNA as did cells with an unreplicated genome, i.e., in the $G_1$ phase. Therefore, the intensity of the signal from the former is twice that of the latter. This technique enables one statistically to determine the status of DNA in a population of cells and at what phase of the cell cycle the culture entered the stationary period.

Analysis of a culture of stationary, zinc-sufficient Euglena shows that most cells are in the $G_1$ phase, whereas zinc-deficient cells are either in S (i.e., in the process of replicating their genome) or $G_2$ phases. In synchronous cultures of zinc-deficient cells blocked in $G_1$ phase, DNA synthesis is initiated following addition of $Zn^{2+}$; the metal induces a transfer of the cells from the $G_1$ to the S phase. The quantity of $Zn^{2+}$ added determines whether cells are blocked in $G_1$, S, or $G_2$ phases. Finally, cells with multiple copies of DNA are present in medium containing $Cd^{2+}$, generally a competitive inhibitor of $Zn^{2+}$ in biosystems.

The foregoing observations indicate that zinc is required during all stages of the cell cycle. This conclusion is further supported by the requirement of DNA and RNA polymerases and other metals involved in nucleic acid metabolism, including DNA-terminal nucleotidyl transferase [153], for zinc (see Table 4-1). Also, the possible role of zinc in microtubule stabilization has been discussed [2,115-118]. Cytochemical studies on zinc have provided useful information. For example, the nucleoli of fertilized sea urchin eggs and starfish oocytes contain a high concentration of zinc [154]. When a nontoxic level of dithizone is added to cell preparations under conditions that selectively favor complexation of $Zn^{2+}$, a purple complex forms. Movement of the zinc-dithizone complex was followed throughout the cell cycle. The stain initially localizes in the nucleolus, passes into the spindle and the chromosomes, and then migrates with the chromosomes into the daughter cells, where it once again localizes in the nucleoli. In prostatic carcinoma cells, the nucleus position corresponding to the nucleolus stains intensely for zinc [155].

To complete this subsection let us consider some interesting and potentially significant observations concerning the ability of zinc selectively to inhibit viral activity in infected animal cells. The addition of 0.1-0.2 mM $Zn^{2+}$ at 3.5 hr postinfection (p.i.) of rabbit kidney cells by herpes simplex virus-1 (HSV-1) inhibited virus-induced giant cell formation [156]. Similarly, 0.2 mM $Zn^{2+}$ added at 3 hr p.i. inhibits the synthesis of viral progeny by 99.8% in HSV-1 infected BSC-1 cells without adversely affecting host cell DNA synthesis or morphology [157]. When added at 7 and 11 hr p.i., zinc has partial inhibitory and no inhibitory effect, respectively, on viral replication. It was suggested that inhibition of HSV-DNA synthesis might result from the selective inactivation of viral DNA polymerase by $Zn^{2+}$ [158]. Although removal of the high zinc culture medium from the infected cells partially restores HSV-DNA polymerase activity, infectious progeny are not synthesized.

The ability of various cations to inhibit viral activity of twelve different picornaviruses following infection of HeLa cells has been tested [159].
Only $Zn^{2+}$ decreases viral activity at concentrations that are not toxic to
the host cells. Of the nine viruses sensitive to this cation, eight are completely inhibited by 0.1 mM $Zn^{2+}$. It has been found that $Zn^{2+}$ prevents
cleavage of polypeptide precursors required for maturation of encephalo-
myocarditis, polio, and human rhino-1A viruses in infected HeLa cells
[160]. Finally, 0.5% (w/v) $ZnSO_4$ is therapeutically effective in the treatment of herpetic keratitis in man [161].


D.  Ribosome Stabilization

The stabilization of ribonucleoproteinaceous complexes in the compact con-
figuration required for protein synthesis is dependent upon the presence of
cationic species [162]. Both $Mg^{2+}$ and polyamines, the principle cations
associated with these organelles, have been reported to be present in
quantities as high as 2.5% of the dry weight of ribosomes. However, prolonged treatment of ribosomes with EDTA yields dissociated intermediates
that do not reassociate to the original conformation upon addition of $Mg^{2+}$
[163]. Tal found that the adverse effect of EDTA on the thermal stability
of ribosomes from E. coli is greatest at neutral pH, the hydrogen ion
concentration at which the transition metals are preferentially complexed
by the chelator [164]. Since these results suggest that metal cations other
than $Mg^{2+}$ and $Ca^{2+}$ might be essential for ribosomal stability, the metal
content of ribosomes was analyzed by emission spectroscopy and X-ray
fluorescence. Zinc, nickel, and iron were found to be present, along with
smaller quantities of manganese, lead, cobalt, chromium, vanadium,
barium, and strontium.
    Dissociated and unfolded ribosomes were dialyzed against 1.0 mM
magnesium, zinc, or nickel salts, and the sedimentation profile and specific viscosity of the reconstituted ribosomes were compared with native
ribosomes [165,166]. Magnesium-reconstituted particles are heterogeneous, characterized by S (Svedberg) values intermediate between unfolded
and native particles and have a relatively high viscosity. In contrast,
zinc- and nickel-reconstituted particles are homogeneous with S values and
viscosities very similar to native particles. Dialysis of ribosomal particles
against manganese, iron, or cobalt also produces homogeneous reconstituted particles with characteristics similar to native ribosomes. Finally, reconstitution of ribosomal subunits from unfolded particles by
dialysis against a mixture of several ions, e.g., $Zn^{2+}$, $Ni^{2+}$, and $Mn^{2+}$,
each at suboptimal concentrations, with subsequent removal of free ions,
produces stable particles with high S values. Since dialysis against transition metals proved to be more effective than against $Mg^{2+}$ in restoring
ribosomal subunits to more normal sedimentation values, Tal concluded

4. ZINC: FUNCTIONS AND TRANSPORT

that in vivo the transition metals have a role in stabilizing the native conformation of the ribosome. Unfortunately, transition metal-reconstituted
particles have not been shown to be functional in an in vitro protein-
synthesizing system [167].

Belitsina et al. [168] reported that the association of 50-S subunits
with both purified and "charged" 30-S subunits decreases in the presence
of $Zn^{2+}$ or $Ni^{2+}$. Spermidine or $Mn^{2+}$ increase the stability of the coupled
subunits, compared with either $Ca^{2+}$- or $Mg^{2+}$-stabilized pairs. The discrepancy between the observations of Tal [165,166] and Belitsina et al.
[168] might be due to the different experimental conditions employed, viz.,
ionic strengths and, in the latter study, the presence of either 9 or 20 mM
$Mg^{2+}$ in reaction mixtures.

More recently, Reisner et al. [169] showed that either 0.2 mM $Zn^{2+}$
or $Ni^{2+}$ preserves the polyribosomal profile from homogenized Paramecium aurelia. Polyribosomes are also stabilized by 50 mM $Mg^{2+}$ or
10 mM $Co^{2+}$, whereas $Fe^{2+}$, $Cu^{2+}$, $Cd^{2+}$, $Sr^{2+}$, and $Ba^{2+}$ fail to preserve
these complexes. However, when the polyribosomes are removed from
the homogenizing solution and placed on sucrose density gradients containing $Zn^{2+}$, aggregation of ribosomal-polyribosomal material occurs.
Furthermore, AAS analysis of purified ribosomes from P. aurelia and
E. coli indicates that only magnesium is tightly bound [169]. The reason
for the discrepancy between these results and those reported by Tal [164]
is not apparent.

Other data support the suggestion that zinc has a role in ribosome
structure in vivo. Ribosome turnover was not observed during growth or
stationary phases in zinc-sufficient cultures of Euglena gracilis [151].
However, zinc-deficient cells lose ribosomes during stationary phase,
without concomitantly increasing the size of the ribosomal subunit pools.
Ribosomes reappear within 2 hr after the addition of $Zn^{2+}$. De novo protein synthesis also is required for the reappearance of cytoplasmic ribosomes. Comparison of ribosomes from zinc-sufficient and zinc-deficient
cells by AAS showed that the former have at least twice as many atoms of
zinc per ribosome as do the latter. It has been suggested that a threshold
quantity of zinc is required for the structural integrity of the ribosome in
this organism [151]. When the zinc level decreases below this limit, the
ribosomes dissociate and unfold; the particles are then attacked by
ribonucleases and proteases.

Cysts of the intestinal parasite Entamoeba invadens condense their
ribosomes into crystalline particles referred to as chromatoid bodies that
are 4-10 μm in length [170,171]. Electron microprobe analysis of cysts
containing visible chromatoid bodies reveal that these cells accumulate up
to 1% zinc by dry weight [172]. Trophozoites and cysts were stained with
dithizone to localize intracellular zinc; the former stain diffusely,
whereas the chromatoid bodies of the latter are selectively stained. Furthermore, the number of zinc atoms associated with the chromatoid body

is approximately the same as the total number of rRNA nucleotides. The authors proposed that the metal may serve to neutralize the negative charge of rRNA. Alternatively, the bound metal could indirectly stabilize these structures by inhibiting localized cytoplasmic RNase activity.

Zinc-deficient cells of the yeast Rhodotorula gracilis have fewer ribosomes and mitochondria than do control cells [131]. Zinc is found in the ribosomes from a number of rat tissues, but the specific content varies from 0.26-4.78 μg per milligram of tissue protein [173]. Addition of 10-25 μM $Zn^{2+}$ increases the incorporation of [$^{14}$C] Leu in a protein-synthesizing reaction mixture containing dog prostate polyribosomes [173]. Finally, zinc-deficient rats have a decreased brain polysomal profile and an increased 80-S ribosomal pool as compared with material from pair-fed and ad libitum control pups [174].

From the above studies it appears that zinc has some role(s) in stabilizing ribosomal subunits, intact ribosomes, and possibly polyribosomes. The specific nature of interactions between the metal and these structures is unknown. The recent report that rat liver protein synthesis elongation factor 1 (equivalent to bacterial elongation factor Tu) is a zinc-dependent enzyme represents the first demonstration of a specific role for zinc in the translation process [175].

E.  Stabilization of Membranes

Chvapil has proposed that one of the general functions of zinc in living matter is its ability to stabilize cellular components and systems [2]. In this subsection, I will consider this hypothesis by discussing the nature and the effects of interactions between zinc and various membrane components. Let us begin by briefly noting some of the earlier observations concerning stabilization of bacterial membranes by divalent cations.

By 1960 it was recognized that the stabilization of osmotically fragile protoplasts and spheroplasts by $Mg^{2+}$ was due to more than the establishment of osmotic equilibrium [176]. Later, $Zn^{2+}$ and $Cu^{2+}$ were shown to be 200 and 1,000 times, respectively, as effective as $Mg^{2+}$ in preventing lysis of psychrophilic marine bacteria following transfer to distilled water; $Ca^{2+}$ and $Mn^{2+}$ are only twice as effective as $Mg^{2+}$ [177]. Also, $Zn^{2+}$ and $Cu^{2+}$ prevent morphological distortion of these cells after transfer to distilled water. These ions protect the cells at concentrations as low as 5.0 mM, a level too low to produce an osmotically favorable environment.

Early insights concerning the nature of the interaction between metals and membranes are perhaps the result of membrane reconstitution studies, such as those performed by Razin and his associates [178-180]. They developed a technique for the reconstitution of Mycoplasma laidlawii membranes from solubilized membrane components that had been obtained by the disruption of the intact structure with detergents. The method requires removal of the detergent, reaggregation of the solubilized lipids and reattachment of released enzymes in the presence of divalent cations. Razin suggested that the cations are required to decrease electrostatic repulsion between lipids and proteins, thereby increasing membrane stability.

In recent years, the direct interaction of metal ions with negatively charged head groups of membrane phospholipids has been studied [181-184]. The obvious effects of cationic binding to both synthetic and natural membranes are a decrease in electrostatic repulsion forces between lipid head groups and a reduction of the net surface charge. A subtle, but perhaps more important, effect is the ability of cations to induce lipid-phase transitions between the ordered and disordered (fluid) states. Träuble and Eibl [185] found that small changes in either pH or the cationic environment markedly alter the structure of artificial membranes. Divalent cations induce fluid-to-ordered phase transitions, while monovalent cations increase membrane fluidity. All alterations occur at constant temperature. Therefore, it is possible that localized changes in the ionic environment have a central role in controlling lipid phase transitions, which in turn affect membrane function. If this is the case, some type of cation specificity is essential. Research on this problem is just beginning: $Ca^{2+}$ has been shown to induce lateral phase separations in mixed lipid membranes, but $Mg^{2+}$ fails to do so [186,187].

Do observations with intact, viable cells also show the importance of interactions between the metal ion and the membrane phospholipids? Multivalent cations protect bacterial membranes from the antibiotic polymyxin; this substance destroys membrane integrity by binding to phospholipids [188,189]. Newton [188] suggested that cations prevent polymyxin from interacting with the membrane components by binding to the negatively charged phosphate residues.

Bacitracin, a cyclic peptide antibiotic, inhibits cell wall biosynthesis by inhibiting the enzymatic dephosphorylation and subsequent recycling of undecaprenyl pyrophosphate [190]. This phosphorylated hydrocarbon functions as a lipid carrier for peptidoglycan precursors. Antibiotic activity is dependent upon the binding of bacitracin to the polar head groups of membrane phospholipids. Bacitracin also requires a divalent cation for activity; $Zn^{2+}$ and $Cd^{2+}$ potentiate this agent more than do other cations [191]. At physiological pH, binding of the antibiotic to pyrophosphate is greatest in the presence of $Zn^{2+}$ [190]. Apparently, zinc complexed to lipid head groups on the membrane surface enhances the action of bacitracin while inhibiting polymyxin.

The effect of the ionic environment on the fluidity of Pseudomonas phaseolicola membranes has been studied using spin-labeled hydrocarbons as membrane probes [192]. As predicted from the studies with artificial membranes, when cells are grown in medium supplemented with 10 mM $Ca^{2+}$, membrane fluidity decreases. Conversely, membrane fluidity increases if the cells are grown in $Na^+$-supplemented medium. Analogous observations have been reported for spin-labeled ghosts of B. subtilis; low concentrations of $Ca^{2+}$, $Mg^{2+}$, and $La^{3+}$ each have an ordering effect on the membrane lipids [193].

Most of the studies to date have used $Mg^{2+}$ and $Ca^{2+}$ as representative divalent cations. However, relatively high concentrations of zinc are present in cell membranes [2]. Some of this metal must be localized on the surfaces, where it is able to interact with the phospholipid head groups. Also, a number of zinc-calcium antagonisms in biological systems are known [194], as we shall see shortly. Are such antagonisms of any importance in regulating lipid phase transitions in the membrane that result in altered membrane function? The binding of zinc to membrane phospholipids needs to be characterized.

The interaction of zinc with membrane proteins is also important in understanding the stabilizing effect of this metal on membranes and membrane-mediated processes. Zinc nonenzymatically protects L-cell membranes during isolation, purification, and storage [195]. DTNB, fluorescein, and mercuric acetate also protect these membranes. Consequently, it was suggested that zinc mediates its effect by binding to free sulfhydryl residues of membrane proteins to form stable mercaptides [195]. Such mercaptides would prevent the formation of free radical-induced disulfide linkages that adversely alter membrane structure and function. Similarly, $Zn^{2+}$ protects lysosomal [2,196-199], macrophage [200], platelet [201], and erythrocyte membranes against physical and chemical substances capable of inducing lipid peroxidation. That zinc inhibits the actual formation of peroxides and other destructive free radicals by binding to and inhibiting enzymes such as NADPH oxidase, lipoxidase, superoxide dismutase, and membrane-bound ATPases has been proposed [2,199,201]. Finally, might the stabilizing effects of the glucocorticoid steroid hormones on lysosome membranes [202] be partially explained by the demonstrated ability of these hormones to increase the quantity of zinc accumulated by certain animal cells [203-205]?

Does experimental evidence exist in support of the foregoing proposals concerning the mechanisms by which zinc stabilizes membranes? The effect of this metal on the respiratory activity of mitochondria, bacteria, yeasts, and animal cells has been investigated. In the mid-1950s, it was reported that less than 10 μM $Zn^{2+}$ or $Cd^{2+}$ induce swelling, inhibit respiration, and uncouple oxidative phosphorylation in animal mitochondria [206,207]. Similar results were obtained with $Ca^{2+}$, $Cu^{2+}$, and $Fe^{2+}$ at tenfold higher concentrations, whereas $Cr^{2+}$, $Mn^{2+}$, $Co^{2+}$, and $Ni^{2+}$ had no such effect on mitochondria, even at 1,000 μM levels. More than a decade later, Skulachev et al. [208] reported that less than 10 μM $Zn^{2+}$ inhibits rabbit liver mitochondrial respiratory activity; $Cd^{2+}$, $Ag^+$, $Hg^{2+}$, $Ni^{2+}$, $Co^{2+}$, $Cu^{2+}$, or $Fe^{3+}$ induce a similar effect, but only at concentrations 50 to 1,000 times greater than that of $Zn^{2+}$. These authors suggested that at physiological levels $Zn^{2+}$, like antimycin A, is capable of blocking electron transfer between cyt b and c.

More recently, Kleiner [209] and Kleiner and von Jagow [210] have shown that in uncoupled rat liver and beef heart mitochondria, $Mg^{2+}$/ATP submitochondrial particles, and partially purified mitochondrial electron transport complex III, less than $10\,\mu M$ $Zn^{2+}$ reversibly inhibits electron transfer between ubiquinone and cyt b. The quantity of reduced ubiquinone increases from 11 to 50% following $Zn^{2+}$ addition. Also, the number of specific binding sites for this metal can be accounted for by ubiquinone, but not by the cytochromes. Complexation of $Zn^{2+}$ by ubiquinone and ubi-hydroquinone was studied, and it was found that the cation binds only to the latter. However, the dissociation constant is too high to suggest that the formation of a zinc-reduced ubiquinone complex mediates the inhibitory effect of $Zn^{2+}$ on electron transfer. Zinc has also been shown to decrease the respiratory activity of yeast mitochondria by blocking the oxidation of NADH and citrate [211]. Unfortunately, a nonphysiological concentration $(3,000\,\mu M)$ of the cation was used.

That the primary site of the inhibitory action of zinc is at the level of substrate rather than cytochrome has received support from observations made independently by two groups studying this phenomenon in bacteria. Addition of $100\,\mu M$ $ZnSO_4$ or $HgCl_2$ produces 100, 99, and 68% inhibition of oxygen consumption by E. coli membrane fractions when succinate, reduced nicotinamide adenine dinucleotide (NADH), and D-lactate, respectively, serve as the energy sources [212]. Since similar results are obtained by the addition of either NEM or pCMB, it was proposed that $Zn^{2+}$ binds to essential sulfhydryl residues of the respiratory chain components. Indeed, subsequent studies revealed that zinc inhibits the activity of succinate dehydrogenase and NADH oxidase by binding to reactive sulfhydryl and histidine residues, respectively [213]. It has also been found that zinc inhibits synthesis of ATP coupled to aerobic glucose metabolism [214].

Finally, the effect of $Zn^{2+}$ on various membrane-bound enzymes in membrane particles prepared from Salmonella typhimurium has been studied in an attempt to localize the site(s) at which $Zn^{2+}$ inhibits electron transfer [215]. Micromolar quantities of the cation markedly reduce the activities of energy-dependent NADH/$NADP^+$ transhydrogenase, succinate, and NADH oxidases, and D-lactate and succinate dehydrogenases; energy-independent transhydrogenase activity is unaffected. Thus, recent studies with mitochondria and bacteria show that $Zn^{2+}$ inhibits substrate oxidation at sites before the cytochromes in the respiratory chain.

Related observations have been made in studies with yeasts and animal cells. The addition of $0.76\,\mu M$ $Zn^{2+}$ and $750\,\mu M$ $Ca^{2+}$ to medium containing an inorganic nitrogen source increases the yield of Saccharomyces pastorianus [216]. When $1.1\,\mu M$ $Zn^{2+}$ is added alone, the respiratory quotient, i.e., the number of moles of oxygen consumed per mole of carbon dioxide produced, increases for both endogenous and exogenous respiration. Similarly, the respiratory adaptation of yeast from an anaerobic to an aerobic environment is inhibited by $2.5\,\mu M$ $Zn^{2+}$ [217]. The cation prevents synthesis of respiratory enzymes, including cytochromes.

The zinc content of spermatozoa and seminal fluid is relatively high
[7]. Spermatozoa have a low respiratory activity in the native milieu
[218]. However, when the cells are washed, the concentration of zinc decreases and respiratory activity significantly increases; increased oxygen
consumption is prevented by addition of $Zn^{2+}$. Likewise, when zinc is
removed from starfish spermatozoa by addition of either histidine,
cysteine, or glycine [219], and from human spermatozoa by histidine,
cysteine, or EDTA [220], oxygen consumption and cell motility both
increase.

Is there a correlation between in vitro observations concerning the
greater respiratory activity following loss of zinc from spermatozoa and
in vivo sperm metabolism? Oxidation of sulfhydryl groups to disulfides is
essential for sperm maturation [221]. Many free sulfhydryl groups are
localized in keratin-like structures in the sperm tail [222]. Much of the
zinc associated with spermatozoa originates in seminal fluid and is bound
to outer surfaces of the cell, rather than being incorporated in intracellular
materials [223]. It has been shown that $Zn^{2+}$ and $Cu^{2+}$ bind to the tail
sulfhydryl groups, thereby blocking maturation of stored spermatozoa
[221]. Furthermore, as sperm transverses the epididymal pathway the
zinc content of the extracellular fluid decreases [224]. In contrast, the
calcium level of the liquid matrix increases, as does the intracellular
concentration of cyclic AMP (adenosine 3':5'-cyclic monophosphate) [225].
The addition of 1 mM $Ca^{2+}$ activates spermatozoa by increasing cyclic
AMP levels [225]. Finally, $Zn^{2+}$ is known to inhibit adenyl cyclase [226].
Thus, it seems that respiratory activity and the development of spermatozoa may be regulated through a complex zinc-calcium antagonism. Sperm
are maintained in a resting state through the binding of zinc to the membrane and tail groups. Activation of the cells prior to ejaculation and
fertilization involves the loss of zinc, maturation of the tail, increased
respiratory activity, accumulation of calcium, and motility. It has been
suggested that the antifertility activity of monochlorohydrin may be due to
its ability to prevent the release of zinc from sperm [227].

Supplementation of mouse diets with low doses of zinc significantly
reduces in vitro cytotoxicity of silica particles for peritoneal macrophages
[200]. Increased viability is correlated with decreased phagocytic activity.
Similarly, the addition of 15 μM $Zn^{2+}$ to macrophages protects the cells
upon in vitro exposure to silica particles [228]. Chvapil [228] suggested
that zinc exerts this protective effect by inhibiting membrane enzymes
required for the normal phagocytic response. Indeed, macrophage
NADPH oxidase and ATPase are inhibited by zinc [228]. The inhibitory
action of $Zn^{2+}$ on membrane-bound ATPases of rat brain microsomal
fractions [229] and pulmonary alveolar macrophages [230] has also been
reported.

The inhibition of respiratory enzymes by zinc has been shown to have
an adverse effect on membrane-mediated processes other than phagocytosis. Galactose and leucine transport by E. coli are inhibited by

5 mM $Zn^{2+}$ [231], and 1 mM $Zn^{2+}$ noncompetitively inhibited glucose uptake and oxidation by <u>Pseudomonas aeruginosa</u> [232].  Moreover, physiological levels of $Zn^{2+}$ (0.015 mM) noncompetitively inhibits succinate-
dependent proline uptake by <u>E. coli</u> membrane vesicles by 50%, whereas
valinomycin-induced proline accumulation by $K^+$-loaded vesicles is not
affected [233].

Although these results are in agreement with the suggestion that $Zn^{2+}$
inhibits the generation of a membrane potential, recent results with intact
cells do not support this explanation.  Proline and leucine accumulation
were studied in a mutant of <u>E. coli</u> lacking succinate dehydrogenase but
having an intact succinate transport system [234].  ATP-dependent leucine
transport is inhibited, since respiratory and oxidative phosphorylating reactions are blocked by the metal.  However, energy-dependent accumulation of proline is stimulated by zinc.  The increased intracellular level of
proline results from selective inhibition of the proline exit-exchange reaction by zinc.  These data indicate that an energized membrane was
generated by endogenous metabolism.

Cell secretory processes require organelle movement and membrane
fusion.  Both platelet aggregation and collagen- or epinephrine-induced
serotonin release are inhibited by 15 $\mu$M $Zn^{2+}$ [201].  Similarly, $Zn^{2+}$
inhibits secretion of histamine from rat peritoneal mast cells in both in
vivo and in vitro systems [235].

Calcium-dependent secretion of biogenic amines, hormones, and
neurotransmitters has been extensively reviewed [236,237].  Recent
work with the ionophore A23187 has shown that increases in intracellular
calcium induces the release of histamine from mast cells [238,239],
lysozyme from human polymorphonuclear leukocytes [240], acetylcholine
from neuromuscular junctions [241], proteins from pancreatic cells [242,
243], and calcitonin from porcine thyroid tissue [244].  As our knowledge
of the functions of calcium and zinc at the molecular and cellular level
increases, it is becoming apparent that they often "catalyze" opposing
reactions.  Calcium is an activator or initiator, while zinc stabilizes or
inhibits.  Have cells evolved mechanisms to regulate the relative levels
of these cations in localized regions of the cell surface and the cell
interior?  Are zinc-calcium antagonisms important in organelle translocation, exo- and endocytotic processes, cell movements, and cell-cell
interactions?  As an example of possible antagonism, we might consider
recent findings on roles of calcium and zinc in the erythrocytes of individuals with sickle cell disease (SCD).

Persons with SCD are often zinc deficient [245].  Initial studies
indicated that the addition of 1.5 mM $Zn^{2+}$ to normal and SCD blood significantly increases the oxygen affinity of erythrocytes [245].  Zinc was
found to interact with hemoglobin and to increase the ability of the protein
to bind oxygen.  However, 0.3 mM $Zn^{2+}$, a level at which the zinc-
hemoglobin interaction is negligible, improves the deformability of

sickled cells [246]. Eaton et al. [247] had reported that sickled cells, especially in the degenerated state, were abnormally permeable to $Na^+$, $K^+$, and $Ca^{2+}$. They proposed that $Ca^{2+}$ cross-linked hemoglobin to the membrane, thereby decreasing membrane plasticity (deformability) and cell filterability. Subsequently, it has been shown that $Zn^{2+}$ inhibits $Ca^{2+}$ incorporation into erythrocytes, decreases the binding of hemoglobin to the cell membrane, and increases the deformability of the cell membrane [246,248].

F.   Presence and Possible Functions
     in the Cell Wall

Repaske originally reported the resuspension of exponentially dividing Gram-negative bacteria in lysozyme-EDTA-Tris buffered solutions induces cell lysis [249,250]. Pseudomonas aeruginosa is especially sensitive; this organism is lysed in the presence of EDTA and Tris buffer alone. Since various cations are capable of protecting organisms against the action of the enzyme-chelator solution, Repaske proposed that the complexation of cations in the bacterial wall by EDTA must be essential for destruction of cell integrity. In 1965, three laboratories independently reported on the mechanism of action of EDTA on Gram negative organisms. Gray and Wilkinson found that the bactericidal activity of the chelating agent on P. aeruginosa and Alcaligenes faecalis is associated with the release of lipopolysaccharides (LPS) from the wall, and the subsequent loss of intracellular components [251,252]. Calcium-, zinc-, and magnesium-EDTA complexes were isolated from hypertonic supernatant fluid following EDTA-induced formation of osmoplasts.

Eagon and Carson showed that lysozyme alone induces "ballooning," i.e., lesions, in the cell wall of P. aeruginosa [253]. Lysis occurs only after EDTA is added to these cells. Ashed cell walls contain magnesium, calcium, and zinc in a ratio of 4:3:2 [254]. When these three metals are simultaneously added in this ratio and at approximately the same concentrations found in the wall, EDTA-induced osmoplasts regain motility, normal morphology, respiratory activity, the ability to synthesize an induced permease system, and the ability to divide [255,256]. These reports of zinc in the cell surface layer confirmed earlier studies of Kozloff and Lute [257] on the presence of tightly bound zinc in the walls of E. coli.

Finally, Lieve developed a procedure in which brief exposure of E. coli to EDTA alters the permeability of the cell without adversely affecting metabolic processes [258,259]. The increased permeability is correlated with the release of 30-50% of LPS from the wall, a phenomenon that by itself alters neither protein nor RNA synthesis nor cell growth [260]. Here, too, the effect of EDTA is prevented by addition of cations; Lieve proposed that the formation of a magnesium-EDTA complex is responsible for LPS release [260,261].

That EDTA selectively removes magnesium, rather than zinc and/or calcium, from cell walls is questionable. Approximately equimolar quantities of magnesium, calcium, and zinc are present in the walls of P. aeruginosa [254]; zinc is also present in walls of E. coli [257]. The stability constant of the zinc-EDTA complex is approximately seven and five log units greater than that of magnesium- and calcium-EDTA complexes, respectively [262]. Thus, it is probable that zinc has an important role in stabilizing or mediating interactions between various components of the cell wall, including the anchoring of LPS.

It may be argued that the metals exist in either different regions or chemical environments of the wall and, therefore, complexation may not merely be a function of the metal-ligand stability constant. Evidence supporting the presence of zinc throughout the wall, including the outer membrane, is available. The quantity of zinc released from isolated walls of Salmonella enteritidis by exposure to EDTA is both pH- and temperature-dependent [263]. Two and three times as much of the cation was released at $37^\circ$C as compared with $22^\circ$C and $4^\circ$C, respectively. It has been suggested that the lipids of the outer membrane undergo a temperature-dependent phase transition and that increased membrane fluidity at higher temperatures increases the availability of zinc for complexation by EDTA. Furthermore, successful invasion of E. coli by bacteriophages T2 and T4 requires the presence of zinc in the cell wall [257]. Removal of zinc does not prevent phage adsorption, but the virus is unable to digest the cell wall. Addition of nanomolar quantities of $Zn^{2+}$ initiates wall digestion by the intact phage. Phage that are altered to expose tail enzyme readily digest the wall, even in the absence of zinc. The authors proposed that a zinc complex with thiolesterase activity is required for unwinding the tail fibers, an event that is a prerequisite for release of digestive viral enzyme. Since phage adsorb to the outermost region of the wall, $Zn^{2+}$ must be present near the cell surface.

Results from other studies support the suggestion that specific divalent cations present in cell walls have essential roles in maintaining structural integrity of this heterogeneous framework. For example, lysozyme-sensitive E. coli with damaged cell walls are protected by $Zn^{2+}$, $Mg^{2+}$, and $Fe^{2+}$ against lysis [264]. Treatment of S. enteritidis with EDTA releases LPS and increases permeability of the cells to actinomycin D [265]; $0.1 \mu M$ quantities of $Zn^{2+}$, $Mg^{2+}$, $Ca^{2+}$, and $Mn^{2+}$ each protect cells against the antibiotic, whereas mono- and trivalent cations, as well as $Co^{2+}$, $Cd^{2+}$, and $Fe^{2+}$, are without effect [266]. Also, the surface protein of the outer membrane of Acinetobacter can be removed from the underlying region of the outer membrane by treatment with urea or EDTA [267]. The latter also releases LPS. Reattachment of the surface protein requires the presence of either $Zn^{2+}$, $Mg^{2+}$, or $Ca^{2+}$. Finally, EDTA enhances the effectiveness of such antibiotics as novobiocin, tobramycin, gentamycin, carbenicillin, and polymyxin [268-270].

Apparently, the chelating agent removes metal ions from the cell wall or the outer surface of the membrane, thereby increasing the ability of the antibiotic to interact with components at the sensitive sites.

Few studies concerning the presence and functions of metal ions in Gram-positive bacterial walls have been done, as compared with Gram-negative cell walls. Cutinelli and Galderio [271] found that the binding affinities of walls isolated from Staphylococcus aureus are equal for various divalent cations. In addition, the walls are capable of binding six times as many divalent as monovalent cations. Baddiley [272] has proposed that one of the principal functions of teichoic acids in Gram-positive bacterial walls is the binding of divalent cations to provide wall enzymes with a proper ionic environment. Selective removal of phosphate from teichoic acids by periodate oxidation drastically reduces the ability of walls to bind divalent cations [273]. Similarly, walls of yeasts, filamentous fungi, and algae are each capable of binding metal ions to anionic groups, and it is probable that some cations have functions in the maintenance of the surface structure. (The nature of this binding is discussed in Sections III. B and III. C.)

The cell wall is attached to the inner membrane at a number of sites and metal ions may mediate or stabilize the linkage [274,275]. Extracellular enzymes are apparently synthesized at the inner surface of the membrane, transferred across the hydrophobic region of the membrane in the primary conformation, and subsequently assume their native structure [276]. Some of these enzymes require metal ions for structural or catalytic purposes. Presumably, the metal is inserted after passage of the linear polypeptide across the membrane. Alkaline phosphatase, amylase, and a number of the "proteolytic" enzymes listed in Table 4-1 are extracellular zinc metalloenzymes.

Another possible role for metals in the cell wall involves the binding of the cell to either inanimate objects or other cells. In animals, $Ca^{2+}$ plays an important function in cell-cell contact phenomena [277]. It is believed that the metal reduces the negative charge of the cell surface and bridges acidic residues of adjacent cells [277]. Alternatively, cations might alter the conformation of surface macromolecules, thereby making interactions with components of other cells more or less favorable. An interesting characteristic of certain malignant cells is that while the net surface charge is increased as compared with normal cells, the ability of the neoplastic cells to bind $Ca^{2+}$ decreases. This suggests that altered cationic binding may have some role in metastasis and perhaps in general cell adhesion and separation processes.

Is it conceivable that zinc may play some role in attachment of cells to other objects? Tzagoloff and Pratt [278] reported that $Zn^{2+}$ prevents initial attachment of male-specific DNA coliphages M13 and f1 to E. coli. Later, it was shown that $Zn^{2+}$ does not affect adsorption of male-specific RNA coliphages R17 and MS2 [279]. It was proposed that $Zn^{2+}$ itself binds to the pili tips and competitively inhibits phage adsorption, since DNA phages attach to the tips of F. pili, whereas RNA phages attach to the sides of pili. This hypothesis is supported by the observation that nontoxic levels of $Zn^{2+}$ block the formation of mating pairs between Hfr and F⁻ strains of E. coli [280]. The cation renders males infertile. Ou also found that treatment of female cells with $Zn^{2+}$ before mating increases the frequency of recombination [281]. Furthermore, $Zn^{2+}$-treated females are capable of mating with Hfr cells that lack F pili. Thus, it is possible that direct wall-to-wall contact may be important in the formation of a mating pair and that $Zn^{2+}$ and/or cations localized on the outer surface of the cell promote such cell-cell attachment.

In conclusion, data show that $Zn^{2+}$ and other cations, principally $Mg^{2+}$ and $Ca^{2+}$, are present in the cell wall of bacteria, and probably of all microorganisms. These metal ions have roles in stabilizing interactions between various components in the wall, the activity of a number of wall or extracellular enzymes, and possibly the attachment-separation forces between cell surfaces and other objects or cells.

## G. Role of Zinc in Iron Metabolism

The sporidia of Ustilago sphaerogena grow rapidly in complex medium as pink budding cells. In defined media, cells are filamentous and nonpigmented [282]. Pigmentation results from high intracellular content of cytochromes, especially cyt c. Zinc was found to be the component in yeast extract responsible for the synthesis of cyt a, b, and c and cyt oxidase. Neilands [283] also reported that the addition of $Zn^{2+}$ to zinc-deficient Ustilago results in a rapid increase in cyt c synthesis, cell growth, and the disappearance of intracellular ferrichrome. Later, it was found that this organism synthesizes greater quantities of ferrichrome in iron-deficient media, and that the presence of $30\,\mu M$ $Co^{2+}$ induces an iron deficiency, characterized by increased ferrichrome synthesis, decreased porphyrin synthesis and decreased σ-aminolevulinic dehydratase activity (σ-ALA [284]). The cobalt effect is completely reversed by the addition of $10\,\mu M$ $Zn^{2+}$. These observations suggest that $Zn^{2+}$ has an essential role in the intracellular metabolism of iron, and that the presence of micromolar quantities of cobalt affects iron metabolism in medium without added zinc. Other studies with Ustilago show that the coproporphyrin content of cells is a function of the $Zn^{2+}$ content of the growth medium and that zinc itself is required for the synthesis of σ-ALA dehydratase [285]. Furthermore, σ-ALA dehydratase is inhibited by o-phenanthroline, and Komai and Neilands suggested that this enzyme requires cobalt for activity [286]. However, more recently it has been found that bovine liver [287,288], rat liver [288], and rat erythrocyte [289] aminolevulinic dehydratases are each zinc metalloenzymes.

Addition of $20 \mu M$ $Zn^{2+}$ to a 24-hr culture of U. sphaerogena caused a sevenfold increase in the net synthetic rate of cyt c within 30 min and a sixfold increase in the incorporation of [$^{14}$C]-Leu into cyt c [149]. Protein, DNA, and RNA synthetic rates are unaffected by exogenous $Zn^{2+}$. When actinomycin D is added simultaneously with $Zn^{2+}$, the zinc-induced cyt c synthesis is inhibited, whereas macromolecular synthesis continues unaltered for up to 6 hr [290]. Thus, the effect of $Zn^{2+}$ on cyt c production appears to be mediated through some type of transcriptional control.

Observations concerning a possible role for zinc in iron metabolism have been made with other organisms. A zinc-coproporphyrin complex has been isolated from the culture medium of Corynebacterium diphtheriae [291]. Zinc is inserted into porphyrins in chromatophores from Rhodopseudomonas spheroides by a zinc-protoporphyrin chelatase [292]. This enzyme is specific for zinc and thus can be readily distinguished from ferrochelatase. Both zinc and iron protoporphyrin chelatase activities have also been found in mitochondria isolated from guinea pig liver and rabbit liver, kidney, and heart [292]. Zinc-protoporphyrin is not an intermediate in Rhodopseudomonas bacteriochlorophyll synthesis, although Schwarz had earlier suggested that the cation might have some role in chlorophyll synthesis by Chlorella [293].

The addition of $61 \mu M$ $Zn^{2+}$ to iron-sufficient cultures of Mycobacterium tuberculosis avium stimulates coproporphyrin synthesis 100-fold, and increases glycerol utilization and cell yield four- and fivefold, respectively [294]. Zinc porphyrins have also been isolated from anaerobically and aerobically cultured yeasts, both for wild type and cytoplasmic mutants deficient in cyt a, $a_3$, and b [295]. The quantity of zinc- and magnesium-protoporphyrins in erythrocytes of patients with $Pb^{2+}$ intoxication and iron-deficiency anemia has been determined by fluorescent spectroscopy and AAS [296]. Zinc-protoporphyrin, present in relatively high concentrations, is bound to globin moieties, probably at heme binding sites. Still unresolved are the issues of whether such incorporation proceeds by enzymatic or nonenzymatic mechanisms and whether or not zinc complexes are normal intermediates in heme biosynthesis that accumulate under various "pathological" conditions in microorganisms and humans. Finally, Ratledge and Hall found that both zinc and manganese are required by M. smegmatis for optimal production of the iron-sequestering agent mycobactin [297].

H.   Morphological Differentiation and the
     Production of Secondary Metabolites

Zinc stimulates conversion of the following fungi from the filamentous to the yeast form: Histoplasma capsulatum [298], U. sphaerogena [282], Mucor rouxii [299] and Candida albicans [125,300]. Zinc-deficient M. smegmatis [128, 129] and N. opaca [129] remain filamentous

after entering the stationary phase, whereas zinc-sufficient organisms of
these bacterial species fragment into short rods. The mechanism of action
of zinc is probably indirect, since zinc-deficient cells of the mycelial bacteria and Candida have significantly lower levels of RNA. A strain of B.
megaterium cultured in "low zinc" medium was unable to sporulate;
vegetative growth was not significantly affected [301]. Zinc is a key element in the production of numerous secondary metabolites of fungal and
actinomycetal origin (see Chapter 7).


III. ZINC TRANSPORT

In contrast to our rapidly increasing knowledge of functions of zinc at the
molecular and cellular levels, the mechanisms of cellular accumulation
and intracellular distribution of this metal remain obscure. A number of
reasons can be cited for this discrepancy. First, extremely small quantities of zinc are required for cellular metabolism and growth. Reliable
data concerning the amounts required are generally unavailable. Chemically defined culture media, prepared from reagent-grade chemicals,
usually contain $0.5-1.0 \mu M$ zinc as an impurity. Most microorganisms
achieve optimal cell yields in such media without $Zn^{2+}$ supplementation,
although some fungi require an additional quantity of the metal [16, 302].
Second, transport studies would be facilitated if the background level of
zinc in the experimental medium could be substantially reduced. However,
elimination of the impurities by solvent extraction techniques or alumina
adsorption have either been unsuccessful or nonspecific. Third, the high
affinity of $Zn^{2+}$ for anionic sites on the microbial surface complicates
studies concerned with the intracellular accumulation of the metal. Finally, specific interest in the functions of zinc at the cellular level is a
relatively recent development. Thus, much less is known about the accumulation of zinc than about the other essential metals discussed elsewhere in this volume, viz., magnesium, calcium, manganese, and iron.
    The available data concerning zinc transport are considered in this
section, along with some of the numerous questions that presently exist.
For the purposes of this discussion the term "accumulation" refers to the
association of zinc with the cell, whether it be surface bound or within the
region bounded by the cytoplasmic membrane. "Uptake" is synonomous
with the influx process, i.e., translocation of the metal from the environment to the cell interior. Finally, "exit-exchange" refers to efflux or the
transfer of metal from within the cell to the external environment.


A. Availability

The uptake and incorporation of zinc by organisms is partially determined
by the physicochemical status of the metal. Some agricultural crops

growing in soils with normal levels of zinc may become zinc deficient [303, 304]. Soil properties such as high pH, levels of carbonate and phosphorus, or low quantities of organic matter decrease the amount of zinc that is available to the plant roots. Likewise, formation of zinc-phytate (inositol hexaphosphate) inhibits zinc uptake by intestinal mucosal cells. Consequently, high levels of phytate in cereals and grains consumed in certain parts of the world have been implicated in the development of zinc deficiency in animals and humans [6,7].

Zinc speciation in the aqueous environment is a complex subject on which there is little agreement. Various authors have proposed that $Zn^{2+}$ [305], $ZnCl^+$ [306], $ZnCO_3$ and $Zn(OH)_2$ [307,308], and zinc-organic complexes [309] each represent the principal species of the metal found in seawater. There is, however, general agreement on the presence of increased levels of either hydrated zinc or $Zn^{2+}$ in acidic environments. Also, in alkaline environments, the tendency for zinc to be present in insoluble precipitates or adsorbed to colloids increases. In unbuffered waters, photosynthetic activity can raise the pH above 9.0 during daylight hours.

It has been proposed that the solubilization of essential metals by organic molecules is a critical factor in determining microbial growth in environments containing "low" levels of such metals, e.g., the sea [310, 311]. Evidence in support of this concept was initially obtained by Johnston [312], who found that algal growth occurs in "bad" waters (samples that do not support growth of plankton) simply by the addition of EDTA-metal mixtures containing iron, manganese, zinc, cobalt, and copper. Moreover, EDTA alone had the same effect. While the results might be due to the removal of toxic metals, rather than solubilization of essential nutrients that are present in the samples, additional experiments have shown that metal chelation by organic molecules promotes growth of plankton in such water [310,313-315].

Lange [316] has recently found that axenic cultures of some cyanobacteria grow well at alkaline pH in the absence of artificial chelators. Filtrates from these cultures support the growth of other blue-green algae that either die or grow slowly in medium without added metal-solubilizing agents. The cyanobacteria do not grow when the medium is deficient in iron, molybdenum, copper, zinc, cobalt, or manganese. These results suggest that organisms differ in their ability to sequester essential trace metals and that such sequestration is mediated by the release of organic substances with metal-binding ability. However, the data fail to differentiate between possible mechanisms for the release of the sequestering agents. Do cells synthesize and secrete complexing agents during metal deficiency, as is known to occur during microbial iron deprivation (see Chapter 5)? If so, what is the specificity of such molecules? Do some cells in the population lyse, releasing intracellular molecules with chelating ability and thereby providing the remaining viable cells with the necessary metal nutrients? Further work in this area is needed.

Do cells secrete zinc-sequestering agents in environments where the metal is present in some form that is not capable of being transported across the permeability barrier? Because of the essentiality of this metal and the ability of microbes to inhabit every known environment, some microorganisms must be capable of producing specific zinc-complexing agents; experimental evidence in support of this hypothesis is not available. The isolation and characterization of natural zinc autosequestering molecules, should they exist, would have numerous medical and industrial uses and would also greatly increase our understanding of an important aspect of the problem of microbial accumulation of zinc.

## B. Accumulation

By the late 1950s, a media campaign by environmentalists had successfully increased the public's awareness that the oceans could not serve as a sink for the radioactive waste materials being introduced from nuclear detonation fallout and the coolant waters of nuclear power plants. Living matter was shown to be capable of accumulating radionuclides by adsorption processes or through food chains. In light of its known essentiality and its relatively long half-life, 245 days, $^{65}Zn$ was studied in some detail. The concentration factors for $^{65}Zn$ by various algae are listed in Table 4-4 [317-328].

The ability of certain algae to accumulate zinc appears to be amazingly high. Almost as striking is the extreme variation among related genera and even among strains of the same species; for example, note Fuscus vesiculosis in Table 4-4. Observations made in a number of the studies from which Table 4-4 is compiled, help to explain the apparent discrepancies. They are as follows: (a) the concentration factor for heat-killed or formalin-treated cells is as great or greater than that for living cells; (b) organisms are also capable of accumulating high quantities of nonessential trace elements such as strontium, yttrium, and cesium; (c) transfer to solutions without radioisotope usually results in a loss of greater than 75% of the associated radioactive counts; and (d) the quantity of zinc bound by the organism decreases if the concentration of either protons or other cations increases. Obviously, much of the cell-associated zinc binds nonspecifically to the cell surface. For this reason the concentration factor fails to differentiate between binding and uptake processes. In an earlier paper it was reported that the quantity of $^{65}Zn$ associated with the cell is directly proportional to the light intensity, suggesting that accumulation requires an energy-dependent process [329]. However, Gutknecht showed that most of the increased accumulation results from increases in pH that occur as a result of photosynthetic activity in unbuffered solutions [321,328]. Moreover, it has been observed that the quantity of zinc associated with macrophytic algae is dependent upon the season of the year, the part of the plant sampled, and the chemical

TABLE 4-4   Accumulation of [65]Zn by Algae

| Organism | CF[a] | References |
|----------|-------|------------|
| Ascophyllum nodosum | 518 | 317 |
|  | 1,400 | 318 |
| Fuscus edentatus | 1,200 | 317 |
| Fuscus serratus | 10,768 | 319 |
|  | 600 | 318 |
| Fuscus vesiculosus | 11,000-64,000[b] | 320 |
|  | 1,100 | 318 |
|  | 1,200 | 317 |
|  | 6,900 | 321 |
| Gomphonema parvalum | 9,600 | 322 |
| Laminaria digitata | 2,455 | 313 |
|  | 725 | 318 |
| Navicula confervacea | 23,000 | 323 |
| Navicula pelliculosa | 21,600 | 322 |
| Navicula seminulum | 16,700 | 322 |
| Nitzschia closterium[c] | 50,000 | 324 |
| Nitzschia species | 42,000 | 323 |
| Pelvetia canaliculata | 1,000 | 318 |
| Phaeodactylum tricornutum | 115 | 325 |
| Phytoplankton (mainly Ceratium sp.) | 15,314 | 326 |
| Platymonas elliptica | 67,800 | 323 |
| Plectonema boryanum[c] | 25,000 | 327 |
| Porphyra umbilicus | 1,200 | 321 |
| Rhodomonas lacustris | 0 | 323 |
| Scenedesmus lubricum | 3,200 | 322 |
| Scenedesmus helveticum | 3,700 | 322 |
| Ulva lactuca | 300 | 328 |
|  | 4,100 | 321 |

characteristics of the environment from which the plant is isolated [313, 320].

The studies with algae, as well as yeasts (discussed in Section III.C.2), show that zinc, like other essential multivalent cations, is accumulated through two distinct processes, namely, adsorption and uptake. Adsorption to the cell surface resembles cation binding to an ion-exchange resin, i.e., the process is rapid, reversible, pH dependent, and relatively non-specific. However, it is also probable that some of the zinc that interacts with surface components is tightly bound. This "incorporated" zinc may have functions in maintaining cell wall and membrane integrity, and possibly as a cofactor for certain enzymes localized in the wall region (see Section II.F). Uptake involves translocation of the metal across the cell membrane by a temperature- and energy-dependent mechanism that exhibits high specificity for the substrate.

C. Uptake

The removal of nonspecifically bound zinc from the cell surface without damaging the membrane is one of the principal problems encountered in determining the quantity of zinc transported into a cell. For radioisotope studies, washing cells with a chilled acidic solution containing nonradioactive $Zn^{2+}$ appears to be the most effective procedure. Both $H^+$ and $Zn^{2+}$ displace surface-bound $^{65}Zn$, and the decrease in temperature minimizes continued uptake and/or exit-exchange. Obviously, the investigator must verify that intracellular zinc is not lost as a result of cold shock.

The characteristics of zinc uptake in microbial and higher cell systems are given in Table 4-5 [330-343]. A discussion of pertinent observations and problems concerning uptake studies with each group of organisms follows.

1. Bacteria

There have been very few studies concerning zinc metabolism or transport in bacteria. Supplementation of defined media with zinc salts fails to increase either cell growth rate or yield. For unknown reasons, the zinc

---

[a] The concentration factor (CF) is defined as the ratio of the radionuclide concentration per gram of material to the concentration per milliliter of solution.

[b] Samples were collected from different locations.

[c] Axenic cultures used in study.

TABLE 4-5 Characteristics of Zinc–Uptake Systems

| System | Energy dependence | Inhibitors | Exit-exchange | Comments | References |
|---|---|---|---|---|---|
| Escherichia coli | + | $Cd^{2+}$ | 25% | $K_m = 20$[a]; $V_{max} = 2.67$[b]; $Q_{10}$[c] $= 10$ | 330 |
| Salmonella enteritidis | | pCMB | 50% 27%[d] | | 114, 263, 265 |
| Candida utilis | + | $CN^-$, CCCP,[d] EDTA, pCMB, NEM, $Cd^{2+}$, phosphate | None | Half saturation =1.3; $V_{max} = 0.21$; pH dependent: 4.5>5.5> 6.2>8.2 | 331 |
| | | $Fe^{2+}$ | | | 332 |
| Neocosmospora vasinfecta | + | $N_3^-$, $UO_2^{2+}$, $Mn^{2+}$, phosphate, $N_2$ | None | $K_m = 200$; protein synthesis not required for uptake; pH dependent: 4.5<5.5 <6.5>7.5 | 333 |
| Saccharomyces cerevisiae | + | | None | $K_m = 1000–1300$; $V_{max} = 0.043$. | 334 |
| | + | $Ca^{2+}$ | | $K_m = 10$; $Co^{2+}$ stimulated | 335 |

| System | | | | | Ref. |
|---|---|---|---|---|---|
| Chlorella fusca | + | $Ca^{2+}$ | | $K_m = 5.7-8.7$; $V_{max} = 0.015$; $Q_{10} = 2.2-2.6$ | 336,337 |
| Dunaliella tertiolecta | ? | | None | DNP and anaerobiasis do not affect uptake; mercurials have no effect. | 338 |
| Beef heart mitochondria | + | Rotenone, CCCP, $Mg^{2-}$, $K^+$ | | | 339 |
| Rat liver mitochondria | + | | | | 340 |
| Rat liver nuclei | | EDTA; ATP | | Albumin, histidine and other amino acids do not inhibit uptake; phosphate, AMP and ADP stimulate | 341 |
| Equine erythrocytes | ? | EDTA; $Fe^{2+}$, $Cu^{2+}$ | | $CN^-$ stimulates; DNP, pCMB, $Cd^{2+}$, and histidine have no effect. | 342 |
| HeLa cells | + | $Cd^{2+}$ | Temperature-dependent exchange | Cyclic AMP and theophylline have no effect; stimulated by prednisolone. | 203-205 |

TABLE 4-5 (continued)

| System | Energy dependence | Inhibitors | Exit-exchange | Comments | References |
|---|---|---|---|---|---|
| 3T3 mouse fibroblasts | | EDTA, histidine, albumin, serum | | Citrate, nitriloacetic acid, and other amino acids have no effect. | 343 |

[a] $K_m$ values are given as micromoles per liter.

[b] $V_{max}$ values given as nanomoles per minute per milligram dry weight.

[c] $Q_{10}$ is the increase in the rate of zinc uptake when the temperature of the system is increased by $10°$ C.

[d] CCCP: m-chlorophenyl carbonylcyanide hydrazone.

requirement for bacteria seems to be tenfold lower than that for algae or fungi [344]. The extreme difficulty in preparing a zinc-deficient medium led Torriani to state, "it is practically impossible to decrease the $Zn^{2+}$ concentration to less than one-tenth the usual amount," which was $2 \mu M$ in her system [114]. Surely, others who have attempted this feat have eventually voiced similar frustrations in language that could not be printed! The most valiant effort is that of Wilson and Reisenauer, who prepared a zinc-deficient medium for four species of Rhizobium [345]. Addition of as little as 0.1 nM $Zn^{2+}$ stimulated growth of all organisms; optimal growth occurred in medium supplemented with 0.1 $\mu M$ $Zn^{2+}$.

Although zinc is present in Gram-negative walls, and probably has structural functions in both Gram-negative and Gram-positive bacteria, binding of this specific cation to the bacterial wall has not been studied. From the few observations given in Table 4-5 it appears that bacteria do possess specific energy-dependent influx and efflux transport systems for zinc.

Uptake of $^{65}Zn$ by Serratia marcescens is linear during exponential growth [346]. The quantity of metal transport is not altered in HEPES-buffered medium[4] with inorganic phosphate levels ranging from 0.3 to 50 mM. Likewise, zinc is accumulated by Bacillus megaterium during exponential growth [301]. During sporulation, the metal binds to the endospore at a level fourfold greater than that associated with the vegetative cell [347]. Deficiency of either manganese, iron, calcium, or zinc causes a decrease in growth and prevents sporulation [301]. Finally, $Cd^{2+}$ inhibits $Zn^{2+}$ uptake by E. coli [330]; pretreatment of cells with $Zn^{2+}$ before exposure to $Cd^{2+}$ decreases the toxicity of the latter; the degree of protection conferred by $Zn^{2+}$ is directly proportional to the length of pre-incubation with this metal [348].

Observations made with several bacteria may facilitate the study of zinc metabolism and transport, in spite of the technical problems already mentioned. In Table 4-3 the effects of zinc deficiency on nucleic acid metabolism in Mycobacterium and Nocardia are listed. In each case zinc-deficient media were readily prepared by treatment with alumina [129, 130]. Obviously, these organisms require greater quantities of zinc than do most other bacteria and, therefore, may lend themselves to zinc transport studies. In addition, the penicillinase plasmids in Staphylococcus aureus often carry determinants for resistance to normally toxic levels of inorganic ions, including $Hg^{2+}$, $Pb^{2+}$, $As^{2+}$, $Bi^{2+}$, $Cd^{2+}$, and $Zn^{2+}$ [349]. Chemically induced loss of resistance to $Cd^{2+}$ is always accompanied by increased sensitivity to $Zn^{2+}$, while resistance to the other metals is not affected. This suggests that genes governing cadmium-zinc resistance map at the same locus (loci). More recently, the genetic characteristics for penicillinase-linked $Cd^{2+}$ resistance have been determined [350]. Such strains might be quite useful in physiological studies concerning zinc transport and metabolism.

---

[4] HEPES = N-2-hydroxyethylpiperazine-N'-2-ethanesulfonic acid.

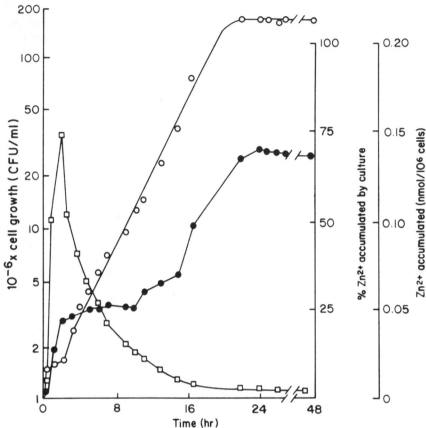

FIG. 4-1   Zinc uptake during growth of <u>Candida utilis</u> in batch culture.
Yeasts from a 24-hr culture were isolated by centrifugation, washed with
PIPES-buffered salts solution, and diluted 200-fold into growth medium;
$Zn^{2+}$ was added to a concentration of $1.1\ \mu M$ ($0.5\ \mu C\ ^{65}Zn$) at 0 hr.   The
quantity of the metal taken up by the cells was determined periodically
throughout the culture cycle.   (See Ref. 331 for experimental details.)
Key: O, colony forming units (CFU)/ml; ●, percentage of $Zn^{2+}$ taken up
by yeasts; □ , quantity of zinc per $10^6$ CFU.   (Modified from Ref. 331.)

2.   Yeasts and Filamentous Fungi

As compared with other microorganisms, mechanisms of zinc accumulation have been studied in some detail for this group.   Several early studies
revealed that cations bind to the yeast surface in a rapid and reversible,
pH-dependent manner [351-353].   Competition experiments yielded the
following affinity series: $UO_2^{2+} > Ba^{2+} > Zn^{2+}, Co^{2+} > Mg^{2+}, Ca^{2+},$

$Sr^{2+} > Mn^{2+}$, $Cu^{2+}$, $Hg^{2+} > Li^+$, $Na^+$, $K^+$, $Rb^+$ [351]. As with algae, accumulation of essential cations by yeasts and filamentous fungi occurs through two distinct processes, viz., energy-independent binding to anionic sites on the cell surface, and energy-dependent, highly specific transfer across the plasma membrane [331,333,334,351].

Radiozinc ($^{65}Zn$) exit-exchange reactions have not been observed in either yeasts or filamentous fungi. Apparently, these organisms regulate zinc influx in such a manner so as to insure that the intracellular environment neither becomes zinc-deficient nor contains toxic quantities of this metal. We have found that Candida utilis takes up zinc in a cyclic manner in batch cultures containing 1.1 μM $Zn^{2+}$ (Fig. 4-1 [331,354]). Uptake occurs during lag phase and the latter half of the growth phase, but neither during early growth nor in the stationary period. Exit-exchange reactions do not occur at any time.

When stationary phase cells are resuspended in growth media containing various concentrations of $Zn^{2+}$ (even as high as 50 μM) the same biphasic, energy-dependent metal uptake process is observed during the culture cycle [354]. Moreover, the maximum intracellular level of zinc that can be accumulated by C. utilis during the lag phase is about 40 nmol per milligram dry weight, regardless of the concentration of $Zn^{2+}$ in the growth medium. When cell division begins, the intracellular level of the metal decreases in each generation until the second period of uptake is initiated. Since the characteristics of zinc uptake are identical during both periods, zinc translocation is apparently mediated by the same "carrier" system.

The ability of the cells to transport zinc varies when the metal is added at different times during the culture cycle (Figs. 4-2 and 4-3). If the metal is added during the earlier periods of cell division, zinc uptake begins immediately. However, once the intracellular level of zinc is equivalent to that present in cells growing in media supplemented with the metal at earlier times, the rate of influx decreases until later in the culture cycle. When $Zn^{2+}$ is not added until late in the growth phase of the culture cycle, both the rate and the quantity of the cation accumulated increase significantly (Figs. 4-2 and 4-3). The initial rate of zinc uptake increases 18-fold [the $V_{max}$ (maximum velocity) increases from 0.2 to 3.6 nmol/min per milligram dry weight], whereas the $K_m$ (Michaelis constant) for uptake by "zinc-deficient" C. utilis is the same as for zinc-sufficient cells [354]. The maximum quantity of zinc accumulated during this period of activated transport is 115-130 nmol per milligram dry weight, or 0.7-0.85% of the dry weight of the yeasts! This high level of zinc is actually located within the cell, since resuspension of yeasts in 1.0 mM solutions of either HCl, EDTA, or nonradioactive $Zn^{2+}$ following activated zinc uptake fails to reduce the quantity of $^{65}Zn$ associated with the microorganisms. Thus, zinc uptake is well regulated; also, it seems that the cells have either an altered capacity or requirement for this metal at different times in the growth cycle.

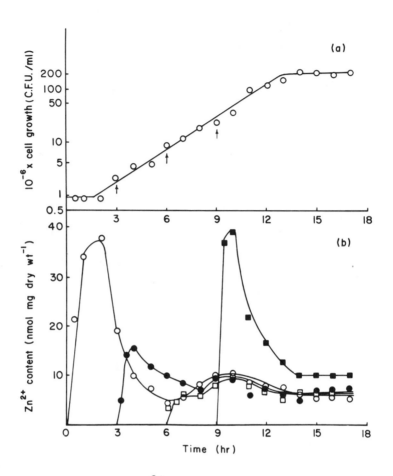

FIG. 4-2  Effect of time of $Zn^{2+}$ addition on zinc uptake by <u>C. utilis</u> growing in batch culture.  (a) Growth curve.  CFU/ml were determined by
standard pour-plate technique and by phase microscopy.  (b) At times indicated by arrows in Fig. 4-2(a), $Zn^{2+}$ was added to cultures of <u>C. utilis</u> to
a concentration of 10 μM (1.0 μC $^{65}Zn$).  The quantity of zinc taken up by
the cells was determined by filtration of 0.5 ml culture aliquots at given
times.  The filters were washed three times with 3.0-ml volumes of
200 μM nonradioactive $ZnCl_2$ solution at pH 3.5.  After air drying, the
filters were counted as described in [331].  Key: O, 0 hr; ●, 3 hr; □ ,
6 hr; ■, 9 hr.  When $Zn^{2+}$ was not added until 9 hr, all the metal was
taken up by the cells within the initial hour.

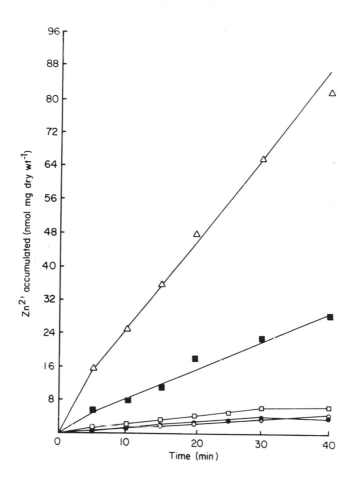

FIG. 4-3  Ability of C. utilis to take up zinc at various times in the growth
cycle.  Yeasts were harvested from growing cultures at cell concentrations given below, washed with PIPES-buffered salts solution and resuspended in fresh growth medium at approximately $10^7$ CFU/ml (equivalent
to 50 µg dry weight per milliliter).  After allowing the cells to equilibrate
in the medium for 5 min, $ZnCl_2$ was added to a concentration of 15 µM
(0.5 µC $^{65}Zn$).  The quantity of cation transported by the cells was determined as described in the legend of Fig. 4-2.  The concentration of yeasts
(CFU/ml) in batch culture at time of harvest are shown as follows:  O,
6 X $10^6$; ●, 1 X $10^7$; □, 2.3 X $10^7$; ■, 5 X $10^7$; Δ, 8 X $10^7$.

Activated zinc uptake is analogous to manganese "hyperactive" transport by manganese-deficient B. subtilis [355]. Following hyperactive
transport, however, the intracellular level of manganese is toxic and bacteria can only initiate growth after 90% of the recently accumulated cation
is "excreted" via the manganese exit-exchange system. Candida utilis
lacks such an efflux system for zinc; remarkably, neither cell viability,
growth rate, nor cell yield are altered in organisms that have accumulated
high levels of zinc by the activated zinc transport process [354]. Apparently, this yeast is able to protect itself from the potentially harmful effects of elevated intracellular levels of zinc. The mechanisms that C. utilis
employs for this phenomenum are presently being investigated. Interestingly, electron-dense regions that probably contain zinc have been observed
near the plasma membrane and within the nucleus of Neocosmospora [333].

## 3.   Algae

As discussed earlier, algal accumulation includes binding to the cell surface and energy-dependent uptake. Theoretical considerations and experimental observations concerning the binding of $^{65}$Zn to the surface of a
lichen, Usnea florida, were recently published [356]. Similarly, a recent
study from Haug's laboratory dealing with both binding and uptake of zinc
by the brown alga Ascophyllum nodosum provides an example of some of the
problems presently being studied in this area [357]. Eutrepia, a single-
celled alga similar to Euglena, has been isolated from marine eutrophic
water with a zinc content of almost 500 μM; much of the metal is localized
in the chloroplasts [358].

As indicated in Section II.C, much of the work at the molecular and
cellular levels concerned with the noncatalytic functions of zinc has been
done with Euglena gracilis (specifically, strain Z). Severe zinc deficiency
can be induced in strain Z simply by decreasing the concentration of the
zinc in the growth medium to 0.1 μM, a level that supports optimal growth
in other strains of Euglena gracilis [359]. Euglena gracilis strain Z
would appear to be an excellent organism for the study of zinc transport
and metabolism. Euglena gracilis strain Klebs rapidly accumulates and
incorporates all the zinc in a medium containing less than 0.1 μM $Zn^{2+}$;
subsequent dilution of the metal with each new generation does not adversely
affect the cell [359].

## 4.   Cell Organelles

Brierly and Knight's work with beef heart mitochondria merits consideration. In addition to the energy-dependent uptake process listed in Table
4-5, an energy-independent, nonsaturatable process that requires phosphate or arsenate was also observed [339]. Although zinc accumulated by

the latter process is bound to the mitochondria, it appears to be present as an inert, insoluble zinc-phosphate complex that can be removed by treatment with EDTA. When respiration begins in mitochondria containing this complex, endogenous $K^+$ and $Mg^{2+}$ are rapidly released. Since concomitant cation efflux is characteristic of energy-dependent zinc uptake, it was proposed that the zinc-phosphate complex is localized close to the mitochondrial inner membrane. It was further suggested that zinc directly interacts with the mitochondrial membrane so as to increase permeability of $K^+$ and $Mg^{2+}$, but not $Na^+$.

## 5. Animal Cells

As for microorganisms, zinc transport in animal cells has not received much attention. One may hope that recent findings on the roles of zinc at the cellular level will generate interest in this problem. Regulation of the uptake process may be extremely complex in light of Cox's observations on the stimulating effect of hormones on zinc uptake in tissue cell cultures ([203,204]; also Cox and Ruckenstein [205]).

## 6. Animal Studies

Many studies on zinc absorption, retention, and excretion in whole animals or with intact organ systems have been performed and this work has been reviewed [6,7,360]. During the past 5 years, mechanisms of zinc transport at the intestinal mucosal/serosal surfaces have been investigated. Suso and Edwards [361] found that zinc is completely complexed by low molecular weight molecules in chicken digesta. Following uptake at the mucosal membrane, the metal is principally associated with a protein (14,000-18,000 daltons) whose affinity for zinc is less than that of a plasma protein (65,000-75,000 daltons), probably albumin. It was therefore suggested that unidirectional transfer of zinc from the lumen to the serosal surface occurs because substances with different binding affinities are present in various regions of the intestine [361].

As zinc deficiency in the rat develops, intestinal and plasma levels of the metal decrease, and the ability of intestinal cells to absorb zinc and transfer it to the plasma increases [362]. Intravenous injection of zinc-deficient rats with $Zn^{2+}$ not only increases the plasma level of the metal but also that of the intestinal mucosal cells. Similarly, when $Zn^{2+}$ is injected prior to feeding zinc-deficient rats a zinc-sufficient diet with $^{65}Zn$, the level of isotope that is taken up from the lumen decreases. These results indicate that the zinc flux is bidirectional; they also suggest that zinc accumulation is regulated by a negative feedback mechanism possibly controlled by the plasma zinc concentration. Kowarski et al. [363] have studied the bidirectional flux of zinc in the rat in both in vivo and in vitro systems. Both mucosal → serosal and serosal → mucosal zinc transfers are Na$^+$-dependent processes requiring the presence of a metabolizable hexose and oxygen, and these processes are inhibited by DNP plus iodoacetate or cyanide plus iodoacetate. $Cd^{2+}$ and $Cu^{2+}$ antagonism of intestinal zinc uptake has also been reported [364].

Fractionation of the soluble components of the mucosal epithelium on Sephadex G75 yields four distinct peaks with zinc-binding ability [365]. In the first three peaks, zinc is associated with high-molecular-weight materials, probably proteins. Hydrolysis and chromatography of the low-molecular-weight fraction, the fourth peak, suggests that zinc is bound to a small peptide.

Additional interest concerning intestinal zinc absorption has been generated by the recent recognition that the disease acrodermatitis enteropathica (AE) is probably the result of malabsorption of zinc from the human intestine; AE is an autosomal, recessive disorder that generally appears in infants that either have not been breast-fed or at weaning. Clinical manifestations include severe skin disruptions and ulceration, loss of hair, diarrhea, alopecia, and irritability [366,367]. Similar characteristics have been reported in laboratory or domestic animals that are zinc deficient [7]. Untreated, the disease leads to death in infancy or early childhood. Previously, therapy with diiodohydroxyquinoline (DIHQ) checked the disease, but continuous use damaged the optic nerve.

Moynahan and Barnes [366] recently reported that DIHQ treatment failed to improve the health of an infant with AE. Analysis of the infant's diet revealed that it was deficient in zinc. Supplementation of the diet with $Zn^{2+}$, in conjunction with DIHQ therapy, returned the child to normal health. Subsequently, AE patients were treated with oral $ZnSO_4$ alone and recovered in a short period [367]. The positive effect from DIHQ is now believed to be due to the formation of a zinc-DIHQ complex that is more readily absorbed by the mucosal cells than is zinc alone [368]. Lombeck et al. [369], reported that three patients with AE absorb significantly less zinc than control subjects. Zinc excretion in these patients is normal, as are histological preparations of mucosal epithelium. Identification of the biochemical lesion(s) await(s) further research.

D.  Competition for Zinc in Microbial-
    Animal Symbiotic Relationships

Visible signs of zinc-deficiency are delayed and less severe in germ-free rats as compared with conventional control animals [370]. The results suggest that germ-free rats have a lower dietary requirement for the metal because microorganisms inhabiting the gut compete with the host for ingested trace metals.

Recently, several investigators have suggested that in addition to humoral and cellular immunity higher animals have another defense against invading parasites [371]. The third system involves withholding essential nutrilites from the parasite and is called "nutritional immunity" (see Chapter 6). Thus far, the phenomenon is mainly concerned with competition for iron. However, competition for zinc may also be fundamentally important in determining the relative success of the invader. Release of an endogenous hormonal factor by polymorphonuclear leukocytes induces hypozincemia, as well as hypoferremia, within a few hours after invasion by a microorganism, a microbial product such as endotoxin, or a virus [372]. Of interest is the observation of Booth and Schubert [373], who found that schistosome worms and eggs accumulate greater quantities of injected $^{65}$Zn than do host tissues. Further work on the effect of infection on host metabolism of zinc is required.

## IV. ZINC STORAGE

Although this aspect of zinc metabolism has not been intensively studied in microorganisms, a relatively large number of investigators have been working on this problem in animals for the past decade. As a result of these efforts, our understanding of the storage and the regulation of the intracellular level of zinc, though still minimal, is rapidly increasing.

Vallee and his associates (see Kagi and Vallee [374]; Margoshes and Vallee [375]; Pulido et al. [376]) originally identified and characterized a family of small, sulfur-rich, soluble proteins from equine and human renal cortex, collectively referred to as metallothionein (MT). These proteins have since been found in various tissues from many organisms, and each has the following properties: (a) 25-35% cysteine residues; (b) no aromatic amino acid residues; (c) a molecular weight of about 10,000 daltons as determined by equilibrium centrifugation and gel filtration; and (d) the ability to bind 6-9 mol of soft metals, including $Zn^{2+}$, $Cd^{2+}$, and $Hg^{2+}$. More recently, a detailed physicochemical analysis has shown that the metal-free polypeptide, thionein, actually has a molecular weight of approximately 6,000 daltons [377]. The earlier higher estimates resulted from the fact that the molecule is ellipsoidal, rather than globular. Also, thionein is actually two very similar polypeptides (see Ref. 377 for additional references). Finally, it had been believed that thionein also complexes $Cu^{2+}$. It was recently found that injected copper induces the formation of chelatin, two proteins very similar to MT but with a smaller molecular weight and containing only 14.6% cysteine residues [378-380]. Chelatin has been isolated from rat, chicken, and rabbit livers, yeast, and the bean plant, Phaseolus aureus.

The biological function of these proteins is controversial. Injection of the salts of group Ib and IIb metals into organisms induces the synthesis of MT in the liver, kidney, and other tissues. This has led to the suggestion that MT protects cells against the cytotoxic effects of these metals by effectively removing the latter through complexation (see Ref. 381 for

additional references). Under normal conditions it is believed that MT
plays a dynamic role in zinc uptake, storage, and metabolism [382–385].
Synthesis of mRNA and protein are required for MT synthesis and $^{65}Zn$
accumulation in the liver [382,383]. Actinomycin D decreases MT synthesis and zinc storage in the intestine of zinc-loaded rats but has no effect
upon the absorption of zinc from the lumen [382]. Therefore, Richards
and Cousins [382] have proposed that body stores of zinc influence MT
synthesis, which in turn affects the level of zinc stored in the intestinal
cells. Interestingly, RNA and protein synthesis are also required for $Zn^{2+}$
uptake by HeLa cells [204].

A soluble, low-molecular-weight, zinc-binding protein with chromatographic characteristics similar to those of MT has been isolated from
yeasts [331,386] and the cyanobacterium Anacystis nidulans [387]. Similarly, chelatin, the copper-binding protein similar to MT, is found in
yeasts [380]. However, cytoplasmic $Cd^{2+}$ in E. coli was reported to be
associated with two proteins of higher molecular weight ( >20,000 daltons)
[348]. Finally, protein synthesis is required for zinc uptake in energy-
starved Candida utilis [331]. While the synthesis of permeases may be
required, it is possible that formation of intracellular "storage" molecules
is necessary before uptake can be initiated. Obviously, additional research
is needed into the problems of intracellular storage, distribution, and
translocation of zinc in microorganisms.


V.  CONCLUSION

The importance of zinc in the structure and activity of numerous enzymes
is well established. Similarly, it is recognized that this metal has essential roles in nucleic acid metabolism and cell division. Studies concerning
the activity of zinc throughout the cell cycle have been initiated. The stabilization of macromolecules, organelles, and membranes seems to be a
critical function for this metal. Recent observations suggest that zinc,
like other cations, may have a significant role in membrane-mediated cell
processes. Additional studies in this area are needed. Likewise, possible
interactions or antagonisms between zinc and calcium within the cell or on
the cell surface merit further investigation.

Our present knowledge of the mechanisms and regulation of zinc transport in microbial and higher cells is minimal. Do microorganisms produce zinc sequestering agents under certain conditions? Is the competition
for limited quantities of zinc a determinant in various symbiotic relationships or in ecological successions? Finally, we have virtually no knowledge concerning the intracellular localization, storage, and translocation
of this very important nutrient.

Surely, advances will continue to be made in relation to all these
questions in the near future, "for zinc is fast becoming a focus of diverse
and powerful interests" [1].

## RECENT DEVELOPMENTS

Various aspects of zinc biochemistry, including the roles of this metal in biochemical and morphological events of the cell cycle of Euglena gracilis, are discussed in a recent review by Riordan [388]. It has been suggested that zinc may have both structural and catalytic roles in yeast alcohol dehydrogenase [389,390]. The amino acid sequence of equine renal metallothionein-1B has been elucidated [391].

Zinc inhibits cleavage of precursor polypeptides of the following viruses: sindbis [392], herpes simplex type 2 [393], avian myeloblastosis [394], and mengoviruses [395]. The metal binds to the precursors, rendering them unsuitable as substrates for proteolysis [394,395]. Also, zinc stabilizes erythrocytes against various lysins of bacterial origin [396]. Mortality following endotoxin-induced shock is reduced by $ZnCl_2$ injection [397]. Presumably the metal mediates such an effect by stabilizing lysosomal membranes.

Weiss et al. [398] have shown that $Cd^{2+}$ accumulation by S. aureus containing plasmid P1258 is an energy-dependent process that competitively inhibits manganese uptake, as well as accelerating loss of intracellular $Mn^{2+}$. Zinc, magnesium, cobalt, and rubidium uptake and exchange are not affected by cadmium. Interestingly, Zn-transferrin, but not Zn-albumin, Zn-ovatransferrin or Zn-acetate, increases nucleic acid synthesis in PHA-stimulated human lymphocytes [399]. Also, these lymphocytes only accumulate zinc when the metal is present as the Zn-transferrin complex. The binding of this complex to specific membrane receptors is inhibited by cyanide, fluoride, and sulfhydryl reagents [400].

An antibacterial system composed of a hexapeptide (3 glutamine-glutamic acid, 2 glycine, and 1 lysine) and 1 g-atom zinc has been isolated from human amniotic fluid [401].

## ACKNOWLEDGMENTS

I wish to thank E. D. Weinberg for initially exposing me to the problem of zinc metabolism, for the countless discussions we have had on mineral metabolism and function, and for his continual guidance throughout the preparation of this chapter. Sincere thanks to J. C. Smith (Trace Element Research Laboratory, Veteran's Administration Hospital, Washington, D.C.) for critically reviewing the manuscript. Also, the patience and secretarial skills of Ferole Minns are appreciated. Finally, the support provided by USPHS Microbiology Traineeship PHS 5T 1-GM 50313 and the aid of a research grant from the National Science Foundation (BMS 75-16753) are gratefully acknowledged.

REFERENCES

1.  Anonymous, Lancet, ii, 351 (1975).
2.  M. Chvapil, Life Sci., 13, 1041 (1973).
3.  B. J. Aylett, in Comprehensive Inorganic Chemistry (J. C. Bailar,
    H. J. Emeleus, S. R. Nyholm, and A. F. Trotman-Dickenson, eds.),
    Pergamon Press, Oxford, 1973, pp. 186-253.
4.  D. Mikac-Devic, Advan. Clin. Chem., 13, 271 (1970).
5.  E. J. Hewlett, in Plant Physiology (F. C. Steward, ed.), Vol. 3,
    Academic Press, New York, 1963, pp. 229-239.
6.  J. A. Halsted, J. C. Smith, and M. I. Irwin, J. Nutr., 104, 345
    (1974).
7.  E. J. Underwood, Trace Elements in Human and Animal Nutrition,
    3rd ed., Academic Press, New York, 1971, pp. 208-252.
8.  B. L. Vallee, Physiol. Rev., 39, 443 (1959).
9.  B. L. Vallee, in Mineral Metabolism (C. L. Comar and F. Bronner,
    eds.), Vol. 2, Pt. B, Academic Press, New York, 1962, pp. 443-
    482.
10. L. S. Hurley and R. E. Shrader, Int. Rev. Neurobiol., Suppl. 1, 7
    (1972).
11. L. E. Sever, Human Ecology, 3, 43 (1975).
12. J. E. Zajic, Microbial Biogeochemistry, Academic Press, New
    York, 1969, pp. 142-155.
13. M. J. Jordan and M. P. Lechevalier, Can. J. Microbiol., 21,
    1855 (1975).
14. O. H. Tuovinen, S. I. Niemela, and H. G. Gyllenberg, Antonie van
    Leeuwenhoek, J. Microbiol. Serol., 37, 489 (1971).
15. J. Raulin, Ann. Sci. Nat. Bot. Biol. Vegetale, 11, 93 (1869).
16. J. W. Foster, Chemical Activities of Fungi, Academic Press, New
    York, 1949, 649 pp.
17. R. E. Feeney, H. D. Lightbody, and J. A. Garibaldi, Arch. Bio-
    chem. Biophys., 15, 13 (1947).
18. D. Keilin and T. Mann, Biochem. J., 34, 1163 (1940).
19. W. E. C. Wacker and B. L. Vallee, J. Biol. Chem., 234, 3257
    (1959).
20. C. A. Price and E. Millar, Plant Physiol., 37, 423 (1962).
21. B. L. Vallee and H. Neurath, J. Amer. Chem. Soc., 76, 5006 (1954).
22. H. Matsubara and J. Feder, in The Enzymes (P. Boyer, ed.), Vol.
    3, Academic Press, New York, 1971, pp. 721-795.
23. A. S. Mildvan and C. M. Grisham, Struc. Bond., 20, 1 (1974).
24. K. Morihara, Advan. Enzymol., 41, 179 (1974).
25. A. F. Parisi and B. L. Vallee, Amer. J. Clin. Nutr., 22, 1222
    (1969).
26. J. F. Riordan and B. L. Vallee, in Protein-Metal Interactions
    (M. Friedman, ed.), Plenum Press, New York, 1973, pp. 33-58.

27.  W. J. Rutter, Fed. Proc., 23, 1248 (1964).

28.  D. D. Ulmer and B. L. Vallee, Advan. Chem. Ser., 100, 187 (1971).

29.  B. L. Vallee, Advan. Protein Chem., 10, 317 (1955).

30.  B. L. Vallee and W. E. C. Wacker, in The Proteins (H. Neurath, ed.), Vol. 5, Academic Press, New York, 1970, 192 pp.

31.  B. L. Vallee, in Metal Ions in Biological Systems (S. K. Dhar, ed.), Plenum Press, New York, 1973, pp. 1-12.

32.  V. Jagannathan, K. Singh, and M. Damodaran, Biochem. J., 63, 94 (1956).

33.  S. Sugimoto and Y. Nosoh, Biochem. Biophys. Acta, 235, 210 (1971).

34.  J. Kowal, T. Cremona, and B. L. Horecker, Arch. Biochem. Biophys., 114, 13 (1966).

35.  N. B. Schwartz and D. S. Feingold, Bioinorg. Chem., 2, 75 (1973).

36.  N. B. Schwartz, D. Abram, and D. S. Feingold, Biochemistry, 13, 1726 (1974).

37.  R. D. Kobes, R. T. Simpson, B. L. Vallee, and W. J. Rutter, Biochemistry, 8, 585 (1969).

38.  C. E. Harris, R. D. Kobes, D. C. Teller, and W. J. Rutter, Biochemistry, 8, 2442 (1969).

39.  F. K. Yoshizumi and J. E. Coleman, Arch. Biochem. Biophys., 160, 255 (1974).

40.  R. T. Simpson, B. L. Vallee, and G. H. Tait, Biochemistry, 7, 4336 (1968).

41.  R. T. Simpson and B. L. Vallee, Biochemistry, 7, 4343 (1968).

42.  W. F. Bosron, F. S. Kennedy, and B. L. Vallee, Biochemistry, 14, 2275 (1975).

43.  E. A. Stein and E. H. Fischer, Biochim. Biophys. Acta, 39, 287 (1962).

44.  T. Isemura and K. Kakiuchi, J. Biochem. (Tokyo), 51, 385 (1955).

45.  M. E. Nelbach, V. P. Pigiet, J. C. Gerhart, and H. K. Schachman, Biochemistry, 11, 315 (1972).

46.  F. L. Hoch and B. L. Vallee, Proc. Nat. Acad. Sci. U.S., 41, 327 (1955).

47.  B. L. Vallee and F. L. Hoch, J. Amer. Chem. Soc., 77, 821 (1955).

48.  B. L. Vallee, F. L. Hoch, S. J. Adelstein, and W. E. C. Wacker, J. Amer. Chem. Soc., 78, 5879 (1956).

49.  T. Keleti, Biochem. Biophys. Res. Commun., 22, 640 (1966).

50.  T. Cremona and T. P. Singer, J. Biol. Chem., 239, 1466 (1964).

51.  C. Gregolin and T. P. Singer, Biochim. Biophys. Acta, 67, 201 (1963).

52.  A. Curdel and F. Labeyrie, Biochem. Biophys. Res. Commun., 4, 175 (1961).

53.  M. Iwatsubo and A. Curdel, Biochem. Biophys. Res. Commun., 6, 385 (1961).

53a. W. H. Taylor, M. L. Taylor, M. E. Balch, and P. S. Gilchrist, J. Bacteriol., 127, 863 (1976).

54. R. B. Davies and E. P. Abraham, Biochem. J., 143, 129 (1974).

55. M. Fugimoto, A. Kuninaka, and H. Yoshino, Agr. Biol. Chem., 38, 785 (1974).

56. M. Fugimoto, A. Kuninaka, and H. Yoshino, Agr. Biol. Chem., 39, 1991 (1975).

57. C. Little and A. Otnass, Biochem. Biophys. Acta, 391, 326 (1975).

58. R. W. Gracy and E. A. Noltmann, J. Biol. Chem., 243, 4109 (1968).

59. R. W. Gracy and E. A. Noltmann, J. Biol. Chem., 243, 5410 (1968).

60. J. P. Slater, A. S. Mildvan, and L. A. Loeb, Biochem. Biophys. Res. Commun., 44, 37 (1971).

61. J. P. Slater, I. Tamin, L. A. Loeb, and A. S. Mildvan, J. Biol. Chem., 247, 6784 (1972).

62. C. F. Springgate, A. S. Mildvan, R. Abramson, J. L. Engle, and L. A. Loeb, J. Biol. Chem., 248, 5987 (1973).

63. M. C. Scrutton, C. W. Wu, and D. A. Goldthwait, Proc. Nat. Acad. Sci. U.S.A., 68, 2497 (1971).

64. K. H. Falchuk, B. Mazus, L. Ulpino, and B. L. Vallee, Biochemistry, 15, 4468 (1976).

64a. H. Lattke and U. Weser, FEBS Letters, 65, 288 (1976).

64b. D. S. Auld, I. Atsuza, C. Campino, and P. Valenzuela, Biochem. Biophys. Res. Commun., 69, 548 (1976).

65. J. E. Coleman, Biochem. Biophys. Res. Commun., 60, 641 (1974).

66. J. L. Oakley, J. A. Pascale, and J. E. Coleman, Biochemistry, 14, 4684 (1975).

67. D. S. Auld, H. Kawaguchi, D. M. Livingston, and B. L. Vallee, Proc. Nat. Acad. Sci. U. S., 71, 2091 (1974).

68. D. S. Auld, H. Kawaguchi, D. M. Livingston, and B. L. Vallee, Biochem. Biophys. Res. Commun., 57, 967 (1974).

69. B. J. Poiesz, G. Seal, and L. A. Loeb, Proc. Nat. Acad. Sci., U. S., 71, 4892 (1974).

69a. M. Nakajima, K. Mizusawa, and F. Yoshida, Eur. J. Biochem., 44, 87 (1974).

69b. G. Metz and K. H. Rohm, Biochim. Biophys. Acta, 429, 933 (1976).

70. J. F. Seber, T. P. Toomey, J. T. Powell, K. Brew, and W. M. Awad, J. Biol. Chem., 251, 204 (1976).

70a. B. A. Chabner and J. R. Bertino, Biochim. Biophys. Acta, 276, 234 (1972).

71. S. Seifter, S. Takahashi, and E. Harper, Biochim. Biophys. Acta, 214, 559 (1970).

72. S. Hayman, J. S. Gatmaitan, and E. K. Patterson, Biochemistry, 13, 4486 (1974).

73. K. Morihara and H. Tsuzuki, Agr. Biol. Chem., 39, 1123 (1975).

74. L. Keay, L. Feder, L. R. Garrett, M. H. Mosely, and N. Cirulis, Biochim. Biophys. Acta, 229, 829 (1971).

75. W. M. Fogarty and P. J. Griffith, Biochem. Soc. Trans., 1, 400 (1973).

76. P.J. Griffith and W. M. Fogarty, Appl. Microbiol., 26, 191 (1973).

77. J. Feder, L. Keay, L. R. Garrett, N. Circulis, M. H. Moseley, and B. S. Wildi, Biochim. Biophys. Acta, 251, 74 (1971).

78. L. Keay and B. S. Wildi, Biotechnol. Bioeng., 12, 179 (1970).

79. J. D. McConn, D. Tsuru, and K. T. Yasumohi, Arch. Biochem. Biophys., 120, 479 (1967).

80. D. Tsuru, H. Kira, T. Yamamoto, and J. Fukumoto, Agr. Biol. Chem., 30, 856 (1966).

81. H. Sekino, Agr. Biol. Chem., 36, 2143 (1972).

82. A. Hiramatsu and T. Ouchi, J. Biochem. (Tokyo), 71, 767 (1972).

83. K. Miyata, K. Tomoda, and M. Isono, Agr. Biol. Chem., 35, 460 (1971).

84. T. B. Griffin and J. M. Prescott, J. Biol. Chem., 245, 1348 (1970).

85. J. M. Prescott and S. H. Wilkes, Arch. Biochem. Biophys., 117, 328 (1966).

86. S. A. Latt, B. Holmquist, and B. L. Vallee, Biochem. Biophys. Res. Commun., 37, 333 (1969).

87. P. I. Levy, M. K. Pangburn, Y. Burnstein, L. H. Ericsson, H. Neurath, and K. A. Walsh, Proc. Nat. Acad. Sci. U.S., 72, 4341 (1975).

88. M. C. Scrutton, M. R. Young, and M. F. Utter, J. Biol. Chem., 245, 6220 (1970).

89. U. Rapp, W. C. Adams, and R. W. Miller, Can. J. Biochem., 51, 158 (1973).

90. H. P. Misra and I. Fridovich, J. Biol. Chem., 247, 3410 (1972).

91. S. A. Goscin and I. Fridovich, Biochim. Biophys. Acta, 289, 276 (1972).

92. D. B. Northrop and H. G. Wood, J. Biol. Chem., 244, 5801 (1969).

93. I. Fridovich, Advan. Enzymol., 41, 35 (1974).

94. I. L. Sun and F. L. Crane, Biochem. Biophys. Res. Commun., 65, 1334 (1975).

95. G. B. Banks, W. K. Holloman, M. V. Kairis, A. Spanos, and G. T. Yarranton, Eur. J. Biochem., 62, 131 (1976).

95a. N. S. Verma, D. Sharma, and K. G. Gollakota, Indian J. Biochem. Biophys., 13, 99 (1976).

96. A. Medina and D. J. D. Nicholas, Biochem. J., 66, 573 (1957).

97. C. A. Price, Biochem. J., 82, 61 (1961).

97a. H. Vachek and J. L. Wood, Biochim. Biophys. Acta, 258, 133 (1972).

98. P. T. Chan and H. Urbanek, Acta Microbiol. Polonica [Sec. B], 6, 21 (1974).

99. H. F. Dvorak and L. A. Heppel, J. Biol. Chem., 243, 2647 (1968).

100. V. Sadasivan, Arch. Biochem. Biophys., 28, 100 (1950).

101. Y. Fujimura, Y. Hasegawa, Y. Kaneko, and S. Doi, Agr. Biol. Chem., 31, 92 (1967).

102. M. Hirose, E. Sugimoto, and H. Chiba, Biochim. Biophys. Acta, 289, 137 (1972).

103. M. V. Isopolatovskaya, in Microbial Toxins (S. Kadis, T. C.
     Montie, and S. J. Ajl, eds.), Vol. 2, Pt. A., Academic Press,
     New York, 1970, pp. 125-126.
104. P. Valenzuela, R. W. Morris, A. Faras, W. Levinson, and W. J.
     Rutter, Biochem. Biophys. Res. Commun., 53, 1036 (1973).
105. D. S. Auld, H. Kawaguchi, D. M. Livingston, and B. L. Vallee,
     Biochem. Biophys. Res. Comm., 62, 296 (1975).
106. A. Nason, N. O. Kaplan, and S. P. Colowick, J. Biol. Chem., 188,
     397 (1951).
107. D. J. D. Nicholas and T. Goodman, J. Exp. Bot., 9, 97 (1958).
108. M. Carsiotis and A. Meyers, Neurospora Newsletter, 8, 3 (1965).
109. D. Bertrand and A. deWolf, C. R. Acad. Sci., Paris, 245, 1179 (1957).
110. F. G. Winder and C. O'Hara, Biochem. J., 90, 122 (1964).
111. J. W. Foster and F. W. Denison, Nature, 166, 833 (1950).
112. F. G. Winder and D. S. Barber, J. Gen. Microbiol., 76, 189 (1973).
113. F. Friedley, Quart. Rev. Biophys., 7, 1 (1974).
114. A. Torriani, J. Bacteriol., 96, 1200 (1968).
115. J. B. O. Olmsted and G. G. Borisey, Ann. Rev. Biochem., 42,
     507 (1973).
116. V. J. Nickolson and H. Veldstra, FEBS Letters, 23, 309 (1972).
117. M. Morisawa and H. Mohri, Exp. Cell Res., 70, 311 (1972).
118. M. Morisawa and H. Mohri, Exp. Cell Res., 83, 87 (1974).
119. H. Kawaguchi and B. L. Vallee, Anal. Chem., 47, 1029 (1974).
120. K. Fuwa, W. E. C. Wacker, R. Druyan, A. F. Bartholmay, and
     B. L. Vallee, Proc. Nat. Acad. Sci. U.S., 46, 1298 (1960).
121. G. L. Eichhorn, N. A. Berger, J. J. Butzow, P. Clark, J. M.
     Rifkind, Y. A. Shin, and E. Tarien, Advan. Chem. Ser., 100, 135
     (1971).
122. G. L. Eichhorn, in Inorganic Biochemistry (G. L. Eichhorn, ed.),
     Vol. 2, Elsevier, Amsterdam, 1973, pp. 1210-1244.
123. U. Weser, Struc. Bond., 5, 41 (1968).
124. C. Zimmer, G. Luck, and H. Triebel, Biopolymers, 13, 425
     (1974).
125. H. Yamaguchi, J. Gen. Microbiol., 86, 370 (1975).
126. W. E. C. Wacker, Biochemistry, 1, 859 (1962).
127. E. S. Schneider and C. A. Price, Biochim. Biophys. Acta, 55,
     406 (1962).
128. A. B. Harris, J. Gen. Microbiol., 56, 27 (1969).
129. F. G. Winder and C. O'Hara, Biochem. J., 82, 98 (1962).
130. D. M. Webley, R. B. Duff, and G. Anderson, J. Gen. Microbiol.,
     29, 179 (1962).
131. M. C. Cocucci and G. Rossi, Arch. Microbiol., 85, 267 (1972).
132. H. Rubin, Proc. Nat. Acad. Sci. U.S., 69, 712 (1972).
133. H. Rubin and T. Koide, J. Cell Biol., 56, 777 (1973).
134. R. Williams and L. Loeb, J. Cell Biol., 58, 594 (1973).

135. J. K. Chesters, Biochem. J., 130, 133 (1972).
136. J. K. Chesters, in Trace Element Metabolism in Animals (W. G. Hoekstra, J. W. Suttie, H. E. Ganther, and W. Mertz, eds.), Vol. 2, University Park Press, Baltimore, Maryland, 1974, pp. 39-49.
137. I. Lieberman and P. Ove, J. Biol. Chem., 237, 1634 (1962).
138. I. Lieberman, R. Abrams, N. Hunt, and P. Ove, J. Biol. Chem., 238, 3955 (1963).
139. M. Fujioke and I. Lieberman, J. Biol. Chem., 239, 1164 (1964).
140. U. Weser, S. Seeber, and P. Warnnecke, Biochim. Biophys. Acta, 179, 422 (1969).
141. T. Yamashita, T. Hidaka, and K. Watanabe, Agr. Biol. Chem., 38, 727 (1974).
142. W. S. Wegener and A. H. Romano, Science, 142, 1669 (1963).
143. H. Rubin and T. Koide, J. Cell Physiol., 81, 387 (1973).
144. H. Rubin and T. Koide, J. Cell Physiol., 86, 47 (1975).
145. H. Rubin, Proc. Nat. Acad. Sci. U.S., 72, 1676 (1975).
146. N. A. Berger and A. M. Skinner, J. Cell Biol., 61, 45 (1974).
147. H. Ruhl, H. Kirchner, and G. Bochet, Proc. Soc. Exp. Biol. Med., 137, 1089 (1971).
148. K. H. Falchuk, D. W. Fawcett, and B. L. Vallee, J. Cell Sci., 17, 57 (1975).
149. K. H. Falchuk, A. Krishan, and B. L. Vallee, Biochemistry, 14, 3439 (1975).
150. L. R. Mendiola and C. A. Price, Amer. J. Clin. Nutr., 22, 1264 (1969).
151. J. A. Praok and D. J. Plocke, Plant Physiol., 48, 150 (1971).
152. C. A. Price and B. L. Vallee, Plant Physiol., 37, 428 (1962).
153. L. M. S. Chang and F. J. Bollum, Proc. Nat. Acad. Sci. U.S., 65, 1041 (1970).
154. T. Fujii, Nature, 174, 1108 (1954).
155. F. Gyorkey, K. Min, J. A. Huff, and P. Gyorkey, Cancer Res., 27, 1348 (1967).
156. B. Falke and G. F. Kahl, Z. Med. Mikrobiol. Immunol., 153, 175 (1967).
157. Y. T. Gordon, Y. Asher, and Y. Becker, Antimicrob. Agents Chemother., 8, 377 (1975).
158. J. Shlomai, Y. Asher, Y. F. Gordon, U. Olshevsky, and Y. Becker, Virology, 66, 330 (1975).
159. B. D. Korant, J. C. Kaner, and B. E. Butterworth, Nature, 248, 588 (1974).
160. B. E. Butterworth and B. D. Korant, J. Virology, 14, 282 (1974).
161. A. deRoeth, Amer. J. Ophthalmol., 56, 729 (1963).
162. A. S. Spirin and L. P. Gavarilova, The Ribosome, Springer-Verlag, New York, 1969, pp. 29-33.
163. R. F. Gesteland, J. Mol. Biol., 18, 356 (1966).

164. M. Tal, Biochemistry, 8, 424 (1969).
165. M. Tal, Biochim. Biophys. Acta, 169, 564 (1968).
166. M. Tal, Biochim. Biophys. Acta, 195, 76 (1969).
167. M. Tal, personal communication (1975).
168. N. V. Belitsina, V. A. Rozenblatt, and A. S. Spirin, Mol. Biol.,
     5, 723 (1971).
169. A. H. Reisner, C. Bucholtz, and B. S. Chandler, Exp. Cell Res.,
     93, 1 (1975).
170. D. C. Barker and K. Deutsch, Exp. Cell Res., 15, 604 (1958).
171. R. S. Morgan and B. G. Uzman, Science, 152, 214 (1966).
172. R. S. Morgan and R. F. Sattilaro, Science, 176, 929 (1972).
173. M. Webb, H. Creed, and S. Atkinson, Biochim. Biophys. Acta, 324,
     143 (1973).
174. G. J. Fosmire, Y. Y. Al-Ubaidi, E. Halas, and H. H. Sandstead,
     in Protein-Metal Interactions (M. Friedman, ed.), Plenum Press,
     New York, 1974, pp. 329-346.
175. P. S. Kotsiopoulos and S. C. Mohr, Biochem. Biophys. Res.
     Commun., 67, 979 (1975).
176. K. McQuillen, in The Bacteria (I. C. Gunsalus and R. Y. Stanier,
     eds.), Vol. 1, Academic Press, New York, 1960, pp. 288-289.
177. R. R. Korngold and D. J. Kushner, Can. J. Microbiol., 14, 253
     (1968).
178. S. Razin, J. Gen. Microbiol., 36, 451 (1964).
179. S. Razin, H. J. Morowitz, and T. M. Terry, Proc. Nat. Acad.
     Sci. U.S., 54, 219 (1965).
180. S. Razin, Biochim. Biophys. Acta, 265, 241 (1972).
181. A. Azzi, Quart. Rev. Biophys., 8, 237 (1975).
182. D. Chapman, Quart. Rev. Biophys., 8, 185 (1975).
183. D. O. Shah, in Effects of Metals on Cells, Subcellular Elements and
     Macromolecules (J. Maniloff, J. R. Coleman, and M. W. Miller,
     eds.), Thomas, Springfield, 1970, pp. 155-189.
184. D. O. Shah, Progr. Surface Sci., 3, 221 (1972).
185. H. Träuble and H. Eibl, Proc. Nat. Acad. Sci. U.S., 71, 214 (1974).
186. T. Ito and S. Ohnishi, Biochim. Biophys. Acta, 352, 29 (1974).
187. K. Jacobson and D. Papahadjopoulos, Biochemistry, 14, 152 (1975).
188. B. A. Newton, Bacteriol. Rev., 20, 14 (1956).
189. L. H. Muschel and L. Gustafson, J. Bacteriol., 95, 2010 (1968).
190. D. R. Storm and J. L. Strominger, J. Biol. Chem., 248, 3940
     (1973).
191. E. D. Weinberg, in Antibiotics: Mechanism of Action (I. D.
     Gottlieb and P. Shaw, eds.), Vol. 1, Springer-Verlag, New York,
     1967, pp. 90-101.
192. J. A. Sands, R. A. Lowlicht, S. C. Cadden, and J. Haneman,
     Can. J. Microbiol., 21, 1287 (1975).

193. M. Ehrstrom, L. E. G. Eriksson, J. Israelachvili, and E. Ehren-berg, Biochem. Biophys. Res. Commun., 55, 396 (1973).
194. W. G. Hoekstra, Fed. Proc., 23, 1068 (1964).
195. L. Warren, M. C. Glick, and M. K. Nass, J. Cell Physiol., 68, 269 (1966).
196. M. Chvapil, J. N. Ryan, and Z. Brada, Biochem. Pharmacol., 21, 1097 (1972).
197. M. Chvapil, J. N. Ryan, and C. F. Zukoski, Proc. Soc. Exp. Biol. Med., 140, 642 (1972).
198. M. Chvapil, J. N. Ryan, and C. F. Zukoski, Proc. Soc. Exp. Biol. Med., 141, 150 (1972).
199. M. Chvapil, A. L. Aronson, and Y. M. Peng, Exp. Mol. Pathol., 20, 216 (1974).
200. L. Karl, M. Chvapil, and C. F. Zukoski, Proc. Soc. Exp. Biol. Med., 142, 1123 (1973).
201. M. Chvapil, D. L. Weldy, L. Stankova, D. S. Clark, and C. F. Zukoski, Life Sci., 16, 561 (1975).
202. G. Melnykovych, Science, 152, 1086 (1966).
203. R. P. Cox, Mol. Pharmacol., 4, 510 (1968).
204. R. P. Cox, Science, 165, 196 (1969).
205. R. P. Cox and A. Ruckenstein, J. Cell Physiol., 77, 71 (1971).
206. F. E. Hunter and L. Ford, J. Biol. Chem., 216, 357 (1955).
207. E. E. Jacobs, M. Jacob, D. R. Sanadi, and L. B. Bradley, J. Biol. Chem., 223, 147 (1956).
208. V. P. Skulachev, V. V. Christyakov, A. A. Jasaitis, and E. S. Mirnova, Biochem. Biophys. Res. Commun., 26, 1 (1967).
209. D. Kleiner, Arch. Biochem. Biophys., 165, 121 (1974).
210. D. Kleiner and G. von Jagow, FEBS Letters, 20, 229 (1972).
211. J. Subik and J. Kolarov, Folia Microbiol., 15, 448 (1970).
212. M. Kasahara and Y. Anraku, J. Biochem. (Tokyo), 72, 777 (1972).
213. M. Kasahara and Y. Anraku, J. Biochem. (Tokyo), 76, 967 (1974).
214. Y. Anraku, E. Kin, and Y. Tanaka, J. Biochem. (Tokyo), 78, 165 (1975).
215. A. P. Singh and P. D. Bragg, FEBS Letters, 40, 200 (1974).
216. L. Lomander, Physiol. Plantar., 18, 968 (1965).
217. L. Ohaniance and P. Chaix, Biochim. Biophys. Acta, 128, 228 (1966).
218. R. E. Eliasson, O. Johnson, and C. Lindholmer, Life Sci., 10, 1317 (1971).
219. T. Fujii, S. Utida, and T. Mizuno, Nature, 176, 1068 (1955).
220. L. Huacuta, A. Sosa, N. Delgado, and A. Rosado, Life Sci., 13, 1383 (1973).
221. H. I. Calvin, C. C. Yu, and J. M. Bedford, Exp. Cell Res., 81, 333 (1973).
222. H. I. Calvin and C. Bleau, Exp. Cell Res., 86, 280 (1974).

223. C. Lindholmer and R. Eliasson, Int. J. Fertil., 19, 56 (1974).

224. S. A. Gunn and T. C. Gould, in The Testis (A. D. Johnson, W. R. Grimes, and N. L. Van Denmark, eds.), Vol. 3, Academic Press, New York, 1970, p. 404.

225. B. Morton, J. Harrigan-Lum, L. Albagli, and T. Jooss, Biochem. Biophys. Res. Commun., 56, 372 (1974).

226. T. W. Rall and E. W. Sutherland, J. Biol. Chem., 248, 3940 (1973).

227. S. A. Gunn and T. C. Gould, Proc. Soc. Exp. Biol. Med., 141, 639 (1972).

228. M. Chvapil, C. F. Zukosi, B. G. Hattler, L. Stankova, D. Montgomery, E. C. Carlson, and J. C. Ludwig, in Trace Elements in Human Health and Disease (A. S. Prasad, ed.), Vol. 1, Academic Press, New York, 1976, pp. 283-294.

229. J. Donaldson, T. St. Pierre, J. Minnich, and A. Barbeau, Can. J. Biochem., 49, 1217 (1971).

230. M. G. Mustafa, C. E. Cross, R. J. Munn, and J. A. Hardie, J. Lab. Clin. Med., 77, 563 (1973).

231. Y. Anraku, J. Biol. Chem., 243, 3128 (1968).

232. R. G. Eagon and M. A. Asbell, J. Bacteriol., 97, 812 (1969).

233. M. Kasahara and Y. Anraku, J. Biochem. (Tokyo), 76, 977 (1974).

234. Y. Anraku, F. Goto, and E. Kin, J. Biochem. (Tokyo), 78, 149 (1975).

235. W. Kazimierczak and C. Malinski, Agents and Actions, 4, 1 (1974).

236. R. P. Rubin, Pharmacol. Rev., 22, 389 (1970).

237. H. Rasmussen, Science, 170, 404 (1970).

238. D. E. Cochrane and W. W. Douglas, Proc. Nat. Acad. Sci. U.S., 71, 408 (1974).

239. J. C. Foreman, J. L. Mongar, and B. D. Gomperts, Nature, 245, 249 (1973).

240. I. M. Goldstein, J. K. Horn, H. B. Kaplan, and G. Weissmann, Biochem. Biophys. Res. Commun., 60, 807 (1974).

241. H. Kita and W. Van Der Kloot, Nature, 250, 658 (1974).

242. S. Eimerl, N. Savion, O. Heichal, and Z. Selinger, J. Biol. Chem., 249, 3991 (1974).

243. J. A. Williams and M. Lee, Biochem. Biophys. Res. Commun., 60, 542 (1974).

244. N. H. Bell and S. Queener, Nature, 248, 343 (1974).

245. F. J. Oelshlegel, G. J. Brewer, A. S. Prasad, C. Knutsen, and E. B. Schoomaker, Biochem. Biophys. Res. Commun., 53, 560 (1973).

246. G. J. Brewer and F. J. Oelshlegel, Biochem. Biophys. Res. Commun., 58, 854 (1974).

247. J. W. Eaton, T. D. Skelton, H. S. Swofford, C. S. Kolpin, and H. S. Jacob, Nature, 246, 105 (1973).

248. S. Dash, G. J. Brewer, and F. J. Oelshlegel, Nature, 250, 251 (1974).

249. R. Repaske, Biochim. Biophys. Acta, 22, 189 (1956).
250. R. Repaske, Biochim. Biophys. Acta, 30, 225 (1958).
251. G. W. Gray and S. G. Wilkinson, J. Appl. Bacteriol., 28, 153 (1965).
252. G. W. Gray and S. G. Wilkinson, J. Gen. Microbiol., 39, 385 (1965).
253. R. G. Eagon and K. J. Carson, Can. J. Microbiol., 11, 193 (1965).
254. R. G. Eagon, G. P. Simmons, and K. J. Carson, Can. J. Microbiol., 11, 1041 (1965).
255. M. A. Asbell and R. G. Eagon, Biochem. Biophys. Res. Commun., 22, 664 (1966).
256. M. A. Asbell and R. G. Eagon, J. Bacteriol., 92, 380 (1966).
257. L. M. Kozloff and M. Lute, J. Biol. Chem., 228, 529 (1957).
258. L. Lieve, Biochem. Biophys. Res. Commun., 18, 13 (1965).
259. L. Lieve, Proc. Nat. Acad. Sci. U.S., 53, 745 (1965).
260. L. Lieve, Biochem. Biophys. Res. Commun., 21, 290 (1965).
261. L. Lieve, J. Biol. Chem., 243, 2373 (1968).
262. S. Chaberek and A. E. Martell, Organic Sequestering Agents, Wiley, New York, 1959, p. 572.
263. J. R. Chipley and H. M. Edwards, Can. J. Microbiol., 18, 509 (1972).
264. P. Hambleton, J. Gen. Microbiol., 69, 81 (1971).
265. J. R. Chipley and H. M. Edwards, Can. J. Microbiol., 18, 1803 (1972).
266. J. R. Chipley, Microbios, 11A, 117 (1974).
267. K. J. I. Thorne, M. J. Thornley, P. Naisbitt, and A. M. Glauert, Biochim. Biophys. Acta, 389, 97 (1975).
268. L. H. Muschel and L. Gustafson, J. Bacteriol., 95, 2010 (1968).
269. S. D. Davis and A. Iannetta, Antimicrob. Agents Chemother., 1, 466 (1972).
270. S. D. Davis and A. Iannetta, Appl. Microbiol., 23, 775 (1972).
271. C. Cutinelli and F. Galdiero, J. Bacteriol., 93, 2022 (1967).
272. J. Baddiley, Acct. Chem. Res., 3, 98 (1970).
273. S. Heptinstall, A. R. Archibald, and J. Baddiley, Nature, 225, 519 (1970).
274. J. W. Costerton, J. M. Ingram, and K. J. Cheng, Bacteriol. Rev., 38, 87 (1974).
275. H. Onishi and D. J. Kushner, J. Bacteriol., 91, 653 (1966).
276. R. L. Sanders and B. K. May, J. Bacteriol., 123, 806 (1975).
277. L. Weiss, in The Cell Periphery, Metastasis and Other Contact Phenomena (A. Neuberger and E. L. Tatum, eds.), North Holland Publ., Amsterdam, 1967, pp. 267-272.
278. H. Tzagoloff and D. Pratt, Virology, 24, 372 (1964).
279. J. T. Ou and T. F. Anderson, J. Virol., 10, 869 (1972).
280. J. T. Ou and T. F. Anderson, J. Bacteriol., 111, 177 (1972).
281. J. T. Ou, J. Bacteriol., 115, 648 (1973).
282. P. W. Grimm and P. J. Allen, Plant Physiol., 29, 369 (1954).

283. J. B. Neilands, J. Biol. Chem., 205, 647 (1953).

284. H. Komai and J. B. Neilands, Science, 153, 751 (1966).

285. H. Komai and J. B. Neilands, Arch. Biochem. Biophys., 124, 456 (1968).

286. H. Komai and J. B. Neilands, Biochim. Biophys. Acta, 171, 311 (1969).

287. A. Chen and J. B. Neilands, Biochem. Biophys. Res. Commun., 55, 1060 (1973).

288. P. E. Gurba, R. E. Sennett, and R. D. Kobes, Arch. Biochem. Biophys., 150, 130 (1972).

289. V. N. Finelli, D. S. Klauder, M. A. Karaffa, and H. G. Petering, Biochem. Biophys. Res. Commun., 65, 303 (1975).

290. D. H. Brown, R. A. Cappellini, and C. A. Price, Plant Physiol., 41, 1543 (1966).

291. C. B. Coulter and F. M. Stone, Proc. Soc. Exp. Biol. Med., 38, 423 (1938).

292. A. Neuberger and G. H. Tait, Biochem. J., 90, 607 (1964).

293. M. Schwarz, Biochim. Biophys. Acta, 22, 463 (1956).

294. D. S. P. Patterson, Tubercule, 41, 191 (1960).

295. T. P. Pretlow and F. Sherman, Biochim. Biophys. Acta, 148, 629 (1967).

296. A. Lamola and T. Yamane, Science, 186, 936 (1974).

297. C. Ratledge and M. J. Hall, J. Bacteriol., 108, 314 (1971).

298. L. Pine and C. L. Peacock, J. Bacteriol., 75, 167 (1958).

299. S. Bartnicki-Garcia and W. J. Nickerson, J. Bacteriol., 84, 841 (1962).

300. A. Widra, Mycopathol. Mycolog. Applicata, 23, 197 (1964).

301. K. Y. Lee and E. D. Weinberg, Microbios., 3, 215 (1971).

302. G. S. Rawla, New Phytol., 68, 941 (1969).

303. W. L. Lindsay, Advan. Agronomy, 24, 147 (1972).

304. R. L. Mitchell, in Chemistry of the Soil (F. E. Bear, ed.), 2nd ed., Reinhold, New York, 1964, pp. 320-368.

305. K. B. Krauskopf, Geochim. Cosmochim. Acta, 9, 1 (1956).

306. E. D. Goldberg, in Chemical Oceanography (J. P. Riley and G. Skirow, eds.), Vol. 1, Academic Press, New York, 1965, pp. 163-196.

307. A. Zirino and M. L. Healy, Limnol. Oceanogr., 15, 956 (1970).

308. A. Zirino and S. Yamamoto, Limnol. Oceanogr., 17, 661 (1972).

309. E. Rona, D. W. Hood, L. Muse, and B. Buglio, Limnol. Oceanogr., 7, 201 (1962).

310. L. Provasoli, in The Sea (M. N. Hill, ed.), Vol. 2, Wiley (Interscience), New York, 1963, pp. 165-219.

311. G. W. Saunders, Bot. Rev., 23, 389 (1957).

312. R. Johnston, J. Mar. Biol. Assoc., U.K., 44, 87 (1964).

313. G. W. Bryan, J. Mar. Biol. Assoc., U.K., 49, 225 (1969).

314. G. E. Fogg, Oceanogr. Mar. Biol. Ann. Rev., 4, 195 (1966).
315. A. Prakash, in Fertility of the Sea (J. D. Costlow, ed.), Vol. 2, Gordon and Breach, New York, 1971, pp. 351-368.
316. W. Lange, Can. J. Microbiol., 20, 1311 (1974).
317. A. H. Mehran and J. L. Tremblay, Rev. Can. Biol., 24, 29 (1965).
318. W. A. P. Black and R. L. Mitchell, J. Mar. Biol. Assoc., U.K., 30, 575 (1952).
319. M. L. Young, J. Mar. Biol. Assoc., U.K., 55, 583 (1975).
320. G. W. Bryan and L. G. Hummerstone, J. Mar. Biol. Assoc., U.K., 53, 705 (1973).
321. J. Gutknecht, Limnol. Oceanogr., 6, 426 (1961).
322. R. S. Harvey and R. Patrick, Biotechnol. Bioeng., 9, 449 (1967).
323. J. B. Lackey and C. F. Bennett, in Radioecology (V. Schultz and A. W. Klement, eds.), Reinhold and AIBS, New York, 1963, pp. 175-178.
324. W. A. Chipman, T. C. Rice, and T. J. Price, U.S. Fish. Wildlife Serv., Fishery Bull., 58, 279 (1958).
325. J. Haywood, J. Mar. Biol. Assoc., U.K., 49, 439 (1969).
326. P. Mayzand and J. L. M. Martin, J. Exp. Mar. Biol. Ecol., 17, 297 (1975).
327. R. S. Harvey, in Proc. Second Nat. Symp. on Radioecology (D. J. Nelson and F. C. Evans, eds.), U. S. Atomic Energy Commission, Oak Ridge, Tennessee, 1967, pp. 266-269.
328. J. Gutknecht, Limnol. Oceanogr., 8, 31 (1963).
329. R. W. Bachmann and E. P. Odum, Limnol. Oceanogr., 5, 349 (1960).
330. F. Bucheder and E. Broda, Eur. J. Biochem., 45, 555 (1974).
331. M. L. Failla, C. D. Benedict, and E. D. Weinberg, J. Gen. Microbiol., 94, 23 (1976).
332. Z. Fencl, V. Zalabak and J. Bene, Folia Microbiol., 19, 489 (1974).
333. W. Paton and K. Budd, J. Gen. Microbiol., 72, 173 (1972).
334. H. Ponta and E. Broda, Planta, 95, 18 (1970).
335. G. F. Fuhrmann and A. Rothstein, Biochim. Biophys. Acta, 163, 325 (1968).
336. G. R. Findenegg, H. Paschinger, and E. Broda, Planta, 99, 163 (1971).
337. S. Matzku and E. Broda, Planta, 92, 29 (1970).
338. G. D. Parry and J. Haywood, J. Mar. Biol. Assoc., U.K., 53, 915 (1973).
339. G. P. Brierly and V. A. Knight, Biochemistry, 6, 3892 (1967).
340. D. P. Garg, D. P. Dubey, and G. S. Gupta, Indian J. Exp. Biol., 12, 169 (1974).
341. U. Weser and H. Brauer, Biochim. Biophys. Acta, 204, 542 (1970).

342. K. Sivarama Sastry, L. Viswannthan, A. Ramaiah, and P. S. Sarma, Biochem. J., 74, 561 (1960).

343. F. J. Schwarz and G. Matrone, Proc. Soc. Exp. Biol. Med., 149, 888 (1975).

344. C. A. Price, Molecular Approaches to Plant Physiology, McGraw-Hill, New York, 1970, p. 207.

345. D. O. Wilson and H. M. Reisenauer, J. Bacteriol., 102, 729 (1970).

346. F. Witney, M. L. Failla, and E. D. Weinberg, Appl. Environ. Microbiol., 33, 1042 (1977).

347. W. H. Crosby, R. A. Greene, and R. A. Slepecky, in Spore Research: Proc. Brit. Spore Group Meeting (A. N. Barker, G. W. Gould, and J. Wolf, eds.), Academic Press, New York, 1971, pp. 143-160.

348. R. S. Mitra, R. H. Gray, B. Chin, and I. A. Berstein, J. Bacteriol., 121, 1180 (1975).

349. R. P. Novick and C. Roth, J. Bacteriol., 95, 1335 (1968).

350. K. Smith and R. P. Novick, J. Bacteriol., 112, 761 (1972).

351. A. Rothstein and A. Hayes, Arch. Biochem. Biophys., 63, 87 (1956).

352. A. Rothstein and C. Larrabee, J. Cell. Comp. Physiol., 32, 247 (1948).

353. R. J. Lowry, A. S. Sussman, and B. von Boventer, Mycologia, 49, 609 (1957).

354. M. L. Failla and E. D. Weinberg, J. Gen. Microbiol., 99, 85 (1977).

355. S. Fisher, L. Buxbaum, K. Toth, E. Eisenstadt, and S. Silver, J. Bacteriol., 113, 1373 (1973).

356. S. J. Wainwright and P. J. Beckett, New Phytol., 75, 91 (1975).

357. O. Skipnes, T. Roald, and A. Haug, Physiol. Plantar., 34, 314 (1975).

358. M. A. Sims, Experientia, 31, 426 (1975).

359. E. S. Kempner and J. H. Miller, J. Protozool., 19, 343 (1972).

360. W. M. Becker and W. G. Hoekstra, in Intestinal Absorption of Metal Ions, Trace Elements and Radionuclides (S. C. Skoryna and D. Waldron-Edward, eds.), Pergamon Press, New York, 1971, pp. 229-256.

361. F. A. Suso and H. M. Edwards, Proc. Soc. Exp. Biol. Med., 138, 157 (1971).

362. G. W. Evans, C. I. Grace, and C. Hahn, Proc. Soc. Exp. Biol. Med., 143, 723 (1973).

363. S. Kowarski, C. S. Blair-Stanek, and D. Schachter, Amer. J. Physiol., 226, 401 (1974).

364. C. J. Hahn and G. W. Evans, Amer. J. Physiol., 228, 1020 (1975).

365. C. Hahn and G. W. Evans, Proc. Soc. Exp. Biol. Med., 144, 793 (1973).

366. E. J. Moynahan and P. M. Barnes, Lancet, i, 676 (1973).
367. E. J. Moynahan, Lancet, ii, 399 (1974).
368. H. T. Delves, J. T. Harries, M. S. Lawson, and J. D. Mitchell, Lancet, ii, 929 (1975).
369. I. Lombeck, H. G. Schnippering, F. Ritzl, L. E. Feinendeden, and H. J. Bremner, Lancet, i, 855 (1975).
370. J. C. Smith, E. G. McDaniel, L. D. McBean, F. S. Doft, and J. A. Halsted, J. Nutr., 102, 711 (1972).
371. E. D. Weinberg, Science, 184, 952 (1974).
372. R. S. Pekarek, R. W. Wannemacher, and W. R. Beisel, Proc. Soc. Exp. Biol. Med., 140, 685 (1972).
373. G. H. Booth and A. R. Schubert, Proc. Soc. Exp. Biol. Med., 127, 700 (1968).
374. J. H. R. Kagi and B. L. Vallee, J. Biol. Chem., 236, 2435 (1961).
375. M. Margoshes and B. L. Vallee, J. Amer. Chem. Soc., 79, 4813 (1957).
376. P. Pulido, J. H. R. Kagi, and B. L. Vallee, Biochemistry, 5, 1768 (1966).
377. J. H. R. Kagi, S. R. Himmelhoch, P. D. Whanger, J. L. Bethune, and B. L. Vallee, J. Biol. Chem., 249, 3537 (1974).
378. D. R. Winge, R. Premakumar, R. D. Wiley, and K. V. Rajagopalan, Arch. Biochem. Biophys., 170, 253 (1975).
379. R. Premakumar, D. R. Winge, R. D. Wiley, and K. V. Rajagopalan, Arch. Biochem. Biophys., 170, 267 (1975).
380. R. Premakumar, D. R. Winge, R. D. Wiley, and K. V. Rajagopalan, Arch. Biochem. Biophys., 170, 278 (1975).
381. K. S. Squibb and R. J. Cousins, Environ. Physiol. Biochem., 4, 24 (1974).
382. M. P. Richards and R. J. Cousins, Biochem. Biophys. Res. Commun., 64, 1215 (1975).
383. M. P. Richards and R. J. Cousins, Bioinorg. Chem., 4, 215 (1975).
384. R. W. Chen, P. D. Whanger, and P. H. Weswig, Biochem. Med., 12, 95 (1975).
385. M. Webb, Biochem. Pharmacol., 21, 2751 (1972).
386. G. W. Evans, personal communication (1975).
387. F. I. MacLean, O. J. Lucis, Z. A. Shaikh, and E. R. Jansey, Fed. Proc., 31, 699a (1972).
388. J. F. Riordan, Med. Clin. N. Amer., 60, 661 (1976).
389. C. J. Dickenson and F. M. Dickinson, Biochem. J., 153, 309 (1975).
390. J. P. Klinman and K. Welsh, Biochem. Biophys. Res. Commun., 70, 878 (1976).
391. T. Kojima, C. Berger, B. L. Vallee, and J. H. R. Kagi, Proc. Nat. Acad. Sci. U.S., 73, 3413 (1976).
392. M. Bracha and M. J. Schlesinger, Virology, 72, 272 (1976).
393. P. Gupta and F. Rapp, Proc. Soc. Exp. Biol. Med., 152, 455 (1976).

214                                          M. L. FAILLA

394. V. M. Vogt, R. Eisenman, and H. Diggelmann, J. Mol. Biol., 96, 918 (1976).
395. B. D. Korant and B. E. Butterworth, J. Virol., 18, 298 (1976).
396. L. S. Avigad and A. W. Bernheimer, Infect. Immun., 13, 1378 (1976).
397. S. L. Snyder and R. I. Walker, Infect. Immun., 13, 998 (1976).
398. A. Weiss, J. Schottel, and S. Silver, Abstr. Ann. Meeting Amer. Soc. Microbiol., No. H34, p. 101 (1976).
399. J. L. Phillips and P. Azari, Cell. Immunol., 10, 31 (1974).
400. J. L. Phillips, Biochem. Biophys. Res. Commun., 72, 634 (1976).
401. P. Schlievert, W. Johnson, and R. P. Galask, Infect. Immun., 14, 1156 (1976).

Chapter 5

MICROBIAL TRANSPORT AND UTILIZATION OF IRON

B. Rowe Byers and Jean E. L. Arceneaux

Department of Microbiology
University of Mississippi Medical Center
Jackson, Mississippi

## I.  INTRODUCTION

### A.  The Problem in Iron Transport

Iron is an essential but elusive metal for Earth's life system.  Planet
Earth is thought to be about one-third iron; among the metals only aluminum
is present at a greater concentration [1].  Iron is catalytic; its two stable
valencies impart considerable range to its chemical reactivities and its
potential for oxidation-reduction.  If it is true that our present self-
replicating system developed from molecules spontaneously generated on
primitive Earth, then iron must have participated in this genesis.  Iron
probably tended to bind with certain groups on organic molecules to form
primeval "enzymes," and thus some restrictions were imposed on its
catalytic functions.  Increasingly advantageous interactions between iron
and developing macromolecules must have further limited the atom's catalytic role, finally achieving combinations with specificity for both substrate and chemical change impressed on the substrate.  (A brief summary
of the functions of iron in metabolism appears in Section IV.)

It is estimated that there may have been a billion years of procaryotic
evolution under reduced atmosphere (in which iron would be in the ferrous
state) before oxygen-evolving blue-green algae created a new selective
pressure for certain biotypes [2].  Although free ferrous ions are of relatively low solubility, the early life form based its utilization of iron on the
availability of ferrous iron.  Appearance of free oxygen switched much of
the iron to the ferric state.  The solubility product constant for ferrous
hydroxide at $25°$ C is $10^{-15}$ [3], whereas the constant for ferric hydroxide
has been calculated as $1.1 \times 10^{-36}$ [4] and $10^{-38.7}$ [5].  Oxygen caused
rapid removal of transportable iron from the environment, and the metal
became even more elusive.  Had a procaryotic intelligence been present it
might have viewed this removal as a disaster, possibly leading to elimination of all life; but being a system with long-term versatility, life was not
eliminated.  Although numerous alterations were induced by free oxygen,
as regards iron metabolism three possible responses can be imagined.
First, the cells could continue prior existence in the remaining anaerobic
niches; descendants of these forms probably still exist as some of the
strict anaerobes.  Second, cells could circumvent most of their iron requirements, using other cofactors for these reactions; some of the lactic
acid bacteria may represent these forms [2].  Finally, the cells could
develop an efficient means for solubilizing and transporting ferric iron,
causing increased flow of iron into the cells; most of the aerobic cells
today may be derived from these forms.  In fact, increased acquisition of
iron probably was a requirement for evolution of aerobic respiration, with
its advantageous use of free oxygen and its heavy dependence upon iron.

B.  The Microbial Solution to the Problem

What mechanism did microorganisms employ to increase their capacity to capture iron?  Current studies of microbial iron transport have revealed that several billion years of evolution produced a transport system in which the cells release high-affinity ferric iron binding (chelating) molecules and then transport the resulting ferric chelates across the lipid-protein boundary by specific membrane-bound transport systems.  Iron is then removed from the chelate for metabolic use by special enzymes.  The major portion of this review is concerned with the molecules which became iron transporters and what is known about the molecular mechanisms of transport of the ferric chelates.  Certain important research, particularly studies of the effects of the iron transport cofactors on growth of microorganisms will not be discussed, although much of this work was the basis for iron transport studies.  Descriptions of this work and citations for earlier papers may be found in Lankford [6], Byers [7], and Golden et al. [8].

II.  MICROBIAL IRON TRANSPORT
     COFACTORS (SIDEROCHROMES)

Iron is found in microbial cells in combination with certain electronegative donor atoms such as oxygen, nitrogen, and sulfur.  The compounds which are known to function as iron transport cofactors commonly are oxygen ligands in which iron is coordinated through an oxygen atom.  This is not surprising in view of the fact that these compounds selectively chelate ferric iron [2].  Microorganisms are known to produce at least two chemical types of all-oxygen ligands, the secondary hydroxamic acids and the phenolic acids (derivatives of 2,3-dihydroxybenzoic acid).  There is evidence that representatives of both chemical categories participate directly in iron transport.  Neilands [9] has proposed that the term siderochromes [10] include both hydroxamate and phenolate type iron transport cofactors. This terminology will be used in the present review.

A.  Hydroxamates

The general chemistry of the hydroxamic acids has been reviewed by Emery [11] and by Neilands [9].  The most outstanding property of these compounds is their capacity to chelate ferric iron; ferrous iron is weakly bound.  This represents an important chemical property in terms of removal of iron from the chelate by a possible enzymatic reductive step following transport of the ferric chelate [12].  The hydroxamate function produces a stable five-membered ring with an iron atom.  At neutral pH three hydroxamate ligands will bind with trivalent iron to yield a neutral

chelate. Many naturally occurring hydroxamic acids contain three hydroxamate groups per molecule, creating a markedly stable chelate. Iron
transport cofactors containing two hydroxamates per molecule have been
isolated; in these chelators oxygen atoms forming other chemical groups
probably participate in chelation. A brief survey of the various families of
microbial hydroxamates with the structure of one representative of each
family is given next.

## 1. Ferrichrome Types

Following Neilands' original isolation of ferrichrome in 1952 [13] a series
of hydroxamates with related structure have been isolated from several
fungal species. The structure of ferrichrome is shown in Figure 5-1. In
the ferrichrome molecule three residues of N-hydroxy-L-ornithine form
the basic structure. Acetylation of the three hydroxyamino groups produces the hydroxamic acid functions. The molecule is constructed to produce a stable ring structure in which all three hydroxamic ligands will be
oriented about a ferric iron atom.

Included in the ferrichrome family is an antibiotic, albomycin, a
structural analog of ferrichrome [11,14,15]. Albomycin is active against
both Gram-negative and Gram-positive bacteria; unfortunately, its lability
precludes clinical application [11]. The activity of albomycin can be
antagonized by certain other siderochromes [10,15].

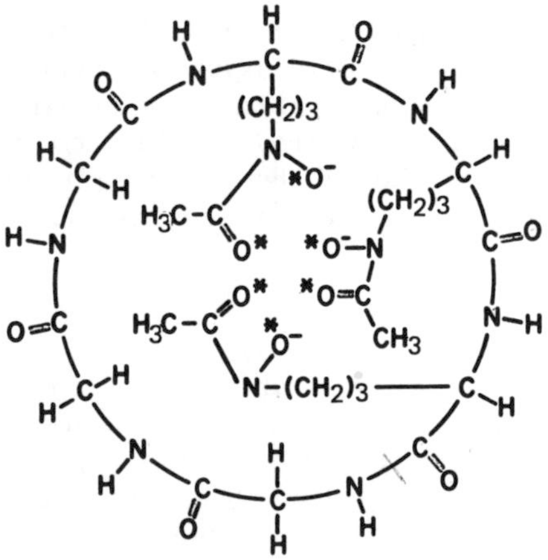

FIG. 5-1  Structure of ferrichrome (deferri) [13]. Iron chelation sites
are indicated by asterisks.

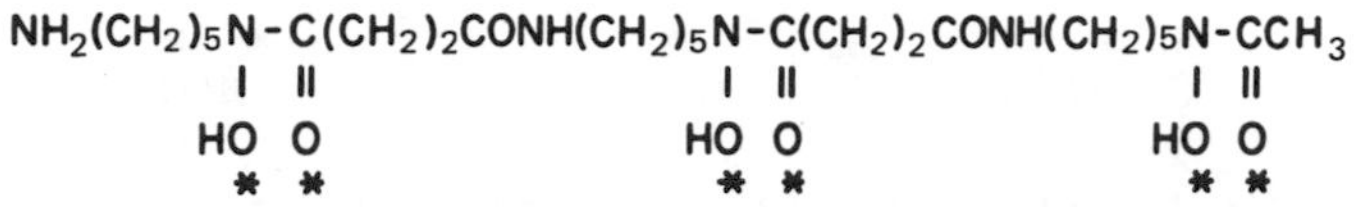

$$NH_2(CH_2)_5N-C(CH_2)_2CONH(CH_2)_5N-C(CH_2)_2CONH(CH_2)_5N-CCH_3$$

FIG. 5-2  Structure of ferrioxamine B (deferri) [16].  Iron chelation sites are indicated by asterisks.

## 2.  Ferrioxamine Types

In contrast to the ferrichrome class of trihydroxamic acids, the ferrioxamine class of compounds contains three hydroxamate groups inserted in a chain.  A typical example, ferrioxamine B [16], is shown in Figure 5-2.  Ferrioxamines have been identified in a number of species of Streptomyces and Nocardia.  The nine atoms separating hydroxamate groups in ferrioxamine B easily allow the molecule to fold around an iron atom accommodating chelation.  Emery [11] has discussed various structural differences found in the ferrioxamines.

Several hydroxamates (structurally related to the ferrioxamines) with antibiotic activity have been described [15].  These substances (for example, ferrimycin, A22765) will compete with certain siderochromes for transport, and many of the mutants resistant to these antibiotics (and to the ferrichrome type, albomycin) have impaired transport mechanisms [15].  Such mutants have been used in some transport studies (described in Section III).

## 3.  Derivatives of Citrate

A related group of dihydroxamates has been isolated from certain bacteria.  The structure of one of these, schizokinen [17], is shown in Figure 5-3.  In these compounds citric acid serves to link hydroxamate functions.  Because the molecules contain only two hydroxamate groups, the free central carboxyl and hydroxyl groups of citrate may participate in the formation of the ferric chelate [17].  These compounds have been identified in Aerobacter aerogenes [18], Arthrobacter pascens [19], and Bacillus megaterium [20].

## 4.  Mycobactins

A series of dihydroxamate compounds has been isolated from the mycobacteria.  The general structure of the mycobactins is shown in Figure 5-4, and Snow [21] has reviewed the chemistry of these compounds.  One of the most significant properties of the mycobactins is their lipid solubility, which contrasts to that of most other hydroxamates.  This property

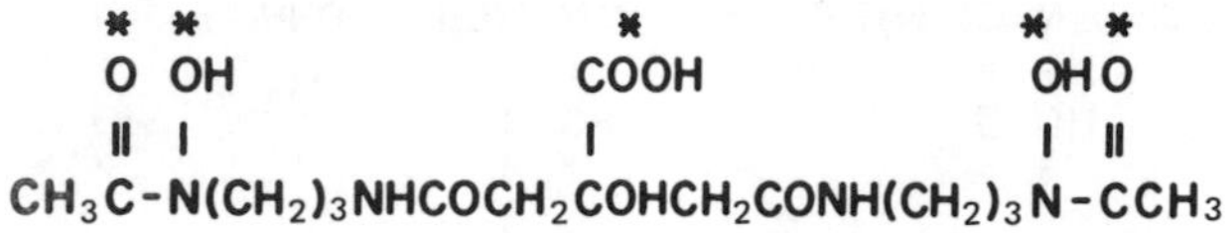

$$\overset{\displaystyle *\ \ *}{\underset{\displaystyle \overset{\|\ \ \ |}{CH_3C-N(CH_2)_3NHCOCH_2COHCH_2CONH(CH_2)_3N-CCH_3}}{O\ \ OH}} \qquad \overset{\displaystyle *}{\underset{\displaystyle |}{COOH}} \qquad \overset{\displaystyle *\ \ *}{\underset{\displaystyle \overset{|\ \ \|}{}}{OH\,O}}$$

FIG. 5-3  Structure of schizokinen (deferri) [17].  Iron chelation sites are indicated by asterisks.

may be related to the high lipid content of mycobacteria.  Like other natural hydroxamates, the mycobactins have strong selectivity for ferric iron.  Lipid-soluble iron-binding compounds termed the nocobactins, which are structurally related to the mycobactins, have been identified in Nocardia species [22].

5.  Rhodotorulic Acid

Rhodotorulic acid is an interesting compound isolated by Atkin and Neilands [23].  Its structure is shown in Figure 5-5.  Rhodotorulic acid is a cyclic dipeptide of N-acetyl-N-hydroxyornithine, a subunit of ferrichrome.  The two hydroxamate functions of this compound cannot both chelate the same iron atom for steric reasons, and rhodotorulic acid forms polynuclear complexes with the metal.  Rhodotorulic acid has been identified in several yeasts [24].

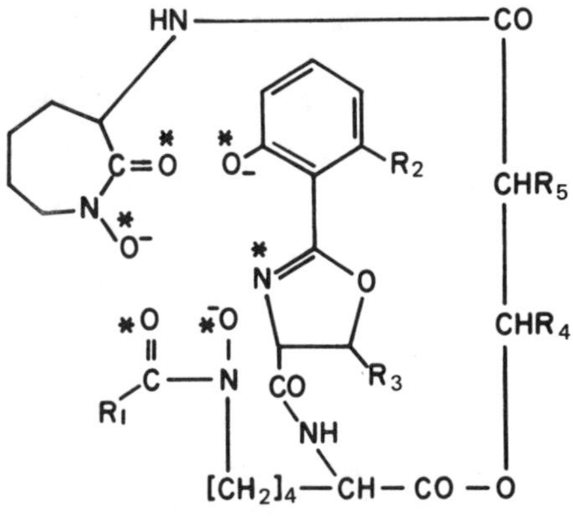

FIG. 5-4  General structure of the mycobactins (deferri) [21].  Variations occur at $R_{1-5}$.  Iron chelation sites are indicated by asterisks.

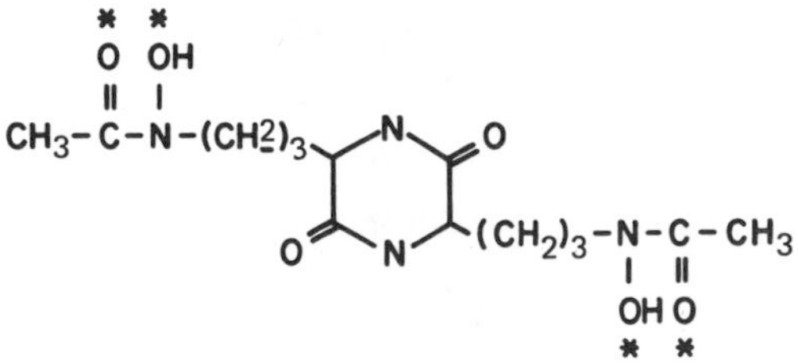

FIG. 5-5  Structure of rhodotorulic acid (deferri) [23]. Iron chelation
sites are indicated by asterisks.

## B.  Phenolic Acids

The second major chemical category of all-oxygen ligands utilized as iron
transport cofactors by microorganisms are amino acid conjugates of 2,3-
dihydroxybenzoic acid.  In A. aerogenes, Escherichia coli, and species of
Salmonella, the cyclic triester of 2,3-dihydroxybenzoylserine (enterobactin
or enterochelin) has been isolated [25,26].  Each enterochelin molecule
complexes with one atom of ferric iron.  A hexadentate complex is formed
between the ferric atom and the six phenolic hydroxy groups of enterochelin.
The structure of enterochelin is shown in Figure 5-6.  A threonine conjugate of 2,3-dihydroxybenzoic acid has been found in cultures of Klebsiella
and E. coli [27].  In the strains of Bacillus subtilis, 2,3-dihydroxybenzoyl-
glycine has been isolated [28]; however, because of its structure, 2,3-
dihydroxybenzoylglycine cannot form a cyclic triester analogous to enterochelin.  A diphenolic acid conjugate of lysine was discovered in low-iron
cultures of Azotobacter vinelandii [29].  Lankford [6] reviewed the isolation of certain other phenolic acid derivatives from bacterial cultures.

## C.  Biosynthesis

Efforts have been made to define biosynthesis of the hydroxamic acids,
particularly the hydroxamate function.  This information has been reviewed recently by Emery [30].  Evidence suggests that enzymatic formation of the hydroxyamino group occurs at the level of the amino acid rather
than after incorporation of the amino acid into peptides or esters.  Biosynthesis of hydroxamates is controlled by the relative iron status of the
producing cell.  Iron deficiency leads to an exaggerated production of the
hydroxamates.  Although some authors have suggested that hydroxamate
synthesis may occur only when the organisms are iron deficient, low-
level synthesis of the hydroxamate schizokinen has been detected in high-
iron cultures of B. megaterium [20].  It is likely that hydroxamate synthesis occurs almost continually and that exhaustion of cellular iron stores

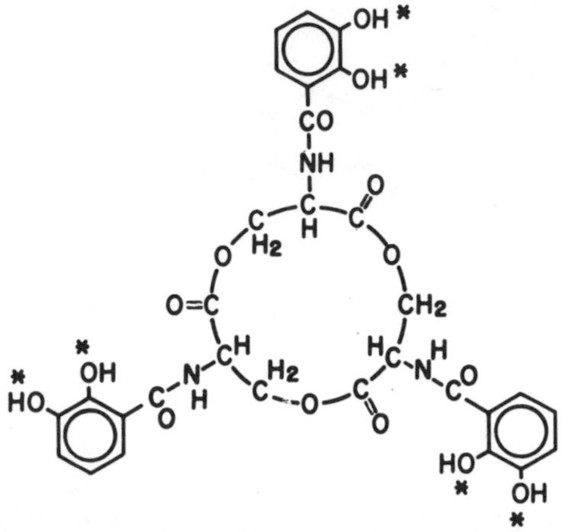

FIG. 5-6  Structure of enterochelin (enterobactin) (deferri) [25,26]. Iron chelation sites are indicated by asterisks.

leads to full release of the biological control mechanisms governing hydroxamate synthesis.  Unfortunately, the enzymes concerned with hydroxamate biosynthesis have not yet been purified sufficiently so that studies of possible feedback inhibition of these enzymes can be undertaken.  It is unknown whether iron represses synthesis of one or more of the enzymes involved in the biosynthetic pathways or if the metal may exert feedback inhibition on one of the steps.  Moreover, the particular form of iron which acts as a negative effector on hydroxamate biosynthesis has not been identified.  The critical need for iron in cell metabolism is shown again by the fact that some organisms divert large amounts of carbon into production of hydroxamates under the pressure of iron deficiency.

Biosynthesis of the phenolates has been carefully studied, and the pathway leading to 2,3-dihydroxybenzoate is well established.  In A. aerogenes, E. coli, and B. subtilis, synthesis of 2,3-dihydroxybenzoate represents a branch from chorismic acid, the major branch-point compound in aromatic biosynthesis [31-33].  Three enzymes have been identified in the pathway [34].  Enterochelin is synthesized from 2,3-dihydroxybenzoate, and L-serine by cell extracts of E. coli [35,36].  There are three proteins involved in synthesis of enterochelin from 2,3-dihydroxybenzoate and serine.  Genes coding for enterochelin biosynthetic enzymes are clustered together on the E. coli chromosome, and recent evidence suggests that these function as an operon [37].

Similar to the hydroxamates, synthesis of the phenols is regulated by iron.  Growth of an organism in high-iron medium leads to repression of synthesis of the enzymes involved in production of enterochelin [32].

Likewise, enzymes converting chorismate to 2,3-dihydroxybenzoate in B. subtilis are repressed by elevated iron concentrations in the culture medium [33,38]. The nature of the corepressors is still unknown.


III.  THE TRANSPORT MECHANISMS

Reviewed in this section are some of the important investigations in various microorganisms that have established a direct role for the siderochromes in transport of iron. These experiments have been conducted largely using radioactive iron ($^{59}$Fe, $^{55}$Fe) as the tracer, although there are important studies in which the organic ligand was labeled with either $^3$H or $^{14}$C. A doubly labeled chelate can be constructed from $^3$H and $^{59}$Fe because the two isotopes can be discriminated by usual counting procedures. The transport process is unfolding as a multistep mechanism involving more than one gene product. It has also become clear that any microorganism may be able to use not only its own endogenously produced siderochrome but also siderochromes produced by other microorganisms. Various siderochromes may also be transported by independent systems within a single organism. It is interesting that an organism should possess the capability to utilize siderochromes of other microorganisms. Maintenance of this capability must be of survival advantage. Presumably, a single organism entering an environment where other organisms have been growing and producing their siderochromes would be unable to grow unless it could both transport and remove iron from these exogenous siderochromes. In evolutionary terms this may represent a curious situation in which several structurally different forms of an essential cofactor exist, so that life forms have developed multiple utilization systems for the various types of the cofactor. The transport story for the siderochromes is by no means complete, and the various molecules present within the cell envelope and within the cytoplasm responsible for effective movement of the chelates and subsequent release of iron are now being isolated and identified. The work summarized in the following subsections supports the prophesy made by Neilands [39] after the isolation of ferrichrome that this compound might function in movement of iron in biological systems.


A.  Bacillus Species

The secondary hydroxamate schizokinen is produced by B. megaterium [20]; see Figure 5-3. A series of mutants unable to produce schizokinen was isolated by Arceneaux and Lankford [40]. These mutants, which retain normal siderochrome transport capacities, as well as mutants isolated from them in which utilization of one of the siderochromes (Desferal) is defective [41], have proved useful in studies of transport and iron assimilation. Loss of the capacity to produce schizokinen did not eliminate

uptake of iron in its ionic form ($^{59}FeCl_3$) in the two mutant strains
B. megaterium SK11 and B. megaterium SK300 [41,42]. In these two
strains rapid uptake of ionic iron was easily observed; however, lack of
schizokinen synthesis prevented strain SK11 from competing for iron with
systems which tend to coprecipitate iron. Thus, addition of concentrations
of aluminum as low as $4 \times 10^{-5}$ M blocked uptake of radioactive iron from
$^{59}FeCl_3$ in strain SK11 [41]. Evidence suggested that aluminum inhibited
uptake of ionic iron through the formation of external complexes with iron,
effectively removing iron from the system. Chelation of radioiron by
schizokinen both prevented aluminum inhibition of iron uptake and stimulated total iron uptake about fourfold in strain SK11. Transport of
schizokinen-iron was rapid, reaching saturation in about 2 min. In the
absence of schizokinen, strain SK11 does not possess an iron-gathering
process of sufficient affinity for acquisition of iron complexed by certain
other systems. These results reveal the essential nature of siderochrome
production in natural situations, although strain SK11 can be grown in the
laboratory in simple medium containing only ionic iron as an iron source.
Stimulation of radioactive iron transport by schizokinen in strain SK300,
which also is unable to produce schizokinen, has been observed [43]. It
is interesting that addition of aluminum to cultures of the hydroxamate-
producing organism B. megaterium ATCC 19213 caused increased schizokinen synthesis [44].

Clearly, then, schizokinen is essential for iron transport in B. megaterium under certain situations. However, the question arises as to
whether schizokinen functions simply to solubilize external iron and deliver
it to a membrane site or whether the schizokinen-iron chelate actually penetrates across the membrane. To resolve this question simultaneous transport of $^3H$ and $^{59}Fe$ from a doubly labeled chelate, [$^3H$]schizokinen[$^{59}Fe$],
was studied in B. megaterium [45]. Kinetic studies showed that at
certain low concentrations of the doubly labeled chelate the picomoles
of radioactive iron transported were nearly equivalent to the picomoles of
[$^3H$]schizokinen taken up by the cells. These results would be expected
if the chelate were translocated as an intact unit. When tested at high
concentrations, uptake of both components of the chelate was rapid; however, release of [$^3H$]schizokinen from the cells after 30 sec of uptake was
evident. Total uptake of radioactive iron during a 10-min testing period
exceeded uptake of [$^3H$]schizokinen by about threefold. These results are
interpreted to mean that the chelate crossed the cell membrane as a single
unit but that iron was rapidly released from the complex and the resulting
deferri [$^3H$]schizokinen was discharged from the cell. Portions of the
doubly labeled chelate may remain as a pool of available iron. Release of
radioactive schizokinen was a temperature-dependent process suggesting
enzyme-mediated removal of iron from the chelate. Chromatographic
examinations of the $^3H$-labeled material released by the cells during the
first 2 min of assay indicated that it was still [$^3H$]schizokinen; therefore,

the organic ligand was not hydrolyzed during the process of transport and iron removal. Similar results, discussed below, with [$^{14}$C]ferrichrome in Ustilago sphaerogena were reported by Emery [30].

Evidence for a specific schizokinen transport system has been obtained in B. megaterium. Like several other organisms, B. megaterium is able to use not only its own hydroxamate but certain other hydroxamates produced by other organisms. Included among those hydroxamates which it can utilize is Desferal (the iron-free form of ferrioxamine B). Aerobactin (a citrate hydroxamate produced by A. aerogenes) cannot be utilized by B. megaterium strain SK11. Determination of transport kinetics in B. megaterium SK11 over a wide range of iron concentrations [46] revealed that rates of iron transport increased sharply as the concentration of [$^{59}$Fe]schizokinen was increased. Similar studies conducted with the [$^{59}$Fe]Desferal chelate showed a difference in the response of strain SK11 to this iron source. Higher concentrations of [$^{59}$Fe]Desferal delivered only about one-fifth of the maximum level of iron transported form [$^{59}$Fe]-schizokinen and did not produce the rapid rates of uptake seen with the endogenous chelate. Thus, B. megaterium SK11 appeared to have a recognition capacity for transport and utilization of hydroxamates based on the chemical structure of the hydroxamate molecule. Such differences might be due to inherent differences in the affinity of a single hydroxamate transport system, or alternatively there may be separate hydroxamate systems for transport of different secondary hydroxamates.

Membrane vesicles prepared by osmotic rupture of B. megaterium protoplasts accumulated the ferric hydroxamates of both schizokinen and Desferal [47-49]. This accumulation was not influenced by temperature and was unchanged by addition of various energy sources. It may represent only binding to possible ferric hydroxamate receptors on the membranes. Such binding was rapid and concentration dependent. Studies of the capacity of a number of different natural ferric hydroxamates to cause dissociation of bound schizokinen-iron or Desferal-iron from the membranes, as well as calculation of binding-affinity constants and maximal binding capacities, revealed that the two chelates probably were bound by independent specific sites. The schizokinen-iron receptors had lower affinity but higher binding capacity than the Desferal-iron binding system showed for its chelate. If it is assumed that this binding was accomplished by components of the transport systems, then the greater number of schizokinen-iron binding sites may explain the increased rate of iron uptake from this chelate at high concentration. Bacillus megaterium may have independent transport processes for these two ferric hydroxamates and for other ferric hydroxamates which it can use. No specific receptors for aerobactin-iron were identified on these vesicles, implying no transport system for this chelate, and explaining lack of aerobactin utilization by B. megaterium.

Further study of the interaction between membrane vesicles of B. megaterium and radiolabeled schizokinen-iron revealed the presence of a schizokinen-iron binding substance present within the membranes [47-50]. The schizokinen-iron binding substance has not been identified chemically; however, its molecular weight is about 50,000-60,000 daltons. There is no direct evidence linking this binding substance with transport of the chelate, although this component may be part of a receptor site or may be functional subsequent to binding of the chelate to the receptor.

Uptake of Desferal was examined in a mutant strain of B. megaterium (strain Ardl) that is unable to grow in the presence of Desferal [51,52]. This strain was isolated on the basis of its resistance to the ferrioxamine-type antibiotic A22765 [41]. Earlier work in Staphylococcus aureus [15] indicated that mutants resistant to A22765 had impaired transport of ferrioxamine B (Desferal-iron). Uptake of radioiron-Desferal by protoplasts of B. megaterium Ardl was observed; however, only the intact Desferal-iron chelate was found in the cell cytoplasmic fraction. No transfer of radioiron to macromolecular components could be noted. Control experiments utilizing schizokinen-iron revealed that by 30 sec of assay a significant amount of the radioiron had been transferred to macromolecular species present in the cytoplasmic fraction, indicating rapid removal of iron from schizokinen. Thus, B. megaterium strain Ardl appears to be able to transport Desferal-iron but cannot remove iron from the chelate for subsequent metabolic use. Careful study of Desferal-iron transport in this organism indicated a facilitated diffusion system, suggesting that ferric hydroxamates are not actively transported. Energy-requiring transfer of iron to macromolecules may drive high-level transport of the chelate. These aspects of ferric hydroxamate transport have been included in a tentative model (Fig. 5-7).

The phenolate-producing organism B. subtilis also has been used in studies of iron transport both from phenolate compounds and from ferric hydroxamates. This organism is known to excrete both 2,3-dihydroxybenzoate and 2,3-dihydroxybenzoylglycine. An increase in total iron uptake in the presence of 2,3-dihydroxybenzoylglycine has been noted [53]. The overall process was temperature sensitive, and total iron uptake was reduced by prior treatment of the cells with sodium azide. Although these latter results suggest that transport was an energy-consuming process, they do not prove that movement of the chelate across the membrane is thermodynamically active. It is not yet clear how 2,3-dihydroxybenzoylglycine might enhance iron transport in B. subtilis. Little significant uptake of radioactive 2,3-dihydroxybenzoylglycine by B. subtilis was noted [54], although rapid release of the iron-free phenol after removal of iron is a possibility. Additional work is required to clarify participation of the phenolate in iron transport in B. subtilis. It is known that the two ferric hydroxamates, Desferal-iron and schizokinen-iron, enhance iron uptake in B. subtilis [33], and ferrichrome also may be a functional iron transport cofactor in this organism [54]. Bacillus subtilis appears to be another example of a microorganism which is able to utilize a number of different siderochromes.

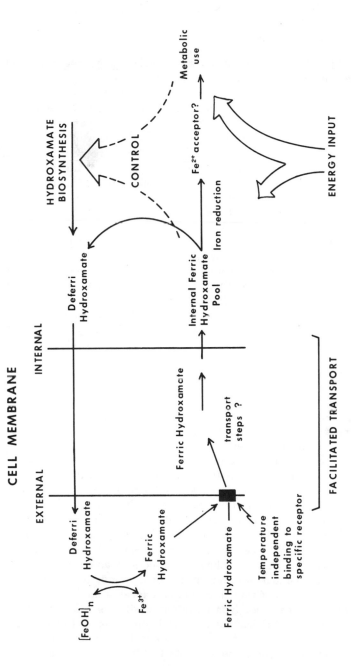

FIG. 5-7  Model of hydroxamate synthesis and ferric hydroxamate transport in B. megaterium.

## B.  Fungal Species

### 1.  Ustilago sphaerogena

Emery [30] has summarized work with Ustilago sphaerogena, which produces the hydroxamate ferrichrome (see Fig. 5-1).  Washed cell suspensions (previously grown under iron-deficient conditions) rapidly incorpo-
rate radioiron-labeled ferrichrome.  Such uptake demonstrated saturation
kinetics and was inhibited by energy poisons or by anaerobic conditions.
Of a number of other hydroxamic acids tested, none was found to be active.
Even ferrichrome A, a structural analog of ferrichrome produced by the
same organism, is not active.  Such recognition must lie in external-
membrane-bound molecules capable of specifically binding ferrichrome.
It is also interesting that chelation of copper by ferrichrome (for which it
has rather low affinity) results in a copper chelate which was not transported by the cells.  On the other hand, chelation of aluminum by ferrichrome results in a chelate which was transported.  With [14]C-labeled ferrichrome, it was noted that iron was not immediately removed following
transport of the chelate.  Eventual release of labeled iron-free ferrichrome
was observed, probably shortly after removal of the metal.  (Exit of [14]C-
labeled ferrichrome which had bound aluminum did not occur, indicating
that aluminum was never removed from the chelate, thereby causing retention of the aluminum-ferrichrome chelate.)  Thus, U. sphaerogena has
a transport system for ferrichrome which has high affinity for this particular ligand.
      Emery [55] has suggested that removal of iron from ferrichrome
might serve as a thermodynamic trap with reference to the direction of
iron flow and has tested this premise in a model transport system.  In this
model, radioiron-labeled ferrichrome was separated by benzyl alcohol-
chloroform from a solution containing a reducing agent (ascorbic acid) and
a ferrous iron chelating agent (ferrozine).  Ferrichrome is soluble in the
organic phase.  At the end of 24 hr over 90% of the radioiron had been
transferred from ferrichrome to ferrozine.  Emery has suggested that
this iron-trapping mechanism provides a thermodynamic drive perhaps
analogous to trapping of iron by its in vivo utilization.
      One of the interesting aspects of siderochrome biosynthesis is found
in U. sphaerogena.  Under iron-deficient conditions this organism excretes
large amounts of iron-free ferrichrome A.  Several workers have been unable to demonstrate that ferrichrome A functions as an iron transport cofactor [30] in a number of different test organisms, including the producing
strain.  The question of why an organism excretes large amounts of an inactive siderochrome under iron-deficient conditions has never been clearly
answered.  Emery [55] has made the following proposal based on his experiments which showed that iron can be slowly transferred from ferrichrome A to ferrichrome.  He suggests that such an exchange at the outer

cell membrane could be driven to completion by subsequent transport of ferrichrome and intracellular removal of the iron from ferrichrome. Ferrichrome A might represent an extracellular iron-solubilizing agent. This hypothesis awaits experimental verification.

## 2. Neurospora crassa and Aspergillus Species

Neurospora crassa produces coprogen, a trihydroxamate of the ferrichrome class. Studies have shown that N. crassa takes up iron from the coprogen-iron chelate and that this organism can use two other ferrichrome-type hydroxamates, ferricrocin and ferrichrysin [56]. Certain other siderochromes (including the ferrioxamines, ferrichrome A and rhodotorulic acid) were ineffective iron transport cofactors for this organism. The siderochrome transport system(s) exhibited saturation kinetics and was inhibited by energy poisons or anaerobiosis.

Members of the genus Aspergillus produce a variety of hydroxamates, usually of the ferrichrome type, although the particular compounds produced often are distinctive of the species. In studies with three species of Aspergillus, Wiebe and Winkelmann [57] noted nearly identical iron transport kinetics with several ferrichrome-type ligands, regardless of the major hydroxamate produced by the species; however, both coprogen and a ferrioxamine-type siderochrome were not utilized. With the hydroxamate fusigen, these investigators observed a $K_m$ value higher than that found with ferrichrome types and concluded that uptake from fusigen may occur by a separate system. These results in N. crassa and Aspergillus species again show the preference of a given organism for certain siderochromes, as well as the capacity to use several different siderochromes. Such recognition must reside in molecules which bind, transport, and remove iron from the siderochromes.

## C. Gram-Negative Bacteria

Iron transport in Gram-negative bacteria has been studied in strains of E. coli, S. typhimurium, and to a very limited extent in A. aerogenes. All these organisms produce a cyclic triester of 2,3-dihydroxybenzoyl-serine (see Fig. 5-6), which was named enterochelin following its isolation from E. coli [25] and enterobactin following its isolation from S. typhimurium [26]. In the present review this siderochrome will be referred to as enterochelin. It is interesting that A. aerogenes strain 62-1 produces both enterochelin and the secondary hydroxamate aerobactin [18, 25]. Certain Gram-negative organisms have the ability to utilize enterochelin and certain of the hydroxamate type siderochromes, and have a relatively low affinity system which transports ionic iron. Moreover,

certain strains of E. coli also possess a transport system for uptake of
iron complexed with citrate. The process of ferric enterochelin uptake and
subsequent release of iron from the chelate has been studied in detail, and
much of this work has been reviewed recently by Rosenberg and Young
[58].

Mutants of both E. coli (ent⁻ strains) and S. typhimurium (enb⁻ strains)
that are unable to produce enterochelin have been isolated [34,35,59].
Using such E. coli mutants it has been shown that enterochelin stimulated
iron uptake and that the ferric enterochelin complex is transported into the
interior of the cell [60]. These conclusions were supported by studies
using $^{14}$C-labeled enterochelin. Once within the cell the enterochelin lig-
and was hydrolyzed to 2,3-dihydroxybenzoylserine, to which iron remains
complexed. Thereafter, iron was released from 2,3-dihydroxybenzoylser-
ine and the iron-free 2,3-dihydroxybenzoylserine was excreted. A model
for this system is shown in Figure 5-8.

Intracellular release of iron from ferric enterochelin was accomplished by the enzyme ferric enterochelin esterase [61]. Evidence supporting the hypothesis that this enzyme functions in the release of iron has been
obtained from studies with mutants of E. coli which lack ferric enterochelin esterase activity (fes⁻ mutants) [62]. The fes⁻ mutants take up
ferric enterochelin, which accumulates unchanged within the cells. At
saturation, the intracellular concentration of ferric enterochelin was fifty
times that of the external fluid, suggesting to these workers an active
transport system (the effect of energy depletion on transport was not reported). It was suggested that the esterase converts the cyclic triester to
iron-2,3-dihydroxybenzoylserine complexes (Fig. 5-8). It has been shown
that the latter complexes will serve as iron sources for heme biosynthesis,
whereas iron bound by ferric enterochelin is not available for such synthesis [63]. Ferric enterochelin esterase is probably composed of at least
two components, designated A and B [61]. The activity of this enzyme,
like that of enzymes involved in the biosynthesis of enterochelin, is repressed by addition of iron to the growth medium, and the gene(s) coding
for the esterase is located on the E. coli chromosome near the other genes
essential for the enterochelin system [62].

Mutants of E. coli defective in enterochelin-mediated iron transport
(designated fep⁻ strains) have been isolated [62]. Only low levels of ferric enterochelin accumulation could be shown in these strains. The fep
gene was located on the E. coli chromosome close to the other genes
coding for both enterochelin synthesis and for the enterochelin esterase
function. Recently, Frost and Rosenberg [64] suggested that the function
coded for by the fep gene may be a component of the inner membrane of
E. coli (Fig. 5-8).

An additional genetic locus, the tonB locus, is critical for enterochelin transport. The products of this locus first were implicated in iron
transport by Wang and Newton [65]. Mutations in the tonB region are

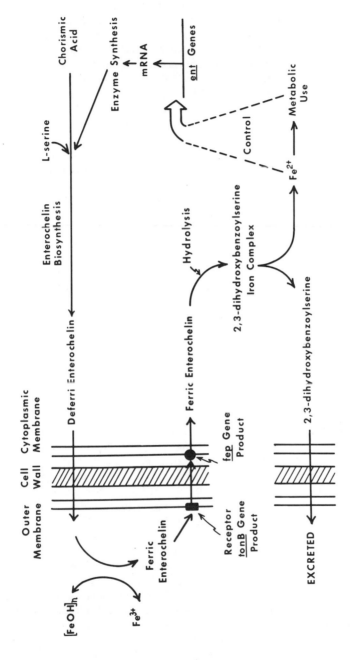

FIG. 5-8  Model of enterochelin synthesis and ferric enterochelin transport in E. coli.  (Adapted from Ref. 64 with permission of the authors and ASM Publications.)

known to impart resistance to bacteriophages $\phi 80$ and T1, and to colicins B, M, V, and I [66-68]. These mutations are thought to involve changes in the receptor present in the outer membrane. Recently tonB⁻ mutants which are resistant to $\phi 80$ have been shown to be unable to transport ferric enterochelin, and several of these mutants have also been found to hyper-excrete enterochelin [64,69]. Cells which are unable to take up iron as the ferric enterochelin complex find themselves in an iron-deficient condition and continue to excrete enterochelin at high levels. This evidence suggests that tonB⁻ strains lack an outer membrane component necessary both for the uptake of enterochelin and for the adsorption of $\phi 80$. In studies with tonB⁻ mutants, Frost and Rosenberg [64] discovered that these mutants were also unable to utilize several hydroxamates (for example, ferrichrome, ferrichrysin, rhodotorulic acid, and coprogen), even though these compounds were functional siderochromes for tonB⁺ parental types. This evidence suggests that the products of the tonB locus may also represent components of hydroxamate transport systems. Of interest also was apparent damage to the citrate-dependent system of iron uptake in their tonB⁻ strains. Certain mutations in the tonB locus appear to affect all the transport systems for iron in E. coli, possibly indicating that they share some common outer membrane components in their transport and/or receptor sites.

The tonB locus is not the only genetic information required for adsorption of bacteriophage $\phi 80$. The related tonA locus is also necessary. Mutants in either the tonA or tonB region are resistant to $\phi 80$; however, tonA mutants have a different pattern of resistance to other bacteriophages and to colicins. Cells of the tonA⁻ phenotype are unable to accumulate iron added as the ferrichrome chelate [70]. Wayne and Neilands [71] obtained evidence that both tonA⁻ and tonB⁻ mutants were unable to transport ferrichrome. In their studies, these authors examined the capacity of ferrichrome to inhibit adsorption of $\phi 80$ to wild-type cells normally sensitive to this bacteriophage. They found a strong competition between ferrichrome and $\phi 80$, apparently for the external binding site functioning both in transport of the hydroxamate and in adsorption of bacteriophage. They were able to show competition between $\phi 80$ and only one other siderochrome tested, rhodotorulic acid. Using partially purified outer membrane phage T5 receptor complexes from E. coli, Luckey et al. [72] showed that ferrichrome competed with phage T5 for binding to this isolated receptor complex. Therefore, particular surface areas containing components coded for by both the tonA and tonB loci are essential in transport of siderochromes in E. coli. The interplay between these two gene products and transport of the chelates is still not clear, but there is convincing evidence of outer membrane receptor sites essential for transport of siderochromes.

The participation of outer membrane binding sites in transport of siderochromes may be necessitated by the molecular weight exclusion limit of 800-900 daltons dictated by the outer cell envelope [71]. Many siderochromes have molecular weights near this exclusion limit and may require special systems to insure their penetration of the outer membrane. Such sites would be essential for survival of the organism; their essential nature is demonstrated by the fact that they have been retained in spite of utilization of these sites for both phage and colicin binding. These latter two destructive entities have capitalized upon a site so essential for cell metabolism that it cannot be modified significantly without serious loss of the capacity to accumulate iron.

In addition to studies in E. coli, S. typhimurium has been used for both transport and genetic studies. This organism produces the phenolate-type siderochrome [59, 73]. Mutants of S. typhimurium which are unable to produce enterochelin (in this case called enterobactin; hence, they are designated enb⁻ strains) require this substance, although at least 15 structurally different hydroxamate-type siderochromes will substitute as growth factors [59]. From an enb⁻ strain, 12 different classes of mutants (termed sid⁻ strains), divided according to their ability to utilize different hydroxamates, were isolated [74]. These mutants were isolated on the basis of their resistance to the antibiotic albomycin. Albomycin is a secondary hydroxamate antibiotic structurally related to ferrichrome. Most of these various classes of mutants had genetic lesions located near panC on the genetic map, although one class of mutants had lesions located near the enterochelin genes. Two of the mutant classes were linked neither to panC nor the enterochelin-producing genes. From these genetic studies it was concluded that S. typhimurium may have several transport systems of high specificity for a variety of different siderochromes produced by other organisms. Unlike E. coli, S. typhimurium is unable to use citrate as an iron transport cofactor [59].

Some, but not all, strains of E. coli do possess a transport system for iron which uses citrate to facilitate movement of the metal [31, 75]. Induction of the system requires growth in the presence of citrate [60]. As mentioned previously, it has now been shown that the tonB chromosomal region also may be involved in use of citrate as an iron carrier [64].

Recently, Frost and Rosenberg [64] have made an interesting observation based on the properties of certain mutants which are both unable to synthesize enterochelin (blocked prior to 2, 3-dihydroxybenzoate) and to transport ferric enterochelin (tonB⁻). These mutants will show a growth response to 2, 3-dihydroxybenzoate, a component of the enterochelin molecule. Because 2, 3-dihydroxybenzoate is itself an iron-chelating agent, it might be predicted that this phenol acted as an iron transporter. This may not be true, and it is likely that 2, 3-dihydroxybenzoate functions only as a precursor for enterochelin synthesis. Using mutant strains that lack the ability to convert 2, 3-dihydroxybenzoate to enterochelin (i.e., containing various genetic lesions after synthesis of 2, 3-dihydroxybenzoate), these investigators showed that enterochelin synthesis was necessary

in order for the organism to show a growth response to 2,3-dihydroxybenzo-
ate.  This represents a complex situation in which enterochelin synthesis
was apparently necessary for these organisms to show a growth response
to 2,3-dihydroxybenzoate in spite of the fact that the organism was unable
to transport ferric enterochelin.  Frost and Rosenberg [64] explained
these findings by suggesting that enterochelin excretion followed by transport of external ferric enterochelin through the outer membrane receptor
may not be necessary in order for the organisms to utilize enterochelin as
an iron transporter.  They suggest that certain amounts of iron may exist
between the cytoplasmic and the outer membranes, perhaps bound to various components in the cell envelope.  Movement of enterochelin across the
cytoplasmic membrane (but not the outer membrane) allows formation of
ferric enterochelin within the cell envelope which can then be transported
back into the cytoplasm.  This hypothesis does not require that enterochelin be released into the environment for that substance to be a functional
siderochrome; if it can be proven, this model will have interesting implications for transport of iron via the siderochromes.  This hypothesis may
also be pertinent to the situation found in the mycobacteria in which their
siderochrome, mycobactin, is almost entirely cell associated (apparently
located in the lipid-rich outer portions of the cell envelope).

A third Gram-negative organism has been used in studies of iron
transport.  Aerobacter aerogenes strain 62-1 produces both enterochelin
and the hydroxamate-type siderochrome aerobactin, as already noted.
Although no iron transport studies with enterochelin in this organism have
been reported, uptake of radioiron bound to aerobactin has been examined,
with the finding that this hydroxamate stimulated iron uptake [45,76].  A
doubly labeled chelate, [3H]aerobactin[ $^{59}$Fe], was transported as an intact
unit, although the labeled iron-free aerobactin was rapidly released from
the cells [45].  It seems likely that A. aerogenes strain 62-1 contains
hydroxamate transport systems similar to those described for B. megaterium.  This organism also appears to contain a citrate-dependent system
of iron transport [31].

D.  Mycobacterium Species

The mycobacteria produce a series of complex ferric chelating secondary
hydroxamates (see Fig. 4-4).  All these substances have the same basic
structure, with variations in the molecules dependent upon the producing
species [21].  The mycobactins have a strong selectivity for ferric iron;
however, in contrast to most other hydroxamates, the mycobactins are
highly soluble in lipid solvents and are almost insoluble in water.  The
hydrophobic nature of the mycobactins may be related to the high lipid
content of the mycobacterial cell.  In culture, most of the mycobactin is
associated with the lipid-rich outer regions of the mycobacterial cells,
and only small amounts of mycobactin may be released into the extracellular

fluid. Lack of mycobactin excretion has caused some controversy among researchers interested in iron transport in this organism (to be discussed shortly).

Mycobactin can serve as a growth factor for various species of Arthrobacter which require hydroxamates, indicating that mycobactin can function as do other hydroxamates [77]. Growth experiments showing that addition of mycobactin will reverse serum inhibition of BCG also suggest a critical role for mycobactin in iron transport [8,78].

In Mycobacterium smegmatis, mycobactin can be extracted from the cells by treatment with ethanol. Therefore, radioiron trapped in the myco-bactin fraction can be separated from total uptake of iron by this extraction procedure. In studies with $^{55}$Fe it was shown that little radioactive iron appeared in the mycobactin fraction during transport of iron; however, in cells starved for an energy source prior to assay, a high percentage of the radioiron was recovered in the mycobactin fraction [79,80]. Mycobactin is a functional component of the iron transport system in this organism. Loading of mycobactin with iron may be a temperature-independent process, whereas release of iron from mycobactin is a temperature-dependent process [79,80].

Actual removal of iron from mycobactin may be an enzymatic event and may require reduced nicotinamide adenine dinucleotide phosphate (NADP) as a cofactor [12,79]. Such a step would involve reduction of ferric iron bound by the mycobactin with subsequent release of iron from the ligand, and it has been proposed that this mechanism may be operational in other hydroxamate iron transport systems [12].

Although it is clear that mycobactin acts as a component of the transport system(s), possibly as an early cellular receptor of iron, there is controversy regarding the role of mycobactin in external solubilization of iron. Ratledge and his associates have suggested that the mycobacteria may require at least two iron-binding molecules, one of these being myco-bactin, which because of its lipid solubility transports iron through the boundary of the bacterial cell. These workers propose the existence of a second iron-binding molecule functioning in the aqueous external environment to solubilize iron for delivery to cellular mycobactin. Previously, they had suggested that excreted salicylic acid might serve as the extracellular chelator [80], but they have now discounted this role for salicylate because it fails to hold iron in solution at physiological pH ranges [81]. Recently, these investigators have detected the existence of certain iron-chelating substances (termed exochelins) in the extracellular fluids of both the BCG strain and M. smegmatis [82,83]. The exochelins produced by the two organisms are apparently of different chemical structure. Both are peptides containing 3 mol of N-hydroxylysine and 1 mol of threonine. The compound produced by M. smegmatis contains β-alanine, whereas exochelin produced by the BCG strain contains salicylic acid. Ratledge et al. concluded that the exochelins probably were neither breakdown

products nor precursors of mycobactin and hence might function as extracellular iron-solubilizing agents.

These conclusions are at variance with the hypothesis of Kochan and
his associates, who suggest that mycobactin is capable of functioning in
extracellular solubilization of iron. They have shown that mycobactin can
remove iron from transferrin and from ferritin, delivering this iron to the
bacterial cell [8, 78 (and papers cited therein)]. These authors have suggested that sufficient amounts of mycobactin may be released from the
cells through the action of serum surface-active agents, and they question
the need of a special extracellular iron chelating agent. Their work is
described in detail elsewhere in this volume.

A major problem in understanding iron transport in the mycobacteria
revolves around the possible necessity of special chelating agents functioning to deliver iron to mycobactin. In any case, the mycobacteria apparently have had to adapt their siderochrome transport system to be functional in their lipid-rich outer layers. Of possible pertinence to this problem is the recent suggestion by Frost and Rosenberg [64] that the siderochrome enterochelin does not have to be excreted in order to be a functional
iron transport cofactor in certain Gram-negative bacteria (as discussed
earlier).

E.   Summary of Transport Systems

Because of the critical need for iron in metabolism (particularly aerobic
metabolism) and because of the tendency of iron to form complexes which
render it unavailable for transport in the ionic form, microorganisms have
evolved special high-affinity systems for acquisition of the metal from the
environment. The special molecules involved in this process are termed
siderochromes and belong to two chemical families, the secondary hydroxamates and the phenolates. Although use of the system may vary among
different microbial types, the general process involves release of the
chelating agent from the cells and subsequent transport of the resulting
external ferric chelate into the inner compartment of the cell. Special
molecules forming receptors and transport components occur at or within
the cell envelope. Once transported the chelate probably remains within
the cell as a pool of available iron; iron can be withdrawn from the chelate
by either its reduction, which causes its release, or by enzymatic hydrolysis of the chelating agent. In Gram-negative organisms a number of genes
have been implicated both in the synthesis of siderochromes and their subsequent transport. Specific receptors also have been identified, both on
the outer membrane of Gram-negative bacteria and on the cytoplasmic
membrane of B. megaterium. These events were shown earlier in two
working models (Figs. 5-7 and 5-8). Some organisms maintain the capacity to transport not only their own siderochrome but siderochromes
produced by other organisms. Such duplication probably represents a

survival mechanism. Microbial siderochrome-mediated transport of iron is a fascinating process about which much remains to be discovered.


## IV.  METABOLIC FUNCTIONS OF IRON

Iron is associated with many varied biological activities of microorganisms. These activities and the roles of iron in them are vital to normal cellular metabolism. The functions of iron in microbial metabolism have been the subject of, or discussed in, several reviews [2, 6, 84-88]. It is not the purpose here extensively to review the literature on the roles of iron in microbial metabolism, but rather to present selected examples in several areas which reflect the diversity of functions affected by iron in microbial cells.

The requirements of iron for growth of microorganisms and some of the parameters which may influence these requirements have been reviewed [6]; a wide range of iron concentrations is required for maximal growth of various microbes. The iron concentration required for maximal growth of an organism may not be that which is necessary or optimal for control of syntheses of microbial products or of other physiological processes [89]. Although it is possible that iron may possess regulatory functions, it is likely that it exerts its greatest influence on microbial cellular metabolism in a catalytic capacity.

Iron deficiency generally results in partial or complete inhibition of growth, or at least in a reduction in growth rate. Iron has been found to influence the cell composition and cellular morphology of many microorganisms. Other microbial processes in which a role for iron has been documented include intermediary metabolism, synthesis of metabolic products, enzymatic activities and microbial virulence and/or pathogenicity. Examples of the effects of iron on microbial metabolism are presented in Table 5-1 at the end of this section.

The intermediary metabolism of microorganisms is an area where the role of iron can be greatly pronounced. Aerobic metabolism of common carbon sources by a variety of microorganisms is influenced by the iron status of the growth medium. Glycolysis and the Krebs cycle-electron transport system are widely distributed in aerobic microorganisms; however, only the oxidative metabolism of the Krebs cycle-electron transport chain is accomplished by enzymes requiring iron [86, 87]. Under conditions of iron limitation, both aerobic and facultative cells can partially satisfy their energy requirements by glycolysis. This shift to an alternate metabolic pathway as a consequence of iron deficiency results in changes in the metabolic end products of glucose and/or other carbon sources supplied for growth. Changes in anaerobic glycolysis during conditions of iron deficiency also are responsible for differences in end products of carbon-containing energy sources supplied for growth.

Biological nitrogen fixation has been reported to require iron salts [90]. Nitrogenase, the enzyme system responsible for the reduction of dinitrogen to ammonia in both aerobic and anaerobic microorganisms, consists of two proteins, both of which contain iron and both of which are essential for activity [91].

Nucleic acid metabolism is another area in which iron has been implicated in microbial cells. Iron-deficient cells of Mycobacterium smegmatis have decreased levels of DNA and RNA, and cells become elongated and filamentous [92]. In a Bacillus subtilis mutant unable to use Desferal-iron, this hydroxamate caused iron starvation and blocked DNA synthesis without immediate interruption of overall RNA and protein synthesis [93]. Structural changes in certain tRNA species during iron deficiency have been documented [94-96]. Although no regulatory function at the translational level has been attributed to these modified tRNAs, certain tRNA species are known to regulate enzyme synthesis [97], and altered tRNA may influence this control system.

A listing of some microbial metabolites whose accumulation has been shown to be increased during conditions of iron limitation can be found in a review by Weinberg [85]. A wide variety of metabolic products has been reported to be dependent on the presence or absence of iron in the growth medium. Porphyrins, toxins, pigments, vitamins, antibiotics, aromatics, and hydroxamates are but some of the metabolic products whose accumulation has been reported to be influenced by iron (see Table 5-1 for examples). The role of iron in control and production of secondary metabolites is discussed elsewhere in this volume. Changes in accumulation of metabolic products as a function of iron status likely reflect alterations in the biosynthetic machinery responsible for these compounds.

The number of enzymes which contain iron or which depend on the presence of iron for activity is extensive. In addition, iron is needed for the synthesis of many enzymes during growth of microorganisms. Table 5-1 lists some enzymes whose synthesis or activity requires iron or which contain iron as an integral part of the protein.

It is apparent when one is reviewing the literature on iron-enzymes that the major roles of physiologically active iron are in oxygen and electron/hydrogen movement. These roles are accomplished by oxygenases, hydroperoxidases, and a large group of proteins referred to collectively as electron transfer proteins. Oxygenases may be of the heme or nonheme type and catalyze the incorporation of molecular oxygen into their substrates; they are the subject of a recent review [98]. The hydroperoxidases (catalase, peroxidase) are heme proteins and function to break down peroxides by catalyzing the transfer of electrons from molecules of substrate to peroxide. Hydroperoxidases have also been reviewed recently

[99]. The number of proteins whose function is the transfer of electrons is large; they may be either heme or nonheme iron proteins. Cytochromes are heme proteins which function in electron and/or hydrogen transport; they occur in all aerobic and in many anaerobic microorganisms and participate primarily in energy-conversion processes. These proteins have been discussed in detail [100]. Nonheme iron electron transfer proteins are redox molecules which are involved in many intracellular pathways including photosynthesis, nitrogen fixation, electron transport, and oxidative phosphorylation; they include nitrogenase, hydrogenase, iron-sulfur proteins, and flavoproteins. Microbial nonheme iron electron transfer proteins are included as part of a review by Malkin and Rabinowitz [101]; recent reviews on nitrogenase [102] and hydrogenase [103] also are available. Iron-sulfur proteins include ferredoxin and rubredoxin, proteins whose primary function is in the transfer of electrons between metabolic enzymes; these proteins have been the subject of several recent reviews [104-106]. The flavoproteins are generally considered to be dehydrogenases; they are responsible for the dehydrogenation of their substrates and utilize a variety of molecules as hydrogen acceptors.

Iron and iron compounds have been implicated in the relationship between pathogenicity/virulence of microorganisms and host resistance to infection. Growth of pathogenic microorganisms in serum has been shown to be restricted by unsaturated transferrin. Evidence for the role of this iron-binding compound as a nonspecific resistance factor to infection by its disruption of normal iron metabolism of pathogens in host-parasite relationships has been reviewed [107]. Excretion of specific iron-binding compounds by microorganisms may be necessary for obtaining iron from transferrin and may be directly related to their degree of virulence. This subject is the topic of recent reviews [108,109] and is also discussed in Chapter 6 of the present volume.


RECENT DEVELOPMENTS

Recent studies [109a] of $^{55}$Fe- and $^3$H-labeled ferrichrome and ferrioxamine B (Desferal) uptake in strains of Salmonella typhimurium and E. coli (which do not produce these hydroxamates) confirm transport of the intact ferric chelates, possibly by a mechanism or mechanisms similar to the model presented here (see Fig. 5-7), prior to removal of iron for metabolic use. However, the evidence suggests that these enteric bacteria also may have an alternate process which involves separation of iron from the ligand at the cell surface, without penetration of the ferric hydroxamate.

TABLE 5-1   Areas of Microbial Metabolism Affected by Iron

| Area | Observed effect | References |
|------|-----------------|------------|
| Cell composition, growth, division and differentiation | Decrease in DNA, RNA, and total lipids and elongation of cells of Mycobacterium smegmatis as a result of iron deficiency | 92 |
| | Decrease in ratio of free to bound lipids in cells of Corynebacterium diphtheriae deprived or iron | 84 |
| | Decrease in RNA/nitrogen and DNA/ nitrogen ratios and prolonged slow growth of Mycobacterium smegmatis in iron-limited cultures | 110 |
| | Growth inhibition of Mycobacterium smegmatis as a result of iron deficiency | 92, 111 |
| | Inhibition of growth rate of Agrobacterium tumefaciens by iron limitation | 112 |
| | Elongation of cells and delay in cell division of Corynebacterium diphtheriae during iron starvation | 84 |
| | Increase in inoculum-dependent lag of Bacillus megaterium as iron concentration in medium was decreased | 6 |
| | Bacterial flagellation in Bacillus megaterium and B. subtilis and in B. cereus, Proteus vulgaris and P. mirabilis suppressed by iron | 89, 113 |
| | Sporulation of Bacillus megaterium required iron | 114 |
| | Latent period of phage infection in Corynebacterium diphtheriae extended during iron limitation | 84 |

TABLE 5-1   (Cont.)

| Area | Observed effect | References |
|------|-----------------|-----------|
| Intermediary metabolism | Shift from predominantly acetic-butyric acid fermentation towards lactic acid fermentation in Clostridium sp. as iron content of medium decreased | 115,116 |
| | Considerable lags in oxidation of Krebs cycle intermediates by Shigella sp. in iron deficient medium | 117 |
| | Altered aerobic metabolism (impaired respiratory activity) of common carbon sources by Staphylococcus aureus in iron-restricted medium | 118 |
| | Decrease in respiratory control ratios and efficiency of conversion of succinate into bacterial mass in Escherichia coli in iron-deficient medium | 119 |
| | Nitrogen fixation by Clostridium pasteurianum required iron salts and biotin | 90 |
| | Modification of certain tRNA species of Escherichia coli and of Bacillus subtilis during iron-deficient growth | 94-96 |
| | Increase in the nucleotide triphosphate-dependent DNA breakdown system of Mycobacterium smegmatis under conditions of iron deficiency | 120 |
| Metabolic products | Increased riboflavin synthesis in Clostridium acetobutylicum during iron deficiency | 121 |
| | Suppression of porphyrin formation and increase in formation of bacteriochlorophyll and heme components in Rhodopseudomonas spheroides by iron salts | 122 |

TABLE 5-1   (Cont.)

| Area | Observed effect | References |
|---|---|---|
| | Accumulation of phenazine pigments by <u>Pseudomonas</u> sp. inhibited by iron | 123 |
| | Inhibition of production of neuro-toxin of <u>Shigella shigae</u> by high iron concentrations | 124 |
| | Maximal neomycin production by <u>Streptomyces fradiae</u> required iron | 125 |
| | Decrease in toxin and coproporphyrin production by <u>Corynebacterium diphtheriae</u> with increasing iron concentrations in the medium | 126 |
| | Accumulation of salicylic acid by <u>Mycobacterium smegmatis</u> and <u>Aerobacter aerogenes</u> was increased during iron deficiency | 127 |
| | Increased production of siderochromes by several microorganisms under conditions of iron deficiency | 87, 128 |
| Enzymes | Lowered levels of $\delta$-aminolevulinate dehydratase during growth of <u>Neurospora crassa</u> under iron deficiency | 129 |
| | Production of glucoside 3-dehydrogen-ase by <u>Agrobacterium tumefaciens</u> inhibited in iron-depleted medium | 112 |
| | Reduction in levels of cytochromes of <u>Aerobacter aerogenes</u> grown in iron-deficient medium | 130 |
| | Synthesis of catalase and peroxidase of <u>Aerobacter indologenes</u> suppressed by iron deficiency | 131 |

TABLE 5-1 (Cont.)

| Area | Observed effect | References |
|------|-----------------|------------|
| | Synthesis and/or assembly of the terminal portion of the steroyl-coenzyme A desaturase system of Mycobacterium phlei required iron | 132 |
| | Activity of aconitase of Bacillus subtilis required ferrous iron | 133 |
| | Maximal activity of phosphotrans-acetylase from Clostridium acidiurici required ferrous iron | 134 |
| | Activity of histidine decarboxylase of Lactobacillus sp. required iron | 135 |
| | Activity of aldolase of Clostridium perfringens required iron | 136 |
| | Dihydroorotic dehydrogenase of Zymobacterium oroticum contained iron | 137 |
| | Ribonucleotide diphosphate reductase subunit B2 of Escherichia coli B contained iron which was essential for enzyme function | 138 |
| | Nitrate reductase of Micrococcus denitrificans contained iron and molybdenum as functional components | 139 |
| | Protocatechuate 4,5-dioxygenase of Pseudomonas sp. contained non-heme iron | 140 |
| Virulence | Reduction in $LD_{50}$ dose of Listeria monocytogenes by iron compounds | 141 |
| | Enhanced virulence (lethality for mice) of nonpigmented strains of Pasteurella pestis by iron | 142 |

TABLE 5-1 (Cont.)

| Area | Observed effect | References |
|------|-----------------|------------|
| | Enhanced virulence (for mice) of *Yersinia pestis* by siderochrome(s) formed under low-iron conditions | 143 |
| | Iron-binding catechols essential for virulence of *Escherichia coli* | 144 |
| | Increased virulence (lethality for chicken embryos) of avirulent T3 and T4 colony types of *Neisseria gonorrhoeae* by addition of iron compounds | 145 |

## ACKNOWLEDGMENTS

The authors' research in *Bacillus* species reported here was supported by U.S. Public Health Service grants CA 11886 from the National Cancer Institute and Research Career Development Award GM 29366 from the National Institute of General Medical Sciences (to B.R.B.). Departmental Chairman Charles C. Randall is thanked for his support.

## REFERENCES

1. J. E. Zajic, *Microbial Biogeochemistry*, Academic Press, New York, 1969.
2. J. B. Neilands, in *Microbial Iron Metabolism* (J. B. Neilands, ed.), Academic Press, New York, 1974, pp. 4-34.
3. D. L. Leussing and I. M. Kolthoff, *J. Amer. Chem. Soc.*, **75**, 2476 (1953).
4. I. M. Kolthoff and P. J. Elving, *Treatise on Analytical Chemistry*, 2nd ed., Thomas, Springfield, Illinois, 1962.
5. G. Biederman and P. Schindler, *Acta Chem. Scand.*, **11**, 731 (1957).
6. C. E. Lankford, *Crit. Rev. Microbiol.*, **3**, 273 (1973).
7. B. R. Byers, in *Microbial Iron Metabolism* (J. B. Neilands, ed.), Academic Press, New York, 1974, pp. 83-105.
8. C. A. Golden, I. Kochan, and D. R. Spriggs, *Infect. Immun.*, **9**, 34 (1974).
9. J. B. Neilands, in *Inorganic Biochemistry* (G. Eichhorn, ed.), Elsevier, Amsterdam, 1973, pp. 167-202.

10. H. Zähner, E. Bachmann, R. Hütter, and J. Nüesch, Pathol. Microbiol., 25, 708 (1962).
11. T. F. Emery, Advan. Enzymol. Relat. Areas Mol. Biol., 35, 135 (1971).
12. K. A. Brown and C. Ratledge, FEBS Letters, 53, 262 (1975).
13. J. B. Neilands, J. Amer. Chem. Soc., 74, 4846 (1952).
14. O. Mikes and J. Turková, Collection Czech. Chem. Comm., 27, 581 (1962).
15. F. Knüsel and W. Zimmerman, in Antibiotics (J. W. Corcoran and F. E. Hahn, eds.), Vol. 3, Springer-Verlag, New York, 1975.
16. H. Bickel, G. E. Hall, W. Keller-Schierlein, V. Prelog, E. Vischer, and A. Wettstein, Helv. Chim. Acta, 43, 2129 (1960), pp. 653-667.
17. K. B. Mullis, J. R. Pollack, and J. B. Neilands, Biochemistry, 10, 1071 (1971).
18. F. Gibson and D. I. Magrath, Biochim. Biophys. Acta, 192, 175 (1969).
19. W. D. Linke, A. Crueger, and H. Diekmann, Arch. Mikrobiol., 85, 44 (1972).
20. B. R. Byers, M. V. Powell, and C. E. Lankford, J. Bacteriol., 93, 286 (1967).
21. G. A. Snow, Bacteriol. Rev., 34, 99 (1970).
22. P. V. Patel and C. Ratledge, Biochem. Soc. Trans., 1, 886 (1973).
23. C. L. Atkin and J. B. Neilands, Biochemistry, 7, 3734 (1968).
24. C. L. Atkin, J. B. Neilands, and H. J. Phaff, J. Bacteriol., 103, 722 (1970).
25. I. G. O'Brien and F. Gibson, Biochim. Biophys. Acta, 215, 393 (1970).
26. J. R. Pollack and J. B. Neilands, Biochem. Biophys. Res. Commun., 38, 989 (1970).
27. H. Korth, C. Spiekermann, and G. Pulverer, Med. Microbiol. Immunol., 70, 297 (1970).
28. T. Ito and J. B. Neilands, J. Amer. Chem. Soc., 80, 4645 (1958).
29. J. L. Corbin and W. A. Bulen, Biochemistry, 8, 757 (1969).
30. T. F. Emery, in Microbial Iron Metabolism (J. B. Neilands, ed.), Academic Press, New York, 1974, pp. 107-123.
31. I. G. Young, G. B. Cox, and F. Gibson, Biochim. Biophys. Acta, 141, 319 (1967).
32. I. G. Young and F. Gibson, Biochim. Biophys. Acta, 177, 401 (1969).
33. D. N. Downer, W. B. Davis, and B. R. Byers, J. Bacteriol., 101, 181 (1970).
34. I. G. Young, L. Langman, R. K. J. Luke, and F. Gibson, J. Bacteriol., 106, 51 (1971).
35. R. K. J. Luke and F. Gibson, J. Bacteriol., 107, 557 (1971).
36. G. F. Bryce and N. Brot, Biochemistry, 11, 1708 (1972).
37. G. C. Woodrow, I. G. Young, and F. Gibson, J. Bacteriol., 124, 1 (1975).

38. B. L. Walsh, W. J. Peters and R. A. J. Warren, Can. J. Microbiol., 17, 53 (1971).
39. J. B. Neilands, Bacteriol. Rev., 21, 101 (1957).
40. J. L. Arceneaux and C. E. Lankford, Biochem. Biophys. Res. Commun., 24, 370 (1966).
41. W. B. Davis and B. R. Byers, J. Bacteriol., 107, 491 (1971).
42. J. E. LeBlanc-Arceneaux and C. E. Lankford, Bacteriol. Proc., p. 131 (1971).
43. K. Todar and C. E. Lankford, Abs. Ann. Mtg. Amer. Soc. Microbiol., p. 171 (1972).
44. W. B. Davis, M. J. McCauley, and B. R. Byers, J. Bacteriol., 105, 589 (1971).
45. J. E. L. Arceneaux, W. B. Davis, D. N. Downer, A. H. Haydon, and B. R. Byers, J. Bacteriol., 115, 919 (1973).
46. A. H. Haydon, W. B. Davis, D. N. Downer, J. E. L. Arceneaux, and B. R. Byers, J. Bacteriol., 115, 912 (1973).
47. A. H. Haydon, Ph.D. dissertation, University of Mississippi Medical Center, Jackson, Mississippi, 1974.
48. J. E. Aswell, J. E. L. Arceneaux, C. A. Dawkins, H. R. Turner, and B. R. Byers, Abs. Ann. Mtg. Amer. Soc. Microbiol., p. 154 (1976).
49. A. H. Haydon, H. R. Turner, J. E. L. Arceneaux, and B. R. Byers, Abs. Ann. Mtg. Amer. Soc. Microbiol., p. 177 (1975).
50. H. R. Turner, Ph.D. dissertation, University of Mississippi Medical Center, Jackson, Mississippi, 1975.
51. C. A. Dawkins, M.S. Thesis, University of Mississippi Medical Center, Jackson, Mississippi, 1975.
52. J. E. L. Arceneaux and B. R. Byers, J. Bacteriol., 127, 1324 (1976).
53. W. J. Peters and R. A. J. Warren, Biochim. Biophys. Acta, 165, 225 (1968).
54. W. J. Peters and R. A. J. Warren, Can. J. Microbiol., 16, 1179 (1970).
55. T. F. Emery, Biochim. Biophys. Acta, 363, 219 (1974).
56. G. Winkelman, Arch. Mikrobiol., 98, 39 (1974).
57. C. Wiebe and G. Winkelmann, J. Bacteriol., 123, 837 (1975).
58. H. Rosenberg and I. G. Young, in Microbial Iron Metabolism (J. B. Neilands, ed.), Academic Press, New York, 1974, pp. 67-82.
59. J. R. Pollack, B. N. Ames, and J. B. Neilands, J. Bacteriol., 104, 635 (1970).
60. G. Frost and H. Rosenberg, Biochim. Biophys. Acta, 330, 90 (1973).
61. I. G. O'Brien, G. B. Cox, and F. Gibson, Biochim. Biophys. Acta, 337, 537 (1971).
62. L. Langman, I. G. Young, G. Frost, H. Rosenberg, and F. Gibson, J. Bacteriol., 112, 1142 (1972).

63. R. J. Porra, L. Langman, I. G. Young, and F. Gibson, Arch. Biochem. Biophys., 153, 74 (1972).
64. G. Frost and H. Rosenberg, J. Bacteriol., 124, 704 (1975).
65. C. C. Wang and A. Newton, J. Biol. Chem., 246, 2147 (1971).
66. N. C. Franklin, W. F. Dove, and C. Yanofsky, Biochem. Biophys.
Res. Commun., 18, 910 (1965).
67. P. Fredericq, Antonie van Leeuwenhoek J. Microbiol. Serol., 17,
103 (1951).
68. A. L. Taylor and C. D. Trotter, Bacteriol. Rev., 36, 504 (1972).
69. S. K. Guterman and L. Dann, J. Bacteriol., 114, 1225 (1973).
70. K. Hantke and V. Braun, FEBS Letters, 49, 301 (1975).
71. R. Wayne and J. B. Neilands, J. Bacteriol., 121, 497 (1975).
72. M. Luckey, R. Wayne, and J. B. Neilands, Biochem. Biophys. Res.
Commun., 64, 687 (1975).
73. T. D. Wilkins and C. E. Lankford, J. Infect. Dis., 121, 129 (1970).
74. M. Luckey, J. R. Pollack, R. Wayne, B. N. Ames, and J. B.
Neilands, J. Bacteriol., 111, 731 (1972).
75. C. C. Wang and A. Newton, J. Bacteriol., 98, 1142 (1969).
76. D. N. Downer, Ph.D. dissertation, University of Mississippi Medical
Center, Jackson, Mississippi, 1972.
77. N. E. Morrison and E. E. Dewbrey, J. Bacteriol., 92, 1848 (1966).
78. I. Kochan, N. Pellis, and C. A. Golden, Infect. Immun., 3, 553
(1971).
79. C. Ratledge, Biochem. Biophys. Res. Commun., 45, 856 (1971).
80. C. Ratledge and B. J. Marshall, Biochim. Biophys. Acta, 279, 58
(1972).
81. C. Ratledge, L. P. Macham, K. A. Brown, and B. J. Marshall,
Biochim. Biophys. Acta, 372, 39 (1974).
82. L. P. Macham and C. Ratledge, J. Gen. Microbiol., 89, 379 (1975).
83. L. P. Macham and C. Ratledge, Infect. Immun., 12, 1242 (1975).
84. L. Barksdale, Bacteriol. Rev., 34, 378 (1970).
85. E. D. Weinberg, Advan. Microb. Physiol., 4, 1 (1970).
86. M. P. Coughlan, Sci. Progr. (Oxford), 59, 1 (1971).
87. J. B. Neilands, Advan. Exptl. Med. Biol., 40, 13 (1973).
88. P. A. Light and R. A. Clegg, in Microbial Iron Metabolism (J. B.
Neilands, ed.), Academic Press, New York, 1974, pp. 35-64.
89. E. D. Weinberg and J. I. Brooks, Nature, 199, 1963.
90. J. E. Carnahan and J. E. Castle, J. Bacteriol., 75, 121 (1958).
91. R. R. Eady and J. R. Postgate, Nature, 249, 805 (1974).
92. F. G. Winder and C. O'Hara, Biochem. J., 82, 98 (1962).
93. J. E. L. Arceneaux and B. R. Byers, J. Bacteriol., in press.
94. F. O. Wettstein and G. S. Stent, J. Mol. Biol., 38, 25 (1968).
95. A. H. Rosenberg and M. L. Gefter, J. Mol. Biol., 46, 581 (1969).
96. M. J. McCauley, J. L. Arceneaux, and B. R. Byers, Abs. Ann.
Mtg. Amer. Soc. Microbiol., p. 173 (1972).

97.  J. E. Burchly and L. S. Williams, Ann. Rev. Microbiol., 29, 251 (1975).

98.  M. Nozaki and Y. Ishimura, in Microbial Iron Metabolism (J. B. Neilands, ed.), Academic Press, New York, 1974, pp. 417-444.

99.  T. Yonetani, in Microbial Iron Metabolism (J. B. Neilands, ed.), Academic Press, New York, 1974, pp. 401-415.

100. T. Yamanaka and K. Okunuki, in Microbial Iron Metabolism (J. B. Neilands, ed.), Academic Press, New York, pp. 349-400.

101. R. Malkin and J. C. Rabinowitz, Ann. Rev. Biochem., 36(1), 113 (1967).

102. R. H. Burris and W. H. Orme-Johnson, in Microbial Iron Metabolism (J. B. Neilands, ed.), Academic Press, New York, 1974, pp. 187-209.

103. L. E. Mortenson and J. S. Chen, in Microbial Iron Metabolism (J. B. Neilands, ed.), Academic Press, New York, 1974, pp. 231-282.

104. R. C. Valentine, Bacteriol. Rev., 28, 497 (1964).

105. B. B. Buchanan and D. I. Arnon, in Advances in Enzymology (F. F. Nord, ed.), Wiley (Interscience), New York, 1970, pp. 119-176.

106. W. Lovenberg, in Microbial Iron Metabolism (J. B. Neilands, ed.), Academic Press, New York, 1974, pp. 161-185.

107. E. D. Weinberg, J. Infect. Dis., 124, 401 (1971).

108. J. J. Bullen, H. J. Rogers, and E. Griffiths, in Microbial Iron Metabolism (J. B. Neilands, ed.), Academic Press, New York, 1974, pp. 517-551.

109. M. Sussman, in Iron in Biochemistry and Medicine (A. Jacobs and M. Worwood, eds.), Academic Press, New York, 1974, pp. 649-679.

109a. J. Leong and J. B. Neilands, J. Bacteriol., 126, 823 (1976).

110. F. G. Winder and M. P. Coughlan, Irish J. Med. Sci., 140, 16 (1971).

111. A. B. Harris, J. Gen. Microbiol., 47, 111 (1967).

112. W. M. Kurowski and S. J. Pirt, J. Gen. Microbiol., 68, 65 (1971).

113. W. T. Sokolski and E. M. Stapert, J. Bacteriol., 85, 718 (1963).

114. B. J. Kolodziej and R. A. Slepecky, J. Bacteriol., 88, 821 (1964).

115. A. M. Pappenheimer, Jr., and E. Shaskan, J. Biol. Chem., 155, 265 (1944).

116. A. M. Hanson and N. E. Rogers, J. Bacteriol., 51, 568 (1946).

117. W. E. van Heyningen, Brit. J. Exp. Pathol., 36, 373 (1955).

118. T. S. Theodore and A. L. Schade, J. Gen. Microbiol., 40, 385 (1965).

119. D. J. Rainnie and P. D. Bragg, J. Gen. Microbiol., 77, 339 (1973).

120. F. G. Winder and M. P. Coughlan, Biochem. J., 111, 679 (1969).

121. R. J. Hickey, Arch. Biochem., 8, 439 (1945).

122. J. Lascelles, Biochem. J., 62, 78 (1956).

123. H. Korth, Arch. Mikrobiol., 77, 59 (1971).

124. W. E. van Heyningen, Brit. J. Exp. Pathol., 36, 381 (1955).
125. M. K. Majumdar and S. K. Majumdar, Appl. Microbiol., 13, 190 (1965).
126. G. D. Clarke, J. Gen. Microbiol., 18, 698 (1958).
127. C. Ratledge, Nature, 203, 428 (1964).
128. J. B. Neilands, Struct. Bond., 1, 59 (1966).
129. S. M. Krishnan, G. Padmanaban, and P. S. Sarma, Biochem. Biophys. Res. Comm., 31, 333 (1968).
130. A. Tissieres, Biochem. J., 50, 279 (1951).
131. W. S. Waring and C. H. Werkman, Arch. Biochem., 4, 75 (1947).
132. Y. Kashiwabara, H. Nakagawa, G. Matsuki, and R. Sato, J. Biochem., 78, 803 (1975).
133. P. Fortnagel and E. Freese, J. Biol. Chem., 248, 5289 (1968).
134. J. R. Robinson and R. D. Sagers, J. Bacteriol., 112, 465 (1972).
135. B. M. Guinand and E. E. Snell, J. Amer. Chem. Soc., 76, 4745 (1954).
136. R. C. Bard and I. C. Gunsalus, J. Bacteriol., 59, 387 (1950).
137. R. W. Miller and V. Massey, J. Biol. Chem., 240, 1453 (1965).
138. N. C. Brown, R. Eliasson, P. Reichard, and L. Thelander, Eur. J. Biochem., 9, 512 (1969).
139. P. Forget and D. V. Dervartanian, Biochim. Biophys. Acta, 256, 600 (1972).
140. K. Ono, M. Nozaki, and O. Hayaishi, Biochim. Biophys. Acta, 220, 224 (1970).
141. C. P. Sword, J. Bacteriol., 92, 536 (1966).
142. S. Jackson and T. W. Burrows, Brit. J. Exp. Pathol., 37, 577 (1956).
143. A. Wake, M. Misawa, and A. Matsui, Infect. Immun., 12, 1211 (1975).
144. H. J. Rogers, Infect. Immun., 7, 445 (1973).
145. S. M. Payne and R. A. Finkelstein, Infect. Immun., 12, 1313 (1975).

Chapter 6

ROLE OF SIDEROPHORES IN NUTRITIONAL
IMMUNITY AND BACTERIAL PARASITISM

Ivan Kochan

Department of Microbiology
Miami University
Oxford, Ohio
and
Wright State University School of Medicine
Dayton, Ohio

I.  INTRODUCTION

The resistance of animals to parasitic microbes is attributed to several
humoral and cellular factors which cause a direct injury to microbial cells.
In addition to these aggressive factors of native immunity, an animal body
possesses a more subtle mechanism for its defense against microbial
parasitism.  Studies in several laboratories indicate that animals possess
an ability to starve parasites by limiting the supply of nutrilites which are
essential for their growth and multiplication.  An acute competition between parasites and their animal hosts for essential nutrilites characterizes
the host-parasite relationship.  In distinction to unrestricted availability
of nutrilites for microbial utilizations in dead tissues, the nutrilites in live
tissues are restricted and frequently unavailable for microbial growth.
    During evolution and coexistence, animals and their microbial para-
sites have developed elaborate mechanisms which they use to satisfy their
requirements for growth-essential nutrilites.  The effectiveness with
which the parasites compete with their hosts for the nutrilites determines
the speed of their growth and, consequently, the progression of an infectious disease.  The effectiveness with which animals can make nutrilites
unavailable for microbial cells contributes to the degree of their resistance
to parasitism.  Experimental findings of the nutritional starvation of microbes in animals led to formulation of the concept of nutritional immunity
[1,2], which expresses the native and acquired ability of an animal to restrict the availability of growth-essential nutrilites for use by parasitic
microbes.  Experimental results indicate that nutritional immunity constitutes one of the principal defense mechanisms of animals against microbial parasites.
    The possibility of nutritional defense of an animal host against micro-
bial parasites has been considered in the past.  Many years ago Louis
Pasteur postulated that exhaustion of some nutrilites by parasitic microbes
in the body of an animal might be the reason for recovery from infection
and for subsequent resistance.  The resistance of the recovered animal
would progressively diminish and disappear as its tissues reconstituted
the depleted nutrilites.  Pasteur believed that the immunity of normal animals to infectious microbes should be attributed not only to native heredi-
tary endowments but also to nutritional and environmental factors.
    The role of hereditary endowment in resistance to infectious microbes
was well documented by the work of Lurie [3].  He found that families of
inbred rabbits possessed various degrees of immunity to tuberculosis.
Very frequently the immunity to infection in nonvaccinated animals of resistant families was stronger than that in vaccinated animals of less
resistant families.  Lurie concluded, therefore, that inborn factors play
a fundamental role in resistance to tuberculosis.  Unfortunately, these
inborn factors were not identified.

The scientific literature records several examples of the dependence of microbial growth upon the presence of essential nutrilites in the body of infected animals. The resistance of animals and mammalian cells to various viral infections can be increased by limiting certain amino acids or vitamins in the diet or in tissue culture media, respectively. Thus, mice can be partially protected against poliomyelitis virus if they are fed a diet deficient in vitamin $B_1$, whereas a diet deficient in folic acid makes mice resistant to the virus of lymphocytic choriomeningitis. A deficiency of para-aminobenzoic acid increases resistance to malaria because the plasmodium needs this nutrilite for its multiplication. In some infections, the nutrilites needed for the growth of parasites have not been identified, but it has been observed that infections occur only in well-fed animals. It is well known that foot-and-mouth disease virus can infect guinea pigs which are well fed and in good health whereas undernourished animals are resistant to this virus. These and similar examples clearly demonstrate that the resistance of animals to infectious microbes can be increased by the withdrawal of essential nutrilites. But little effort has been made so far to determine the effects of nutritional depletion on the development and progression of more than a few microbial infections.

Metals participate in many biological reactions which are essential for the life of animals, plants, and microbes. Consequently, these inorganic nutrilites must exert an important influence on the host-parasite relationship. To survive in an animal body, a parasite must be able to obtain metals that are crucial for its metabolism and its growth. Since the essential metals in the animal body are usually bound to biological carriers or deposited within various cells, the effectiveness with which a microbe can establish itself in tissues of an animal depends upon its ability to mobilize inorganic nutrilites for its own purposes. Of all the essential metals so far investigated, iron appears to be most critical in determining whether a parasite is to be permitted to grow in the tissues of an animal [1,4,5]. This chapter discusses the competition for iron between iron-binding molecules of animal and bacterial origin; it describes how animals withhold iron from bacterial invaders and how bacteria overcome this native or acquired nutritional resistance of the host by the use of strong iron-chelating substances.


## II.  IRON STARVATION OF MICROBES IN TISSUES
    AND FLUIDS OF NORMAL ANIMALS

The fluids and tissues of animals possess nutrients and nutrilites which can support luxuriant growth of most microbial parasites. These microbial growth-supporting qualities of animal tissues are observed most clearly after the death of an animal; in dead and disintegrating tissues, parasitic and saprophytic microbes grow without any apparent hindrance.

On the basis of this observation, animal tissues were generally considered
to have the makings of microbial growth-supporting media, able to sustain
parasitic growth during life as well as after death.

In the past, the resistance of live tissues against microbial parasites
has been attributed exclusively to the cell-damaging activity of aggressive
antimicrobial factors of the host. Recent studies in several laboratories
have shown that antimicrobial activity in the living host may be largely due
to the unavailability of growth-essential nutrilites for utilization by microbial cells. It has been found that the exposure of various parasites to the
isolated tissues and fluids of an animal stops their multiplication, slows
down their metabolism, and induces rather rapidly the death by starvation.
Usually, supplying the unavailable nutrilites alleviates the microbiostasis
and induces microbial growth without any evidence for the activity of aggressive antimicrobial factors.

Of the many essential nutrilites needed for the growth of microbes in
an animal host, the requirement for iron seems to be defined most clearly.
In order to survive and to multiply in the living animal, parasitic microbes
have to be able to obtain iron which is not readily available because of its
association with organic carriers. The effectiveness with which the parasite can remove iron from these carriers determines the speed of its
growth and is an indicator of its virulence.

The competition for iron between biological iron carriers of the host
and parasite takes place in the fluids and tissues of the host. Since iron
carriers of the host become ineffective in dead tissues, the retention of
physiological conditions in isolated fluids or tissues is of crucial importance in studies of the microbial quest for iron. The effectiveness of
biological iron carriers in binding the metal and the stability of iron-
carrier complexes are influenced by hydrogen ion concentration (pH), temperature, the presence of other iron-binding materials or exogenous iron,
and the enzymatic disintegration of the carriers. Usually, the maintenance
of excised tissues or blood at $3^\circ$ C during the preparation of cell lysates or
the separation of serum, and the subsequent adjustment of the processed
materials to pH 7.5, preserve the iron-binding activity of iron carriers
and the stability of their complexes with the metal. It is much more difficult to avoid contamination of the processed materials with exogenous
iron. Traces of iron are present on glassware, in water, and in various
inorganic or organic compounds used for the preparation of microbial
media. The presence of exogenous iron interferes with the measurement
of microbiostatic activity of fluids and tissues and obliterates the competition for iron between iron-chelating substances of host and microbial
origin. There is no easy method to remove ionic iron from media; it has
been done with some success by repeated extractions of the medium with
8-hydroxyquinoline and chloroform [6].

The investigation of the competition for iron between iron-binding
substances of animal and microbial origin has been facilitated by the use

of serum and iron-poor Dubos medium (0.09 μg of iron/ml) mixtures. Although the minute amount of iron present in iron-poor Dubos medium is sufficient to support microbial growth, the medium becomes microbiostatic when mammalian serum is added to it in a dilution as high as 1:8. We have found that the iron-free transferrin (Tr) of serum binds iron of the medium, and in the absence of ionic iron this medium-serum mixture cannot support the growth of various microbes.

In most of our studies we have used an agar-plate diffusion test [7,8] to study the effects of restricted amounts of iron on the growth of tubercle bacilli in various materials derived from the animal. This test is being used effectively in a study of the competition for iron between animal and bacterial iron-transport products, i.e., siderophores. The test is performed in plastic petri dishes of 5.5-cm diameter filled with 12 ml of iron-poor Dubos agar medium. A well, 1 cm in diameter, is made in the agar by placing a glass cylinder in the center of each plate. The surface of the medium is inoculated with washed bacteria previously adjusted to a desired concentration. The wells are filled with test substances which diffuse into the agar, bind iron of the medium, and inhibit microbial growth around the wells. In experiments devised to measure the iron-supplying activity of siderophores, the plates are filled with iron-poor Dubos medium-Tr mixtures, and the wells are charged with various siderophores. The plates inoculated with slow-growing tubercle bacilli are incubated for about 3 weeks at 37° C and 70% humidity; mycobactin, the hydrophobic siderophore of tubercle bacilli, does not leave the lipoid bacterial wall and, therefore, does not obliterate tuberculostasis in zones containing Tr-iron complexes. The plates inoculated with fast-growing bacteria are incubated at 37° C; these plates are examined periodically during 48 hr of incubation. Usually after 35 hr of incubation these plates show little or no microbiostasis, because water-soluble siderophores (low-molecular-weight phenolates or hydroxamates) diffuse rapidly into zones of microbiostasis and alleviate it by supplying bacteria with iron, removed by them from Tr-iron complexes. After an appropriate period of incubation, the growth-inhibiting and growth-supporting activities of various iron-binding substances are determined by measuring the widths of growth-void and growth-filled zones around wells charged with iron-restricting and iron-donating substances, respectively. Because of its sensitivity and simplicity, the agar-plate diffusion test can be used effectively not only to measure the competition for iron between various biological and chemical compounds but also to screen human sera for the degree of iron-associated microbiostasis.

A. Microbiostasis in Mammalian Sera

Several vaguely defined substances in sera of normal animals are credited with antimicrobial activities [9]. In most cases, the in vivo determinations of their value to the host defense systems have been rather disappointing. It was mainly the failure to exert a direct microbicidal effect that discredited the value of normal mammalian sera in antimicrobial defense. In fact, the superior microbial growth in serum-growth medium mixtures suggested that such sera have excellent growth-supporting qualities. Because of these findings, the antimicrobial effect of serum in vivo is attributed largely to the facilitation of phagocytosis rather than to a direct microbicidal activity.

Studies in several laboratories have shown that, in spite of the excellent growth-supporting quality for most microbes of serum-medium mixtures, serum by itself is microbiostatic because of its failure to provide microbes with the iron essential for their growth. In 1946, Schade and Caroline described the antimicrobial iron-binding protein Tr in human plasma and suggested that this ligand withholds the metal from microbial invaders [10]. Every Tr molecule can chelate two atoms of ferric iron with the formation of a salmon-pink colored complex. The color of the complex remains stable on the alkaline side of pH 7, but on the acid side the color disappears as the complexes dissociate. We have found that at pH values lower than 6 Tr fails to bind iron at all.

Transferrin binds ferric iron by means of tyrosyl and histidyl residues with a stability constant of more than $10^{30}$ [11,12]: it serves as a carrier of iron in the mammalian body; 100 ml of human serum contains 200–300 mg of Tr and only 110 µg of iron (SI value). Since 1 mg of Tr can bind 1.25 µg of iron [13], Tr is usually saturated only to one-third of its total iron-binding capacity (TIBC value). Considering the loss of less than 0.1 mg of iron per day in urine, the concentration of free iron in plasma must be much less than 0.01 µg/ml [14] and free iron concentrations lower than 0.01 µg/ml are insufficient to support the growth of commonly studied bacteria [15].

The extremely strong binding of iron to Tr voids the serum of free iron and, as suggested by Schade and Caroline [10], the lack of free iron in serum determines its antimicrobial effect. In subsequent investigations Schade and Caroline showed that the inhibitory activity of human serum against Staphyloccus albus and Candida albicans can be neutralized by saturating the serum Tr with iron [16,17]. The high level of Tr saturation with iron in splenectomized patients with acute leukemia or thalassemia major seems to predispose them to candidiasis [18,19]. Also, it has been observed that children with kwashiorkor, in whom Tr is at a low level and is fully saturated with iron, succumb much more frequently to bacterial infection than do normal children; the administration of iron to such patients seems to predispose them to overwhelming infections [20].

The growth of microbes in sera, or in patients with iron-saturated Tr, can be explained in two ways: (a) the presence of growth-supporting levels of free iron; or (b) the neutralization by iron of a direct antimicrobial activity of iron-free Tr. Some findings have suggested a direct antifungal activity of Tr for C. albicans [21].

The question of mechanism of antimicrobial activity of Tr has been answered in the case of tubercle bacilli exposed to tuberculostatic solutions of Tr. If Tr were to possess a direct static effect on tubercle bacilli, then (a) it should be adsorbed to bacterial cells; (b) the effect should be neutralized not only by iron but also by zinc and cupric copper, which, in the absence of iron, combine with Tr; and (c) its antimycobacterial activity should not be duplicated by other iron-binding substances. It has been found that the mixture of Tr solution with tubercle bacilli (ratio of 1 mg Tr to 1 mg of the bacilli) failed to remove its tuberculostatic activity (Fig. 6-1), and that treatment of Tr with large amounts of zinc and cupric copper did not decrease its antibacterial activity [22]. Further investigation showed that the growth of tubercle bacilli can be inhibited by iron-binding Desferal [desferrioxamine B methane sulfate (Ciba Pharmaceutical Products, Inc., Summit, New Jersey)]. Since tubercle bacilli exposed to Tr or Desferal grew when plated on suitable medium, it has been concluded that Tr, as well as Desferal, do not inhibit tubercle bacilli by a direct effect on bacterial cells, but prevent bacterial multiplication by depriving the bacteria of iron [23].

The effect of Tr-iron interplay on the fate of tubercle bacilli has been investigated by adding ferric chloride and iron-free Tr in various proportions to tuberculostatic human serum [22]. The results showed that the addition of iron vitiated the tuberculostasis of serum; between 0.5 and 1.0 μg of iron neutralized the tuberculostatic activity present in 1 ml of 1:4 diluted serum. Furthermore, the addition of Tr to iron-neutralized serum reconstituted tuberculostasis (Table 6-1). The data in this table show quite clearly that the amounts of Tr which were effective in the reconstitution of tuberculostasis in iron-neutralized serum correlated closely with the known iron-binding capacity of this protein.

The correlation between the percentage of iron-saturated Tr and the strength of serum tuberculostasis was investigated by determinations of SI, TIBC, and the tuberculostasis of various mammalian sera. Table 6-2 presents the results of this study. The mammalian sera tested, and especially human and guinea pig sera, possess similar amounts of Tr (expressed by TIBC values). Pronounced differences were observed in the amounts of SI. Sera of the cow, rabbit, and mouse contained twice and of guinea pig thrice as much iron as was present in human serum. Consequently, the percentage saturation of Tr in guinea pig serum was much higher than in human, rabbit, mouse, and bovine sera. The data in Table 6-2 also show that the fate of tubercle bacilli in mammalian serum depends upon the degree of saturation of Tr with iron: human and bovine sera with about 30% saturated Tr were strongly tuberculostatic; rabbit and mouse sera with about 60% saturated Tr were weakly tuberculostatic; and guinea pig serum with 84% saturated Tr supported luxuriant growth of tubercle bacilli. These and some other results [23] have shown that the higher the percentage of iron-saturated Tr, the less efficient is the serum in inhibiting microbial growth.

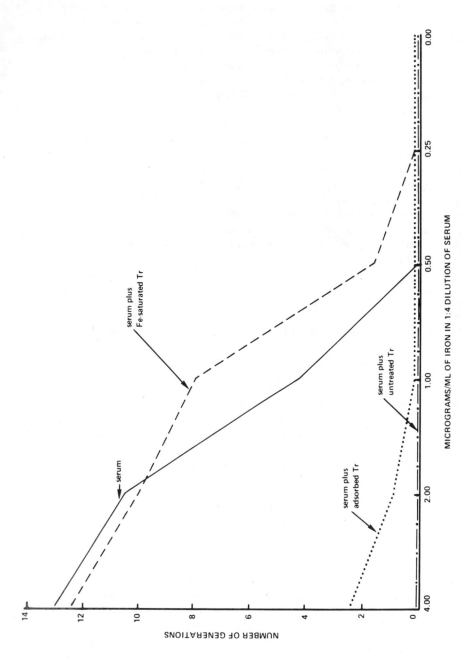

FIG. 6-1  The growth of tubercle bacilli in serum-iron mixtures containing 5 mg of unadsorbed, BCG-adsorbed, and iron-saturated transferrin (Tr) per milliliter of the solution.  (From Ref. 22.)

TABLE 6-1   Iron Neutralization of Tuberculostasis in 1:4 Dilution of Human Serum and the Reconstitution of Tuberculostasis in Iron-Neutralized Serum by Addition of Transferrin [a]

| Transferrin | Generations[b] in serum with iron[c] ($\mu$g/ml) | | | | | | |
|---|---|---|---|---|---|---|---|
| (mg/ml) | 0.0 | 0.5 | 1.0 | 2.0 | 4.0 | 8.0 | 16.0 |
| 0.00 | 0.0 | 2.3 | 11.0 | 12.5 | 12.9 | 13.2 | 12.7 |
| 1.25 | 0.0 | 0.0 | 2.7 | 9.6 | 9.1 | 12.2 | 13.0 |
| 2.50 | 0.0 | 0.0 | 0.5 | 4.5 | 10.2 | 13.1 | 12.1 |
| 5.00 | 0.0 | 0.0 | 0.0 | 0.3 | 2.6 | 9.4 | 12.3 |
| 10.00 | 0.0 | 0.0 | 0.0 | 0.0 | 0.0 | 5.0 | 10.7 |
| 20.00 | 0.0 | 0.0 | 0.0 | 0.0 | 0.0 | 0.0 | 0.0 |

[a] These data were taken from Ref. 22.

[b] The growth of BCG was determined after a 2-week incubation period at $37^\circ$ C.

[c] Iron added as $FeCl_3 \cdot 6H_2O$ salt.


Further study of serum tuberculostasis showed that sera with high iron-saturated Tr require less iron for the neutralization of antibacterial activity than do sera with low iron-saturated Tr (Fig. 6-2). Two bovine sera, one with 21% saturated Tr (serum A) and the other with 60% saturated Tr (serum B prepared from serum A with exogenous iron), were tested by the agar-plate diffusion test for the amount of iron required to neutralize their tuberculostatic effects. The extent of bacterial growth around iron-filled wells demonstrated that the percentage of unsaturated Tr determines the degree of tuberculostasis and the amount of exogenous iron necessary to neutralize it. Similar results were obtained with normal human sera [7]. Human sera with relatively high iron saturation of Tr (45%) showed larger area of bacillary growth around iron-charged wells than did the more typical human sera with only 30% saturated Tr.

The stimulating effect of iron compounds on the growth of various fast-growing bacteria exposed to antibacterial mammalian sera has received much attention in recent years. Bullen and Rogers found that the antibacterial activities of normal rabbit serum against Escherichia coli can be abolished by saturating Tr with iron [24]. Iron and hemoglobin neutralized the antibacterial activity of human serum against a strain of E. coli isolated from a patient with pyelonephritis [25]. Iron compounds exerted

TABLE. 6-2   Correlation between Levels of Iron-Saturated Tr and Degrees of Bacillary Growth in Mammalian Sera [a]

| Source of serum | No. of tests | Amount of iron ($\mu$g) [b] | | Tr saturation [e] (%) | Tuberculostasis in serum [f] |
|---|---|---|---|---|---|
| | | TIBC | SI [d] | | |
| Man | 10 | 327 | 97 | 30.0 | Present |
| Cow | 4 | 490 | 191 | 39.0 | Present |
| Rabbit | 8 | 317 | 204 | 64.3 | Limited |
| Mouse | 10 | 382 | 230 | 60.2 | Limited |
| Guinea Pig | 20 | 323 | 273 | 84.4 | Absent |

[a] These data were taken from Ref. 23.

[b] Individual determinations of TIBC and SI fell within 10% variation of the mean value shown in the table.

[c] TIBC value shows the mean of the amount of iron present in 100 ml of iron-saturated serum.

[d] SI value shows the mean of the amount of iron present in 100 ml of untreated serum.

[e] Percentage of iron-saturated Tr in serum sample equals (SI/TIBC) X 100.

[f] Tuberculostasis was scored as "present" when bacillary growth was less than one generation, "limited" when it was less than five generations, and "absent" when it varied between 5 and 14 generations during a 2-week incubation period.

pronounced neutralizing effects on bacteriostatic activities of sera for Yersinia septica and Y. pestis. It has been found that virulent and avirulent strains of Y. pestis grew much better in human serum when it was supplemented with exogenous iron [26]. Experiments with Y. septica showed that the presence of unsaturated Tr was essential for the inhibition of bacterial multiplication since the saturation of sera with iron neutralized the inhibitory effect [27].

The growth-supporting effect of iron for E. coli exposed to mammalian sera was shown quite clearly by the use of the agar-plate diffusion test [28]. Wells in the growth-supporting iron-poor Dubos agar medium were charged either with 0.4 ml of mammalian sera or 0.4 ml solution of Tr

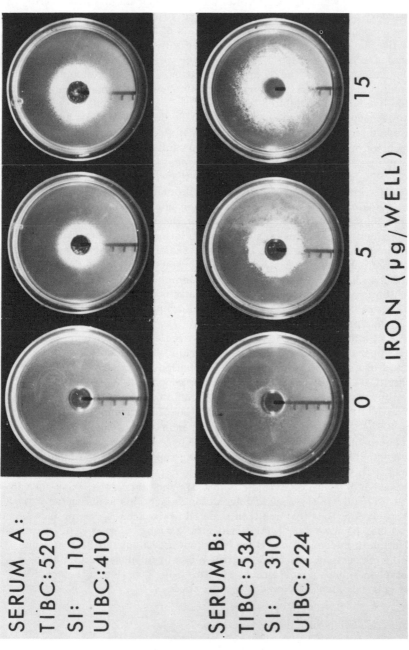

FIG. 6-2 Bacillary growth around wells charged with 0, 5, and 15 μg of iron on Dubos agar medium containing 1:4 dilution of a strongly tuberculostatic serum A or a weakly tuberculostatic serum B. (From Ref. 7.)

SERUM A:
TIBC: 520
SI: 110
UIBC: 410

SERUM B:
TIBC: 534
SI: 310
UIBC: 224

IRON (μg/WELL)

0    5    15

(10 mg/ml), and wells in the inhibitory medium-serum mixtures were
charged with either 0.4 ml saline containing 40 μg of ferric iron or 0.4 ml
of E. coli spent medium.   After the diffusion of the materials from the
wells, the plates were uniformly inoculated with 1:100 dilution of E. coli
suspension adjusted to Klett 1 value.   Plates were incubated at 37°C and
examined after 15, 25, and 35 hr of incubation.   The results obtained in
this study are presented in Figure 6-3; they show that human serum and its
purified iron-free Tr inhibited the growth of E. coli around wells made in
the growth-supporting medium.   Iron-free Tr diffused into the agar medium, tied up iron, and made the metal unavailable for bacterial utilization.
Although not presented in Figure 6-3, the results also showed that guinea
pig serum was not inhibitory for E. coli because it does not possess enough
iron-free Tr to bind iron in the vicinity of the wells.   Iron or spent medium
of E. coli supported the growth around wells made in growth-inhibitory
serum-medium mixture; the addition of iron in excess of the saturating capacity of Tr satisfied the bacterial need for the metal, and the addition of
spent medium provided preformed iron-chelating compound which removed
iron from Tr and supplied it to bacteria.   These and previous results indicate that effects of sera on E. coli and Mycobacterium bovis or M. tuberculosis are identical; these microbes are inhibited by the lack of iron in
mammalian sera or in media in which iron is bound to Tr.   This antimicrobial effect can be alleviated by bacterial siderophores which remove iron
from Tr and provide it to bacteria.

The results presented in Figure 6-3 revealed a phenomenon which we
had not observed before in the study of serum tuberculostasis.   The examination of bacterial growth during the incubation period showed that the
peripheral growth of E. coli on the growth-supporting medium spread
gradually towards the wells charged with inhibitory serum or Tr; however,
the growth of the tubercle bacillus on similar plates remained constant
during 3 to 4 weeks of incubation.   This phenomenon is due to water-soluble
enterochelin, produced by E. coli, that diffuses into the area of Tr-induced
microbiostasis, removes iron from Tr, and supplies it to the inhibited
bacteria.   Tubercle bacilli produce hydrophobic mycobactin (M) which remains associated with their lipoid cell walls and, therefore, does not supply iron to bacilli in the zone of tuberculostasis.   This assistance by growing bacteria for those that are iron starved in the area of Tr- or lactoferrin-
induced inhibition we call "nutritional vitalization."   The role of nutritional
vitalization in microbial infections is unknown; the possibility that iron-
chelating products of intestinal microbes can predispose an animal to
microbial infections poses a new and interesting question in the consideration of host resistance to infectious microbes.

B.   Microbiostasis in Milk

The resistance of colostrum-fed babies, calves, and newborn pigs to
gastroenteritis caused by E. coli suggested that milk possesses antibacterial

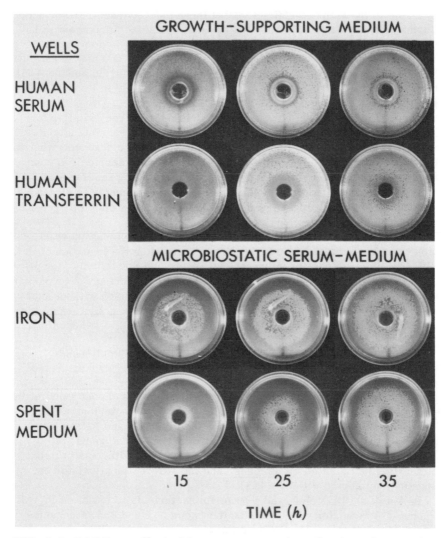

FIG. 6-3  Inhibitory effect of human serum or transferrin and promoting effect of iron or spent medium on the growth of E. coli around wells made in growth-supporting and growth-inhibiting media, respectively.

factors [29-31]. The search for these in colostrum or milk led to the discovery of an iron-binding protein, lactoferrin, which inhibits bacterial growth [32,33]. It has been found that lactoferrin is not limited to milk; the protein is found in other secretions and tissues of the animal body [34], including granules of polymorphonuclear leukocytes [35] where it may

play a role in the antibacterial activity of these phagocytes. The amounts
of lactoferrin and Tr in milk vary with animal species. In contrast to human milk with a large quantity of lactoferrin (2-6 mg/ml) and a small
quantity of Tr (0.05 mg/ml), the milk of the cow, goat, and sow is deficient
in these iron-binding proteins (0.02-0.2 mg/ml of lactoferrin and Tr) [36].
Since the iron in milk (about 1.5 μg/ml) is present mainly in its lipid fraction [37], lactoferrin and Tr exist in the iron-free state.

With regard to iron-binding activity, lactoferrin is very similar to Tr.
Both proteins bind two atoms of iron with similar metal-binding sites.
However, the stabilities of their complexes with iron are different; whereas
the Tr-iron complexes dissociate around pH 5, the lactoferrin-iron complexes remain stable until the pH is below 4 [38]. This property of lactoferrin to bind iron at acidic pH values may be of considerable importance
in the depletion of the metal in phagocytic vacuoles into which the secondary
granules of leukocytes discharge their content [39].

In order to exert antibacterial activity in the gastrointestinal tract,
lactoferrin should retain its iron-binding activity after the ingestion of
milk. The stomach content in breast-fed infants has a relatively high pH
(6.0-6.5) and shows no evidence of protein hydrolysis during a 2-hr period
after feeding. Most of the ingested milk leaves the stomach without any
digestion of proteins [40]. After passing the stomach in active form,
lactoferrin could effectively bind iron in the small intestine, where the pH
is slightly alkaline, and deprive bacteria of iron.

The antibacterial property of lactoferrin has been established by
Masson et al. [32] in bronchial mucus. A purified, 27% iron-saturated
lactoferrin of cow milk was found to be bacteriostatic for Bacillus stearothermophilus and B. subtilis in nutrient broth; human and goat lactoferrins
were equally effective bacterial inhibitors [33]. The bacteriostatic action
of lactoferrin was not limited to vegetative cells; Oram and Reiter [33]
have shown that the protein inhibits the germination of the spores of
B. stearothermophilus on nutrient agar. Human milk with unsaturated
lactoferrin also inhibited C. albicans, and this effect was alleviated by
saturating lactoferrin with iron [41].

Bullen et al. [42] observed a powerful bacteriostatic effect on E. coli
in human milk samples which contained between 56% and 89% of unsaturated iron-binding capacity. This antibacterial effect of milk was abolished
by the addition of exogenous iron, and potentiated by the addition of anti-
E. coli antibodies which act "in concert" with iron-binding proteins.
Bullen and his associates observed a pronounced sensitivity of the bacteriostatic system to pH changes; human milk adjusted to pH 6.95 or below had
no effect on the growth of E. coli, whereas at pH 7.20 the milk exerted a
powerful antibacterial activity. Since the stability of lactoferrin-iron complexes to acid pH has been well documented, the loss of antibacterial
activity in the acidified milk cannot be explained on the basis of the dissociation of lactoferrin-iron complexes. This is indicated by our recent results obtained in the study of the competition for iron between Tr or

lactoferrin and enterochelin of E. coli. We found that, at acidic pH, enterochelin obtains iron from Tr-iron complexes much more effectively than from lactoferrin-iron complexes [28]. These pH effects on milk or serum bacteriostasis could very likely be attributed to changes in the pH-dependent efficiency with which enterochelin removes iron from lactoferrin-iron or Tr-iron complexes and, less likely, to the pH-determined stability of the protein-iron complexes. In either case, these findings suggest that pH exerts a pronounced effect on the competition for iron between animal and microbial iron-binding compounds.

## C. Microbiostasis in Mammalian Cells

It is well documented by in vitro and in vivo experiments that phagocytic cells of animals engulf most kinds of microbes without much difficulty. Some microorganisms are killed within phagocytic cells, whereas others, especially those belonging to the group of facultative intracellular parasites, have the ability to survive and grow within such cells. Factors which determine bacterial killing or multiplication in phagocytic cells are under investigation in several laboratories.

It seems that polymorphonuclear leukocytes and macrophages of normal animals possess different antimicrobial mechanisms. The mechanism by which leukocytes kill intracellular bacteria involves the phagocytosis-induced generation of hydrogen peroxide [43]. Considerable attention has been given to an antibacterial system composed of myeloperoxidase, hydrogen peroxide, and halide [44], but the precise mechanism by which this system exerts its lethal activity has not yet been identified. Knowledge of the microbicidal mechanisms of macrophages is much more deficient. Although enzymatic differences between macrophages of bacteria-resistant and bacteria-susceptible animals have been identified [45], there is no evidence that these differences determine the fate of intracellular bacteria. Recent studies suggest that the primary agent in an antibacterial system of macrophages is hydrogen peroxide [46,47], but the generation of this substance by macrophages has not as yet been clearly demonstrated.

Little is known about the nutrition of microbes within phagocytic cells. Differences in the predisposition of certain tissues of the same host to invasion by tubercle bacilli [3] suggest that tissues differ in the availability of growth-supporting nutrients. It has been found that the growth of tubercle bacilli within macrophages of normal guinea pigs was partially suppressed, while the growth in guinea pig serum was uninhibited [48]. This limitation of bacterial growth has been attributed to the competition for nutrients between animal and bacterial cells. The nutritional starvation of intracellular microbes has been defined by the work of Hanks and his associates [49,50]. They showed that seclusion within tissue cells does not isolate microbes from external nutrients which the cells cannot provide. External supplies of factors such as mycobactin or iron stimulated the growth of Mycobacterium paratuberculosis within macrophages; each factor alone converted indolent intracellular infections into fulminating ones. The dramatic stimulation of intracellular growth of bacteria by mycobactin, the iron-binding product of tubercle bacilli, constitutes the first evidence for the iron starvation of bacteria within macrophages. It will be shown later (Section IV) that mycobactin can remove iron deposited in ferritin molecules of mammalian cells and provide the metal to tubercle bacilli.

One of the major proteins present in neutrophilic leukocytes is an iron-binding lactoferrin [35,51]. Although defensive activity of this protein against intracellular parasites has not been shown, recent results suggest that it can strengthen antimicrobial resistance by inducing hypoferremia in acute inflammatory processes. At the acidic pH of inflammation, iron is removed from the "loose" iron-Tr complexes by lactoferrin of disintegrated leukocytes, and the stable lactoferrin-iron complexes are taken up by cells of the reticuloendothelial system [52]. The lactoferrin-induced depletion of iron in inflammatory processes may increase the resistance of the host against extracellular and intracellular parasites.

## III.   IRON STARVATION OF MICROBES IN TISSUES AND FLUIDS OF IMMUNE ANIMALS

The study of iron metabolism suggested that nutritional immunity can be acquired by an animal in the response to bacterial infections. The acquisition and potentiation of nutritional immunity could come about in the infected or immunized host either by increased synthesis of Tr and lactoferrin or by loss of cellular or humoral iron. No evidence exists for an increase in amounts of Tr in infected or vaccinated animals. There is good evidence, however, for the development of hypoferremia in the blood of animals during infection or after treatment with bacteria or their endotoxins.

### A.   Acquired Hypoferremia in Vaccinated Animals

Low iron levels in human blood were observed by clinicians to be associated with infectious diseases [53]. A decrease in plasma iron levels was induced in mice infected with Listeria monocytogenes [54] and in humans infected with Francisella tularensis [55]. The fall in serum iron during infections probably is elicited by bacterial endotoxins [56], which stimulate leukocytes to release an endogenous mediator of hypoferremia [57, 58]. The treatment of animals with this low-molecular-weight mediator of hypoferremia protected them from lethal salmonellosis [59]. Although the levels of hypoferremia or degrees of iron-saturation of Tr were not examined in these studies, various experiments clearly indicate that the fall in serum iron determines an enhancement of resistance to bacterial infections.

The relationship between hypoferremia and microbiostasis in sera of vaccinated or endotoxin-treated guinea pigs has been demonstrated by work of Kochan and his associates [23,60]. This study originated from the finding of a difference in antimycobacterial activity of normal and immune sera [61]; in contrast to the growth-supporting nature of normal guinea pig serum, serum of immune animals exerted a pronounced suppression on the growth of tubercle bacilli. This initial observation of acquired anti-mycobacterial activity of immune sera developed later into the study of the relationship between iron-free Tr and the tuberculostatic activity in sera of normal, BCG-vaccinated, E. coli LPS(lipopolysaccharide)-treated, or TCW(tuberculous cell wall)-treated guinea pigs [23]. The results showed that the sera of vaccinated or bacterial product-treated animals gained antimicrobial activity in direct proportion to the loss of iron and the decrease in iron saturation of Tr (Table 6-3). The induced tuberculostasis in sera of treated animals can be readily neutralized by the addition of exogenous iron. Since the hypoferremic response transforms animal plasma from a bacterial growth-supporting medium to one that exerts microbiostasis on a wide spectrum of microbes, it should be considered as a protective response of the animal body to parasitic invasion.

B. Antimicrobial Effect of Transferrin-
   Antibody System

The effective neutralization of bacterial toxins by specific antibodies has been regarded as strong evidence for a crucial role of antibodies in protection to microbial infections. Although antibodies neutralize bacterial toxicity, they only infrequently exert a direct antibacterial activity; bacterial lysis by specific antibody and a complement is limited to a few bacterial species only, and other antibody activities do not damage the bacterial cell.

Recent reports by Bullen and his associates have suggested that an antibody may play a role in the antibacterial effects of immune serum. These investigators observed that a specific antibody to Y. septica potentiates the antibacterial effect of Tr [62]. Further study revealed that the bacteriostasis is attributable to a system composed of Tr, a specific antibody, and complement [27]. Since the antibacterial effect of the system is neutralized by iron or heme compounds, which do not affect the activities of antibody and complement, the antibacterial effect has been attributed to the iron-binding activity of Tr. This conclusion has been supported by observations which showed that at pH 6.85, at which the antibody and complement are fully active but the association constant between Tr and iron is reduced, the bacteriostasis is alleviated. Although the role of the antibody and complement in the iron-starvation of bacteria is not clear,

TABLE 6-3   Iron Neutralization of Tuberculosis in Sera of LPS- or TCW-Treated and BCG-Vaccinated Guinea Pigs [a]

| Days after treatments [b] | Tr saturation (%) | Generations [c] in serum with iron (μg/ml) | | | | |
|---|---|---|---|---|---|---|
| | | 0 | 1 | 2 | 4 | 8 |
| LPS-1 | 17.5 | 0.0 | 0.0 | 0.4 | 8.0 | 9.2 |
| 2 | 42.0 | 0.0 | 10.6 | 11.3 | 10.7 | 12.2 |
| 3 | 60.2 | 0.0 | 11.8 | 11.9 | 9.5 | 11.1 |
| 5 | 74.7 | 8.7 | 12.0 | 11.4 | 10.7 | 11.3 |
| 10 | 93.8 | 12.5 | 11.6 | 12.0 | 12.0 | 11.3 |
| TCW-1 | 26.5 | 0.0 | 5.1 | 9.9 | 9.5 | 9.5 |
| 2 | 42.1 | 0.5 | 10.2 | 10.3 | 10.0 | 10.6 |
| 3 | 59.3 | 1.2 | 9.7 | 9.3 | 10.7 | 11.0 |
| 5 | 66.3 | 0.7 | 7.3 | 12.0 | 11.3 | 12.3 |
| 14 | 78.8 | 11.8 | 12.6 | 12.1 | 11.9 | 12.7 |
| BCG-3 | 86.2 | 10.5 | 11.5 | 11.9 | 11.0 | 11.2 |
| 7 | 79.4 | 9.1 | 10.6 | 11.3 | 12.1 | 11.7 |
| 14 | 75.0 | 4.3 | 9.5 | 10.2 | 12.3 | 11.1 |
| 21 | 75.3 | 0.4 | 9.1 | 10.3 | 9.9 | 10.7 |
| 28 | 68.6 | 0.0 | 9.7 | 10.2 | 10.9 | 11.3 |
| Saline-1 | 85.6 | 11.5 | 11.8 | 11.6 | 11.5 | 12.1 |

[a] These data were taken from Ref. 23.

[b] On day 0, animals were injected intraperitoneally with 0.05 mg of LPS, 1 mg of TCW preparation, or 1 mg of BCG cells per 100 g of body weight.

[c] Tests for the presence and neutralization of tuberculostasis were performed in a 1:4 dilution of serum samples. Bacillary fate was determined after a 14-day incubation period.

it has been suggested that they could block the release of bacterial siderophores.

Another explanation for the activity of a specific antibody in relation to the antibacterial activity of Tr has been suggested by the work of Rogers [63]. The investigation of iron-transport in E. coli showed that traces of a specific antibody inhibited the synthesis of diffusable iron-transporting compounds. This activity of the antibody would induce an increased level of antibacterial resistance by lessening the ability of bacterial parasites to compete for the Tr-bound iron.

Using the agar-plate diffusion test for the study of bacterial iron metabolism in sera of normal animals or in solutions of purified Tr, we observed that the iron-neutralizable antibacterial activity of Tr does not
depend upon the presence of a specific antibody or complement (Table 6-1
and Fig. 6-3). In the absence of any serum in the growth-supporting medium, purified Tr inhibits the growth not only of tubercle bacillus but also of
E. coli, Salmonella typhimurium, and L. monocytogenes [28]. It is possible that specific antibodies could potentiate the antibacterial effect of Tr,
but they are not required for its antimicrobial activity. The possibility
exists of a concerted antibacterial activity of specific antibody and Tr and
may be a very rewarding study for students of host-parasite relationships.

C. Antimicrobial Effect in Cells
   of Vaccinated Animals

It has been discussed previously (Section II. C) that the fate of microbes in
normal macrophages is influenced by substances in and conditions of the
external environment. The presence of iron or mycobactin in tissue
culture media stimulated the luxuriant intracellular growth of tubercle
bacilli in peritoneal macrophages [49]. This finding shows quite clearly
that the growth of intracellular parasites is dependent to a large extent on
the extracellular supply of iron. Human serum with a high level of iron-
free Tr added to cultures of normal and immune guinea pig macrophages
inhibited the intracellular growth of tubercle bacilli much more in the immune than in the normal macrophages [48]. It seems that intracellular
iron starvation can be induced by Tr and lactoferrin much more effectively
in immune than in normal macrophages; perhaps increased pinocytosis and
hypersensitivity-induced increased permeability of immune macrophages
[48,64] facilitate the passage of Tr from the extracellular medium into
the cell. These results suggest that Tr exerts its iron-depleting antibacterial activity not only in humoral fluids but in the phagocytic cells of normal and, especially, of immune animals.

Considering the iron requirements of the tubercle bacillus, we investigated the availability of this metal for bacillary utilization inside
macrophages. Iron is stored in macrophages mainly as insoluble ferric
hydroxide within large ferritin molecules. The ability of tubercle bacilli
to utilize ferritin iron has been investigated by the agar plate diffusion
test [65]. The results showed that ferritin, in which iron constitutes 32%
of its dry weight, does not neutralize serum tuberculostasis; iron of ferritin becomes available for bacillary utilization only in the presence of
minute amounts of mycobactin (Fig. 6-4). It has been concluded, therefore, that iron stored in ferritin molecules of host cells is as unavailable
for bacterial utilization as iron bound to Tr. Mycobactin can remove
iron from Tr or ferritin to promote extracellular and intracellular growth
of tubercle bacilli.

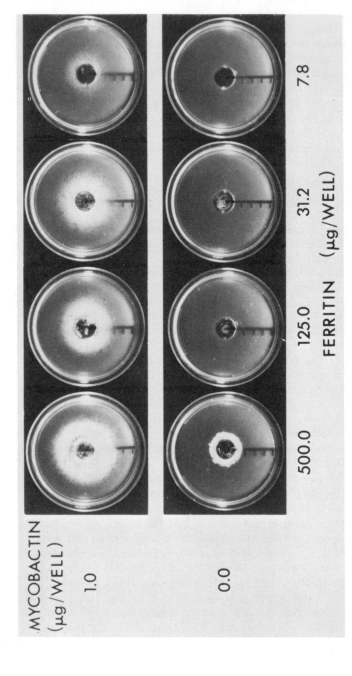

FIG. 6-4  Neutralization of serum tuberculostasis around wells charged with various concentrations of ferritin and ferritin–mycobactin solutions.  (From Ref. 65.)

This study does not explain how facultative intracellular bacterial parasites obtain iron within normal macrophages exposed to microbiostatic serum. Our attempts to reveal an iron carrier in serum or in macrophages with an ability to supply the metal of Tr or of ferritin to tubercle bacilli were unsuccessful. The possibility exists that acid pH in infected cells favors the dissociation of iron from ferritin, as it does from Tr, and then that these minute amounts of iron initiate bacterial metabolism. The intracellular iron-deficient environment stimulates increased production of mycobactin [7], which could provide bacilli with enough metal for unhindered multiplication.

In addition to the iron-depleting effect of Tr, and possibly lactoferrin, on the growth of bacteria in immune macrophages, these cells are armed by an additional antibacterial mechanism. It has been found that lysates of macrophages collected from immunologically stimulated guinea pigs (activated macrophages) exert a strong bactericidal activity even in iron-rich media [66-68]. The antimycobacterial effect was found to be mainly associated with macrophage membranes which produce fatty acids with a wide spectrum of antibacterial activity. It has been found that toxic free fatty acids are produced during the hydrolysis of phospholipids by the activity of membrane-bound lipases. During a severe immunological reaction, 1 to 5 days after intravenous challenge of BCG-vaccinated animals with specific antigen, free fatty acids were also present in the blood of stimulated animals. Serum levels of free fatty acids are dependent upon the number of vaccinations; sera of guinea pigs vaccinated three times contained much larger amounts of free fatty acids and exerted much stronger antimycobacterial activity than did sera of animals vaccinated only one time or untreated [60]. A similar correlation between amounts of free fatty acids and antimycobacterial activity has been observed in lysates or in the heptanic-extractable fractions prepared from activated macrophages [68]. In the light of therapeutic use of BCG in antitumor treatment and the known antitumor activity of free fatty acids, it is possible that the mechanisms of the antimycobacterial and antitumor activity in BCG-treated animals are the same; both activities could be attributed to the formation of toxic levels of fatty acids by macrophages of immunologically activated animals.


IV.   COMPETITION FOR IRON BETWEEN
      IRON-BINDING COMPOUNDS
      OF ANIMAL AND MICROBIAL ORIGIN

In Sections II and III we described the iron starvation of microbes in normal and immune animals. Various experimental observations have shown that microbes cannot utilize iron in host tissues and fluids because the metal is bound to chelating proteins. Thus, in the absence of free iron in animals, one would expect them to be absolutely resistant to infectious microbes. Obviously this is not the case; although animals manifest different susceptibilities to microbial invasion, which correlate closely with the levels of iron-free Tr [23], they all succumb to disease when challenged with a heavy dose of pathogenic bacteria. The growth of microbes in animals with high levels of iron-free Tr suggests that the parasites possess a mechanism by which they can obtain iron needed for their growth. An attempt will be made in this section to describe the in vitro and in vivo experiments which illustrate the molecular competition for iron between microbes and their animal hosts.

A.   In Vitro Competition for Iron

It has been suggested by Neilands that microbes can be divided into two types on the basis of their ability to secrete iron-binding compounds [69]. The great majority of bacterial species (autosequesteric) produce iron-binding compounds which transport the metal into bacterial cells. A few species (anautosequesteric) lack the ability to manufacture their own siderophores and depend on other related organisms for iron-providing compounds. The best known example is the dependence of M. paratuberculosis on the presence of mycobactin, an iron-binding compound which is produced by many mycobacteria but not by the organism itself. The literature records several microbial products which chelate iron and promote bacterial growth [70]. In most cases, these products were considered as growth factors and only occasionally was their iron-binding activity considered in terms of the microbial iron-transport mechanism [71].

During the last few years a large amount of information has been obtained about the chemical nature of several microbial iron-chelating compounds. Studies as to whether these compounds help microbes to overcome iron-starvation imposed by Tr, lactoferrin, or ferritin have been done only infrequently. The development of the agar-plate diffusion test in our laboratory helped in defining the biological role of siderophores in the host-parasite relationship. Originally, our studies defined the role of mycobactin in the life of the tubercle bacillus; more recently, we have investigated the role of enterochelin, the cyclic trimer of 2,3-dihydroxybenzoyl serine, which was isolated from spent media of S. typhimurium [72], E. coli, and Klebsiella aerogenes [73].

Results presented in Section II.A showed that the addition of exogenous iron to tuberculostatic sera lessens tuberculostasis. Results presented in Figure 6-5 show that a similar neutralization of serum tuberculostasis can be achieved by the addition of purified, iron-free mycobactin. These results suggest that mycobactin promotes bacillary growth by removing iron from Tr-iron complexes and making it available for tubercle bacilli, or by acting as a growth factor which lessens the bacterial need for iron. The growth of BCG cells around wells charged with iron or mycobactin was

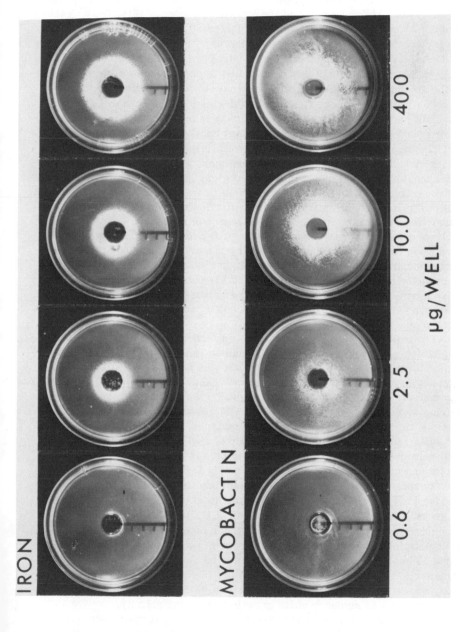

FIG. 6-5 Bacillary growth around wells charged with 0.6, 2.5, 10.0 and 40.0 μg of iron or mycobactin on tuberculostatic bovine serum–Dubos agar medium. (From Ref. 7.)

different: around iron wells most of the colonies were not corded and the growth was rather sharply defined, whereas around mycobactin wells the colonies were highly corded and the growth was spread (Fig. 6-5). Other studies showed that in the presence of Tween 80, virulent tubercle bacilli ($H_{37}Rv$ strain) lose hydrophobic mycobactin from their cell walls and grow as does the avirulent, uncorded $H_{37}Ra$ strain [7]. These observations suggest that the cording of tubercle bacilli is controlled to a large extent by the amount of hydrophobic mycobactin on bacillary cell walls. Since mycobactin is bound more firmly to lipid-rich virulent than to lipid-poor avirulent bacilli [65], virulent bacilli grow in cords on Dubos medium containing 0.05% Tween 80 whereas avirulent bacilli are not corded.

The biochemical function of mycobactin has been clarified by experiments in which this substance in various quantities was added to samples of tuberculostatic serum which received in addition quantities of iron-free or iron-saturated Tr [74]. The fate of BCG organisms in tuberculostatic serum-Tr-mycobactin mixtures was determined by plating on growth-supporting medium after a 2-week incubation period (Fig. 6-6). Since at each concentration of mycobactin its growth-promoting effect varied according to the amount of unsaturated Tr, it has been concluded that mycobactin does not serve as a growth factor but as a carrier of iron which mycobactin (as contrasted to Tr) provides to tubercle bacilli. This conclusion has been supported subsequently by work of Ratledge and Marshall [75], who showed that mycobactin was indeed acting as the carrier of iron for Mycobacterium smegmatis.

The exposure of tubercle bacilli to an iron-poor medium stimulates production of increased amounts of mycobactin [74] and salicylic acid [76]. Since both substances bind iron, it was postulated that salicylate may act as a specific donor of iron to cell-bound mycobactin [75]. Using two mutants of M. smegmatis, one requiring salicylate and the other requiring mycobactin, Ratledge and Hall [77] observed that the addition of mycobactin to the salicylate-requiring mutant did not alleviate the need for salicylate. They suggested, therefore, that salicylate performs a dual role in the life of M. smegmatis; it serves as a precursor of mycobactin and mobilizes iron in the environment for bacillary utilization. In view of these findings, recent results obtained in our laboratory are disturbing [78]. Results showed that the addition of salicylate to growth-supporting media inhibited the growth of BCG, and this stasis could not be alleviated by the addition of large quantities of iron or mycobactin. Salicylate fails to promote bacterial growth in tuberculostatic sera. We found no basis for considering this compound as an iron mobilizer; on the contrary, salicylate does not stimulate but inhibits bacillary growth. Our study shows that the adsorption of salicylate to mycobacteria should not be interpreted as evidence of its growth-supporting activity.

The production of siderophores is influenced by temperature; the production of enterochelin by E. coli at various temperatures is presented

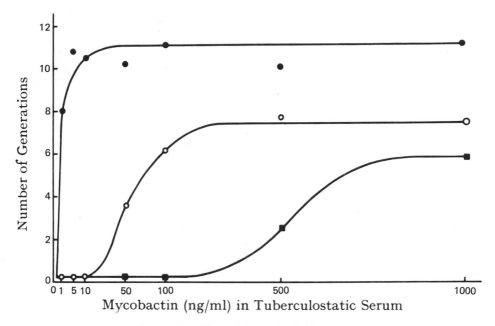

FIG. 6-6 Growth of BCG bacilli in tuberculostatic serum containing 5 mg of transferrin and various concentrations of mycobactin per milliliter of the solution. Key: ●, iron-saturated transferrin; ■ , iron-free transferrin; O, no transferrin. (From Ref. 74.)

in Table 6-4 [79]. The amount of enterochelin in spent medium of bacteria grown at 26° C was found to be larger than in spent medium of bacteria grown at 37° C; E. coli grown at 41° C and 44° C produced little or no enterochelin. These experiments revealed also the phenomenon of nutritional vitalization; the iron-supplied E. coli growing around wells charged with spent medium (enterochelin) produced siderophore which diffused into the area of microbiostasis and promoted the growth of inhibited bacteria by supplying them with iron obtained from Tr-iron complexes. Therefore, the area covered by bacterial growth broadens with time of incubation at 37° C.

The foregoing results confirm and extend the findings of Garibaldi [80], who showed that the biosynthesis of enterochelin by S. typhimurium is progressively reduced when the temperature is raised from 31° C to 40° C. The decrease in siderophore production at high temperatures suggests that fever associated with infectious diseases may be beneficial to the host because it promotes iron starvation of microbial parasites. Experimental results suggest that the animal body reacts to microbial invasion

TABLE 6-4   Bacterial Growth on Microbiostatic Serum-Medium Mixture around Wells Charged with Spent and Unused Media [a]

| Wells charged with | Width (mm) of E. coli growth around wells at: | | | |
|---|---|---|---|---|
| | 24 hr | 30 hr | 40 hr | 48 hr |
| 26°C-spent medium | 9 | 13 | 18 | 20 |
| 33°C-spent medium | 6 | 9 | 12 | 15 |
| 37°C-spent medium | 5 | 8 | 12 | 15 |
| 41°C-spent medium | 0 | 0 | 3 | 3 |
| 44°C-spent medium | 0 | 0 | 3 | 3 |
| Unused medium | 0 | 0 | 0 | 1 |

[a] Spent media were prepared by growing E. coli at various temperatures for 48 hr in iron-poor Dubos medium. After the growth-period, the cultures were adjusted to the same turbidity, bacteria were removed by filtration, and spent media were tested for the ability to alleviate the microbiostasis in bovine serum-agar medium mixture.

with two responses which favor iron-starvation of parasites: (a) hypoferremia (Section III.A), limiting the availability of iron for parasitic utilization; and (b) fever, which suppresses production of siderophores.

There is little information in the literature about the ability of enterochelin or other iron-binding microbial products to remove iron from Tr or lactoferrin. It has been found that antibacterial effect of human serum for S. typhimurium can be neutralized by the addition of enterochelin produced by the organism growing in a low-iron medium [81]. A bacteriostatic effect of Tr in the presence of serum albumin for Clostridium welchii has been described by Rogers and his associates [82]. The bacteriostatic properties of whole serum for C. welchii was not attributed entirely to Tr, because only the mixture of Tr and β- or γ-globulin imitated the bacteriostasis of the whole serum. The concerted action of Tr and antibodies on bacteria has been discussed in Section III.B.

The iron-providing effect of bacterial iron-binding compounds for bacterial growth is being investigated in our laboratory by the use of the agar-plate diffusion test. The plates are filled with 12 ml of bacteriostatic medium consisting of 9 ml of iron-poor Dubos agar medium and 3 ml of mammalian serum. The wells are charged with 0.4 ml of bacteria-free spent media prepared by growing bacteria in iron-poor Dubos liquid medium for 48 hr. After the diffusion of spent and unused media from the wells, the plates are inoculated uniformly with a standard suspension (10,000 microorganisms) of homologous or heterologous bacteria. The inoculated plates are examined for bacterial growth around spent-medium-charged wells after a 20-hr incubation at 37°C. Preliminary results have shown that the spent medium of E. coli and S. typhimurium neutralized serum tuberculostasis not only for the homologous but also for heterologous bacteria. In distinction to the growth-promoting activity of spent media, unused medium did not promote bacterial multiplication on the microbiostatic serum-medium mixture. Since widths of bacterial growth around spent-medium-charged wells were nearly the same, it was concluded that E. coli and S. typhimurium produce similar quantities of enterochelin, equally effective in providing iron for the growth of homologous as for heterologous bacteria.

## B. In Vivo Competition for Iron

The enhancing effect of iron on the progression of infection with Y. pestis was observed in 1956 by Jackson and Burrows [83]. Since that time many investigators have made similar observations. It has been reported that the administration of iron or iron-containing compounds such as hematin, hemin, or hemoglobin promoted infections with Klebsiella pneumoniae, Pseudomonas aeruginosa [84], S. typhimurium [85], L. monocytogenes [54], Y. septica [62,86], E. coli [87], C. welchii [88], and M. tuberculosis [5].

Several suggestions have been made to explain the infectious disease-promoting activity of iron. It has been suggested that iron promotes the growth of Candida albicans in the presence of iron-free Tr because the metal neutralizes the antifungal activity of Tr [21]. Other findings have suggested that iron neutralizes the bactericidal action of a basic protein extract obtained from polymorphonuclear leukocytes [89]. The predominance of experimental results suggest, however, that exogenous iron promotes the development of microbial infections by becoming available for microbial utilization. Studying the role of iron in the host-parasite relationship, Burrows found that iron compounds do not interfere with phagocytosis, intracellular digestion, or the production of antibodies [90]. Similar findings were reported by Bullen and his associates, who observed that iron interfered neither with phagocytosis nor with the fixation of complement by antibody-antigen complexes [88]. The blockage of the reticuloendothelial system in mice infected with attenuated plague bacilli did not stimulate infection, whereas iron treatment of such mice promoted bacterial growth and the development of disease [91]. These findings suggest that iron does not interfere with the defensive activity of the reticuloendothelial system.

Neither is there evidence to suggest that iron impairs some other defense mechanisms of the host. Burrows observed that iron injected even
in large quantities into animals infected with an avirulent strain of Y. pestis
failed to establish an infection, whereas much smaller quantities of iron
enhanced the pathogenicity of virulent bacteria [90]. Iron treatment of
animals infected with a virulent or an attenuated strain of tubercle bacilli
significantly favored the multiplication of the former but not of the latter
strain [1]. If the activity of iron were to neutralize the defense mechanism
of the host, iron-treated animals might be expected to become susceptible
to infections with avirulent bacteria.

The consequences of high iron saturation of Tr in patients with
kwashiorkor [20] and in splenectomized patients with acute leukemia or
thalassemia major [18,19] on susceptibility to microbial infections suggest that a close correlation exists between the hyperferremic state and
the development of infectious diseases. Since hyperferremia seems to be
advantageous to parasites, then hypoferremia should be beneficial to the
host. This has been found to be the case: when hypoferremia was induced
experimentally with endotoxins, the hypoferremic animals became more
resistant to bacterial infections than untreated animals [84,92]. The
treatment of animals with iron-binding compounds such as desferrioxamine
B or iron-free Tr induced hypoferremia which suppressed the growth of
various bacteria [54,84]. These findings indicate that any treatment that
increases the level of iron-free Tr makes iron less available for microbial
utilization and, therefore, increases resistance to microbial parasites.

The documentation of the importance of iron-binding lactoferrin to the
antimicrobial resistance of newborn animals is quite limited. It has been
suggested that colostrum-deprived calves tend to develop infections with
E. coli more frequently than do colostrum-fed animals [93]. The count of
E. coli in the feces of bottle-fed babies tends to be much higher than in
breast-fed babies [94]. The evidence that milk actually suppresses the
growth of E. coli was provided by experiments in which colostrum-fed
guinea pigs were dosed orally with the bacteria and hematin [42]. Results
showed that the presence of hematin in the small intestine enhanced the
growth of E. coli by a factor of over 10,000-fold. It seems that lactoferrin
in milk may play a protective role in the resistance of newborn animals to
enteritis caused by E. coli.

In Section IV.A evidence was presented that the antimicrobial effects
of mammalian sera can be diminished not only by the addition of exogenous
iron but also by the addition of iron-binding microbial products. Therefore, it should be possible to enhance the development of infections not
only with iron but also with siderophores isolated from microbes or their
spent media. The growth-promoting effect of enterochelin isolated from
spent medium of E. coli was tested in mice infected intraperitoneally with
E. coli [63]. Treatment of mice with the purified catechol produced no
observable effect in uninfected mice, but in infected mice bacteria grew

logarithmically and the animals died within 18 hr. This iron-binding compound, which can remove iron from Tr and promote bacterial growth in inhibitory sera, apparently facilitated the development of the overwhelming infection by performing the same activity in infected mice. This study also showed that traces of a specific antibody inhibited the synthesis of catechol by E. coli. If so, then this would be one of the most effective defense mechanisms a host could develop against iron-dependent parasites.


V. RELATIONSHIP BETWEEN SIDEROPHORES
AND MICROBIAL VIRULENCE

The efficiency with which certain microbes invade and grow in the fluids and tissues of an animal determine their virulence. Since exogenous iron promotes microbial growth in microbiostatic tissues and fluids, this metal can be considered as an enhancing factor of microbial virulence. Knowledge of the mechanism by which iron stimulates microbial growth is rather deficient. It is known that microbes respond to the lack of iron very rapidly with stasis and, therefore, the iron deficiency may inhibit some essential metabolic reactions. There is some evidence to support this possibility. The first observable effect of iron depletion in Y. septica is the inhibition of net RNA synthesis [95,96]. In the absence of iron, cells of E. coli produce abnormal bacterial phenylalanyl-tRNA and cease their multiplication; the addition of iron restores their ability to produce normal phenylalanyl-tRNA and to multiply [97,98]. Also, there is some evidence that iron depletion causes a reduction or a complete absence of certain iron-containing enzymes [99]. At the present time, it is not possible to decide which iron-dependent activity in iron-depleted microbes is responsible for microbiostasis. Most of the available evidence deals with the availability of iron as being crucial to microbial growth and to the development of microbial infections.

If iron were the only factor determining microbial virulence, then the metal should promote, with the same effectiveness, the multiplication of virulent and avirulent strains of bacteria. Recent experimental results show that bacteria differ in the ability to acquire iron in the body of the host. Thus, the ability of Pseudomonas aeruginosa to utilize iron in mice was significantly increased after 16 serial passages of the bacteria in these animals [100]. Since unpassaged bacteria required much more iron for the killing of mice than did bacteria of the passaged culture, it was concluded that the virulence of Ps. aeruginosa increased together with its ability to acquire iron in the infected host by means of siderophores. This leads to the question whether or not bacterial effectiveness in the production of siderophores is synonymous with the degree of virulence. Rogers [63] found that there is a direct relationship between the ability to synthesize catechols and the virulence of various strains of E. coli for mice. This work showed, also, that catechols behaved as virulence factors; the injection of the purified catechol together with E. coli promoted bacterial multiplication and development of disease in mice. Although siderophores per se cause no harm to an animal, they behave as true virulence factors in encouraging microbial growth in an otherwise microbiostatic environment of animal tissues and fluids.

The study of the virulence of M. tuberculosis showed that virulent and avirulent bacilli contain similar amounts of mycobactin on their surfaces [65]. The hydrophobic nature of mycobactin and its close association with the bacterial wall are factors which may account partly for the difference between the findings of Rogers and of ourselves. Experiments with various surfactants which remove mycobactin from the bacterial surface showed that virulent and avirulent mycobacteria differ in the firmness with which mycobactin is associated with the mycobactin, and their growth is inhibited at much lower concentrations of the surfactant than is the growth of virulent bacilli. Our results suggested that the retention of mycobactin by virulent cells is attributable to the high content of lipid present in the walls of the bacteria. These findings indicate that the resistance of virulent bacilli to the mycobactin-removing activity of the surfactants is an indicator of the ability of the bacilli to multiply in surfactant-containing fluids and cells of the infected host. In distinction to a direct correlation between the production of water-soluble siderophores and virulence in fast-growing bacteria, there is no correlation between the amount of cell-wall-associated mycobactin and virulence in slow-growing tubercle bacilli; in the case of these bacteria, the virulence seems to be expressed by the firmness of the association between mycobactin and the wax of bacterial cell wall.

VI.   ROLE OF IRON IN NUTRITIONAL IMMUNITY
        AND MICROBIAL PARASITISM

The multiplication of microbial parasites in animals depends upon their ability to obtain iron, which is an essential metal in metabolic processes of living cells. Although limited amounts of iron are present in the fluids and tissues of animals, the metal is unavailable for microbial utilization because it is firmly bound to iron-binding proteins of the host. Whether or not an infection can develop at the portal of microbial entry depends in part upon the ability of parasites to utilize iron bound to Tr, lactoferrin, and ferritin. Various studies have shown that microbes exposed to protein-iron complexes at the physiological pH values of animal fluids cannot remove the metal from the complexes, and hence they cease their multiplication. Thus, microbes cannot overcome the iron-deficiency in the animal body unless they create changes or produce substances which could make the iron of Tr-iron complexes available to them.

The iron starvation of microbes in animal tissues can be alleviated by the localized area of inflammation and the production of microbial iron-

binding compounds, i.e., siderophores. The formation of organic acids in
a toxin- or endotoxin-induced site of inflammation lowers the pH to a point
where the bonds between and Tr and iron are weakened, and some iron is
transferred to iron-binding organic acids. These organic acid-iron complexes provide microbes with a small amount of the metal which stimulates
production of large amounts of siderophores. These strong iron-binding
microbial products diffuse into surrounding tissues, remove iron from
iron-containing proteins, and provide the metal for utilization of the invading parasites. It seems that siderophores have some degree of microbial
specificity; while some can provide iron to some heterologous microbes,
others are restricted to the homologous organisms only. Experimental
evidence suggests that effectiveness in the production of siderophores is
related to microbial virulence; if so, then the rate of siderophore production could be used as a measure of microbial ability to invade animal tissues and to cause infectious disease.

In the presence of siderophore-provided iron, microbes multiply and,
if not inhibited by host responses to the infection, invade tissues around
the focus of original entry. Once microbes reach large numbers in host
tissues, then their products, especially endotoxins, elicit fever and localized and generalized hypoferremia. Although bacteria grow in the presence
of iron at temperatures of $40^\circ$ to $42^\circ C$, they produce little or no siderophores. The absence of siderophores stops the microbial invasion of surrounding tissues. The granules of dead leukocytes in microbial inflammation
liberate lactoferrin, which binds iron even at the existing acidic pH, transports the metal into monocytes, and induces localized hypoferremia. The
generalized hypoferremia in humoral fluids of infected animals is elicited
by bacterial endotoxins; the exact mechanism responsible for this is not
known, but the iron-neutralizable resistance of endotoxin-treated animals
to microbial infections indicates that the hypoferremia should be considered
as a protective response of the animal body to microbial invasion.

The native or acquired abilities of the host to limit the availability of
iron for microbial utilization has been defined as nutritional immunity. It
is not known whether other essential metals or organic nutrilites play a
role in nutritional immunity, but the dependence of host and parasite on the
supply of the same growth-essential compounds suggests that the competition for iron is not an exception but rather an example of the very complicated
nutritional relationship which exists between hosts and their parasites.

VII.  RECENT DEVELOPMENTS
      AND CONCLUDING COMMENTS

During the past two years, most attention in our laboratory was given to
study the mechanism of the acquisition of iron by serum-exposed avirulent

and virulent bacteria. Several investigators suggested the existence of a
close correlation between bacterial virulence and bacterial ability to acquire iron. It has been observed that bacteria subjected to many passages
through animals become virulent and require less iron for the initiation of
infection than do avirulent bacteria [100]. Bacterial virulence has been
attributed to prolific production of siderophores [63] and to the ability of
virulent cells to acquire iron in vivo [101]. Since siderophores are products of growing bacteria, their production may help in the spread of the
infection but not in its initiation. We have found that highly virulent bacteria can grow in mammalian sera without neutralizing them for the growth
of avirulent cells [28]. This finding demonstrates that the ability of virulent bacteria to overcome antibacterial effect of serum should be attributed
to the properties of the bacterial cell itself. A similar conclusion has been
made by Miles and Khimji [102] who decided that "the capacity to produce
chelators is necessary for virulence, but does not determine it."

The most likely possibility that could explain the ability of virulent
bacteria to grow in mammalian sera is the possession of large stores of
intracellular iron in virulent but not in avirulent cells. Determinations of
iron in bacterial ashes showed that avirulent and virulent bacteria contain
similar quantities of iron [28,103]. Also, neither virulent nor avirulent
bacteria can multiply in egg white which is void of iron; both kinds of cells
grow in iron-supplemented egg white. These findings suggested that virulent bacteria have the ability to acquire iron in serum, and therefore the
role of bacterial outer membranes in the acquisition of iron by serum-
exposed bacteria has been investigated in our laboratory. This study
showed the virulent bacteria can obtain iron directly from iron-saturated
Tr by the use of LPS [28]. Trichloroacetic acid-extracted LPS was able
to remove iron from Tr and promote bacterial growth on antibacterial
serum-agar medium. Virulent cells possess larger amounts of much more
effective LPS than do avirulent cells.

The study of bacterial virulence and bacterial ability to grow in serum
has been facilitated by the use of Salmonella typhimurium strains which
are genetically identical except for the depth of lesions in LPS-sugar
chain [104]. Using these strains Lyman et al. [105] showed that decreasing amounts of sugar in LPS are associated with the progressive loss of
bacterial virulence for mice. Results in our laboratory showed that the
virulence of smooth and LPS-sugar rich rough strains can be increased by
the injection of the infected mice with iron; however, iron treatment cannot
promote infection in mice injected with LPS-sugar deficient rough strains
[106]. In addition, these results showed that pronounced differences exist
among LPS-deficient strains in the utilization of enterochelin-iron complexes. Mutants containing decreasing amounts of LPS-sugar progressively
lost the ability to use enterochelin-iron complexes and, finally, LPS-sugar
lacking mutants failed to grow around wells made in antibacterial serum-
agar medium and charged with enterochelin (Fig. 6-7). The growth of

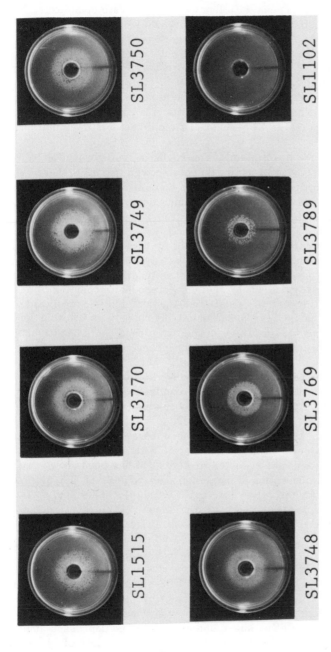

FIG. 6-7  The growth of smooth (SL1515 and SL3770) and rough LPS-defective strains of S. typhimurium around wells made in antibacterial serum-agar medium and charged with 2 μg of enterochelin. Rough strains are arranged according to decreasing amount of sugar in LPS (strain SL1102 has no LPS-sugar).

serum-exposed smooth and rough strains with superficial or deep LPS
lesions was promoted with equal efficiency by iron-citrate complexes.
This difference between the use of iron-citrate and iron-enterochelin complexes indicates that sugar in LPS molecule controls the absorption of
siderophore-iron complexes.  The failure of enterochelin-iron complexes
to promote the growth of LPS-sugar deficient mutants is supported by observations which showed that the treatment of S. typhimurium-infected
mice with enterobactin or iron-binding pacifarin significantly increased
the survivorship of the animals [6,107].  Thus, the presence of siderophores may be detrimental to LPS-sugar-deficient bacteria in animal body
or in serum.

Further study of the mechanism of enterochelin-iron absorption by
S. typhimurium strains with LPS core defects showed that LPS-sugar-
lacking bacteria can utilize siderophore-iron complexes in the absence but
not in the presence of Tr.  Various experiments demonstrated that enterochelin does not remove iron from iron-saturated Tr but combines with it
and forms Tr-iron-enterochelin complex [106]; this complex can be used
as a source of iron by LPS-sugar-rich but not by LPS-sugar-poor strains.
Preliminary results showed that heat-killed bacteria rich in LPS-sugar
adsorb more of Tr-iron-enterochelin complexes than LPS-sugar poor bacteria.  It is very unlikely that whole Tr-iron-enterochelin complex is taken
into the bacterial cell; possibility exists that the complex is hydrolyzed on
the bacterial surface and only iron-saturated siderophore is transported
into the cell.  If this is so, then the bacterial virulence could be attributed
to an enzyme which hydrolyzes Tr-iron-siderophore complexes on the
virulent but not on the avirulent bacterial cell.  Much has to be learned
about the role of these complexes in animal body.  The possibility that
mammalian cells might be unable to obtain iron from Tr-iron-siderophore
complexes may open a new approach not only to study of the acquisition of
iron by normal cells but also by tumor cells.

This discussion of the acquisition of siderophore-iron complexes by
virulent and avirulent cells would be incomplete without the consideration
of the role of outer membrane receptor proteins.  It has been shown that
siderophores prevent adsorption of bacteriophages to cells of S. typhimurium [108] and of colicins to cells of E. coli [109,110].  This implies that
siderophores, phages, and colicins have a common cell envelope receptor;
there is some evidence that this receptor is sensitive to protease treatment and was called protein 7 [111].  These findings suggest that the
receptor of enterochelin-iron complexes on cells of S. typhimurium is a
protein and not LPS-sugar.  In view of our results with the LPS-sugar
deficient strains, the possibility exists that LPS-sugar controls the placing
or amount of protein 7 in the outer membrane of bacterial cell.  That this
may be the case is suggested by findings which showed the existence of
close correlation between short sugar chains in LPS and depressed amounts
of outer membrane proteins [112].  Irrespective of the nature of

siderophore-iron receptors, the results of our recent study demonstrate that LPS of outer membranes of virulent bacteria facilitates the growth of serum-exposed bacteria by providing them with iron of Tr-iron complexes. In vivo and in vitro experiments demonstrated the importance of LPS-sugar in the acquisition of iron from Tr-iron-siderophore complexes. These properties of the bacterial outer membrane facilitate the growth of bacteria in animal tissues, which is the first and main prerequisite for bacterial virulence.

## ACKNOWLEDGMENT

I thank Dr. Sidney Raffel for his critical review of this manuscript.

## REFERENCES

1. I. Kochan, Curr. Top. Microbiol. Immunol., 60, 1 (1973).
2. E. D. Weinberg, JAMA, 231, 39 (1975).
3. M. B. Lurie, Resistance to Tuberculosis, Harvard University Press, Cambridge, Massachusetts, 1964, Chap. 6.
4. E. D. Weinberg, Bact. Rev., 30, 136 (1966).
5. I. Kochan, in Microbiology — 1974 (D. Schlessinger, ed.), American Society for Microbiology, Washington, D. C., 1975, pp. 273-288.
6. E. J. Wawszkiewicz and H. A. Schneider, Infect. Immun., 11, 69 (1975).
7. I. Kochan, D. L. Cahall, and C. A. Golden, Infect. Immun., 4, 130 (1971).
8. I. Kochan and C. A. Golden, Infect. Immun., 8, 388 (1973).
9. M. Shilo, Ann. Rev. Microbiol., 13, 255 (1959).
10. A. L. Schade and L. Caroline, Science, 104, 340 (1946).
11. R. Aasa, B. G. Malmstrom, P. Saltman, and T. Vanngard, Biochim. Biophys. Acta, 75, 203 (1963).
12. S. K. Komatsu and R. E. Feeney, Biochemistry, 6, 1136 (1967).
13. A. G. Bearn and W. C. Parker, in Iron Metabolism (F. Gross, ed.), Springer-Verlag, Berlin, 1964.
14. C. B. Laurell, in The Plasma Proteins (F. W. Putnam, ed.), Vol. 1, Academic Press, New York, 1960, Chap. 10.
15. E. D. Weinberg, J. Infect. Dis., 124, 401 (1971).
16. A. L. Schade, Biochem. Z., 338, 140 (1963).
17. L. Caroline, C. L. Taschdjian, P. J. Kozinn, and A. L. Schade, J. Invest. Dermatol., 42, 415 (1964).
18. L. Caroline, F. Rosner, and P. J. Kozinn, Blood, 34, 441 (1969).
19. L. Caroline, P. J. Kozinn, F. Feldman, F. H. Stiefel, and H. Lichtman, Ann. N.Y. Acad. Sci., 165, 148 (1969).

20. H. McFarlane, S. Reddy, K. J. Adcock, H. Adeshina, A. R. Cooke, and J. Akene, Brit. Med. J., 4, 268 (1970).

21. J. W. Landou, N. Dabrowa, M. D. Newcomer, and J. R. Rowe, J. Invest. Derm., 43, 473 (1964).

22. I. Kochan, J. Infect. Dis., 119, 11 (1969).

23. I. Kochan, C. A. Golden, and J. A. Bukovic, J. Bacteriol., 100, 64 (1969).

24. J. J. Bullen and H. J. Rogers, Nature, 224, 380 (1969).

25. J. Fletcher, Immunol., 20, 493 (1971).

26. S. Jackson and B. C. Morris, Brit. J. Exp. Pathol., 42, 363 (1961).

27. J. J. Bullen, H. J. Rogers, and J. E. Lewin, Immunol., 20, 391 (1971).

28. I. Kochan, J. T. Kvach, and T. I. Wiles, J. Infect. Dis., in press (April 1977).

29. D. Levi, Brit. Med. J., 1, 963 (1941).

30. M. B. Alexander, Brit. Med. J., 2, 973 (1948).

31. H. W. Smith, J. Pathol. Bacteriol., 84, 147 (1962).

32. P. L. Masson, J. F. Heremans, J. J. Prignot, and G. Wauters, Thorax, 21, 538 (1966).

33. J. D. Oram and B. Reiter, Biochim. Biophys. Acta, 170, 351 (1968).

34. P. L. Masson, J. F. Heremans, and C. Dive, Clin. Chim. Acta, 14, 735 (1966).

35. P. L. Masson, J. F. Heremans, and E. Schonne, J. Exp. Med., 130, 643 (1969).

36. P. J. Masson and J. F. Héremans, Comp. Biochem. Physiol., 39B, 119 (1971).

37. R. L. King, J. R. Luick, I. I. Litman, W. G. Jennings, and W. L. Dunkley, J. Dairy Sci., 42, 780 (1959).

38. M. L. Groves, J. Am. Chem. Soc., 82, 3345 (1960).

39. M. Baggiolini, C. de Duve, P. L. Masson, and J. F. Heremans, J. Exp. Med., 131, 559 (1970).

40. S. Mason, Arch. Dis. Child., 37, 387 (1962).

41. C. H. Kirkpatrick, I. Green, R. R. Rich, and A. L. Schade, J. Infect. Dis., 124, 539 (1971).

42. J. J. Bullen, H. J. Rogers, and L. C. Leigh, Brit. Med. J., 1, 69 (1972).

43. B. B. Paul and A. J. Sbarra, Biochim. Biophys. Acta, 156, 168 (1968).

44. B. B. Paul, A. A. Jacobs, R. R. Strauss, and A. J. Sbarra, Infect. Immun., 2, 414 (1970).

45. M. J. Allison, P. Zappasodi, and M. B. Lurie, Amer. Rev. Resp. Dis., 84, 364 (1961).

46. M. J. Cline, Infect. Immun., 2, 156 (1970).

47. T. E. Miller, Infect. Immun., 3, 390 (1971).

48. I. Kochan and L. Smith, J. Immunol., 94, 220 (1965).

49. W. C. Wheeler and J. H. Hanks, J. Bacteriol., 89, 889 (1965).
50. J. H. Hanks, Bacteriol. Rev., 30, 114 (1966).
51. M. S. Leffell and J. K. Spitznagel, Infect. Immun., 6, 761 (1972).
52. J. L. Van Snick, P. L. Masson, and J. F. Heremans, J. Exp. Med., 140, 1068 (1974).
53. G. E. Cartwright, M. A. Lauritsen, and P. J. Jones, J. Clin. Invest., 25, 65 (1946).
54. C. P. Sword, J. Bacteriol., 92, 536 (1966).
55. R. S. Pekarek, K. A. Bostian, P. J. Bartellony, F. M. Calia, and W. R. Beisel, Am. J. Med. Sci., 258, 14 (1969).
56. P. J. Baker and J. B. Wilson, J. Bacteriol., 90, 903 (1965).
57. R. F. Kampschmidt and H. Upchurch, Am. J. Physiol., 216, 1287 (1969).
58. R. S. Pekarek, R. W. Wannemacher, Jr., and W. R. Beisel, Proc. Soc. Exp. Biol. Med., 140, 685 (1972).
59. R. F. Kampschmidt, Ann. Okla. Acad. Sci., 4, 62 (1974).
60. I. Kochan and M. Berendt, J. Infect. Dis., 129, 696 (1974).
61. I. Kochan and S. Raffel, J. Immunol., 84, 374 (1960).
62. J. J. Bullen, A. B. Wilson, G. H. Cushnie, and H. J. Rogers, Immunol., 14, 889 (1968).
63. H. J. Rogers, Infect. Immun., 7, 445 (1973).
64. R. J. Patterson and G. P. Youmans, Infect. Immun., 1, 30 (1970).
65. C. A. Golden, I. Kochan, and D. R. Spriggs, Infect. Immun., 9, 34 (1974).
66. I. Kochan, N. R. Pellis, and D. G. Pfohl, Infect. Immun., 6, 142 (1972).
67. I. Kochan and C. A. Golden, Infect. Immun., 8, 388 (1973).
68. I. Kochan and C. A. Golden, Infect. Immun., 9, 249 (1974).
69. J. B. Neilands, Bacteriol. Rev., 21, 101 (1957).
70. C. E. Lankford, CRC Crit. Rev. Microbiol., p. 273 (March 1973).
71. G. A. Snow, Bacteriol. Rev., 34, 99 (1970).
72. J. R. Pollack and J. B. Neilands, Biochem. Biophys. Res. Commun., 38, 635 (1970).
73. I. G. O'Brien and F. Gibson, Biochim. Biophys. Acta, 215, 393 (1970).
74. I. Kochan, N. R. Pellis, and C. A. Golden, Infect. Immun., 3, 553 (1971).
75. C. Ratledge and B. J. Marshall, Biochim. Biophys. Acta, 279, 58 (1972).
76. A. D. Antoine and N. E. Morrison, J. Bacteriol., 95, 245 (1968).
77. C. Ratledge and M. J. Hall, J. Gen. Microbiol., 72, 143 (1972).
78. I. Kochan and J. T. Kvach, unpublished results (1975).
79. I. Kochan and C. B. Rogerson, unpublished results (1975).
80. J. A. Garibaldi, J. Bacteriol., 110, 262 (1972).
81. T. D. Wilkins and C. E. Lankford, J. Infect. Dis., 121, 129 (1970).

82. H. J. Rogers, J. J. Bullen, and G. H. Cushnie, Immunol., 19, 521 (1970).

83. S. Jackson and T. W. Burrows, Brit. J. Exp. Path., 37, 577 (1956).

84. C. M. Martin, J. H. Handl, and M. Finland, J. Infect. Dis., 112, 158 (1963).

85. G. C. Chandlee and G. M. Fukui, Bacteriol. Proc., p. 45 (1965).

86. A. Wake, H. Morita, and M. Yamamoto, Japan J. Med. Sci. Biol., 25, 75 (1972).

87. J. J. Bullen, L. C. Leigh, and H. J. Rogers, Immunol., 15, 581 (1968).

88. J. J. Bullen, G. H. Cushnie, and H. J. Rogers, Immunol., 12, 303 (1967).

89. G. P. Gladstone and E. Walton, Nature, 177, 526 (1956).

90. T. W. Burrows, Curr. Top. Microbiol. Immunol., 37, 59 (1963).

91. A. Wake, M. Yamamoto, and H. Morita, Japan. J. Med. Sci. Biol., 27, 229 (1974).

92. D. J. Purifoy and C. E. Lankford, Texas Rep. Biol. Med., 23, 637 (1965).

93. P. L. Ingram and R. Lowell, Vet. Record, 72, 1183 (1960).

94. C. L. Bullen and A. T. Willis, Brit. Med. J., 3, 338 (1971).

95. E. Griffiths, Nature New Biol., 232, 89 (1971).

96. E. Griffiths, Eur. J. Biochem., 23, 69 (1971).

97. F. O. Wettstein and G. S. Stent, J. Mol. Biol., 38, 25 (1968).

98. A. H. Rosenberg and M. L. Gefter, J. Mol. Biol., 46, 581 (1969).

99. W. S. Waring and C. H. Werkman, Arch. Biochem., 4, 75 (1944).

100. C. M. Forsberg and J. J. Bullen, J. Clin. Pathol., 25, 65 (1972).

101. S. M. Payne and R. A. Finkelstein, Infect. Immun., 12, 1313 (1975).

102. A. A. Miles and P. L. Khimji, J. Med. Microbiol., 8, 477 (1975).

103. I. Kochan, Advan. Chem. Ser. Amer. Chem. Soc., Washington, D. C. (1977).

104. T. T. Kuo and B. A. D. Stocker, J. Bacteriol., 112, 48 (1972).

105. M. B. Lyman, J. P. Steward, and R. J. Roantree, Infect. Immun., 23, 1539 (1976).

106. I. Kochan, M. Mellencamp, J. T. Kvach, and T. I. Wiles, J. Infect. Dis., 135, 623 (1977).

107. E. J. Wawszkiewicz, in Microbiology — 1974 (D. Schlessinger, ed.), Amer. Soc. Microbiol., Washington, D.C., 1975, pp. 299-305.

108. M. Luckey and J. B. Neilands, J. Bacteriol., 127, 1036 (1976).

109. S. K. Guterman, J. Bacteriol., 114, 1217 (1973).

110. R. Wayne, K. Frick, and J. B. Neilands, J. Bacteriol., 126, 7 (1976).

111. M. Inouye in Membrane Biogenesis (A. Tzagoloff, ed.), Plenum Press, New York, 1975, pp. 351-391.

112. G. F. Ames, E. N. Spudich, and H. Nikaido, J. Bacteriol., 117, 406 (1974).

Chapter 7

MINERAL ELEMENT CONTROL OF
MICROBIAL SECONDARY METABOLISM

Eugene D. Weinberg

Department of Microbiology and Program in Medical Sciences
Indiana University
Bloomington, Indiana

So narrow is the zone (of iron concentration) in which (diphtheria)
toxin is obtained and so sharp the peak of maximal production
that this single uncontrolled factor must have played a greater
role in any previous experiments than specific conditions sup-
posedly under investigation. [From Ref. 1.]

Without a doubt, it is possible to alter the pigment excretion of
any fungus seriously by trace element nutrition. [From Ref. 2,
p. 270.]

Sporulation effects are among the most conspicuous actions of trace elements. [From Ref. 2, p. 268].

## I.  INTRODUCTION

Microbial cells that stop dividing produce synthetases that catalyze transformation of unused substances into biochemically bizarre molecules called secondary metabolites [3,4].  The process of secondary metabolism occurs also in plants and, to a lesser extent, in animals [5].

Substrates of secondary metabolism consist of unused pools of such primary metabolites as acetate, pyruvate, malonate, mevalonate, shikimate, prephenate, amino acids, and purines.  These are packaged by the synthetases into a very wide diversity of molecules.  The latter are highly restricted in taxonomic distribution; however, because of variable yields in strains of taxonomic units, they are not dependable markers in classification schemes.  Examples of classes of organic compounds that contain secondary metabolites are listed in Table 7-1.  In addition to these low-molecular-weight materials, bacterial protein toxins possess some attributes of secondary substances [4].

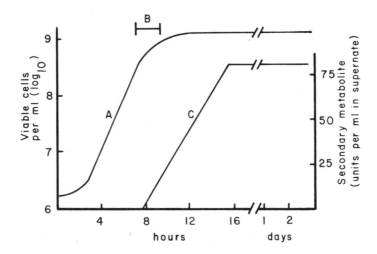

FIG. 7-1  Diagrammatic representation of kinetics of growth and secondary metabolism in a bacterial batch culture:  (A) viable cells; (B) approximate period of transcription and translation of secondary metabolic synthetases; (C) secondary metabolite.

TABLE 7-1 Examples of Classes of Organic Compounds in Which Secondary Metabolites are Found

| | | |
|---|---|---|
| Amino sugars | Lactones | Pyridines |
| Anthocyanins | Macrolides | Pyrones |
| Anthraquinones | Naphthalenes | Pyrroles |
| Aziridines | Naphthaquinones | Pyrrolidones |
| Benzoquinones | Nitriles | Pyrrolines |
| Coumarins | Nucleosides | Pyrrolizines |
| Diazines | Oligopeptides | Quinolines |
| Epoxides | Perylenes | Quinolinols |
| Ergoline alkaloids | Phenazines | Quinones |
| Flavonoids | Phenoxazinones | Salicylates |
| Furans | Phthaldehydes | Terpenoids |
| Glutaramides | Piperazines | Tetracyclines |
| Glycopeptides | Polyacetylenes | Tetronic acids |
| Glycosides | Polyenes | Triazines |
| Hydroxyamines | Polypeptides | Tropolones |
| Indole derivatives | Pyrazines | |

The temporal relationship between the kinetics of cell growth and secondary metabolism in batch cultures is represented diagrammatically in Figure 7-1. The overall process of secondary metabolism includes (a) derepression of the appropriate portion of the genome to obtain transcription and translation of specific synthetases and (b) subsequent activity of the latter for a brief time in the life of the culture. Note that synthetase formation occurs during the period of transition from the exponential to the stationary phase of the growth cycle. The products of secondary metabolism then begin to accumulate at a nonexponential linear rate. The duration of the accumulation phase in cell cultures ranges between one-half to slightly more than twice the time required for growth. Kinetic data are not yet available for the process in plant or animal cells contained in tissues.

In continuous cell cultures, secondary metabolism cannot occur unless the rate of dilution is lowered so as to enable some of the cells to stop dividing. In natural environments, there occur complex series of events that possess features of both batch and continuous cultures. There is abundant evidence that secondary metabolites are produced in nature, and it is assumed that the cells responsible for such syntheses are those that have recently stopped increasing in number.

In batch cultures, the duration of production of secondary metabolites is considerably less than the longevity of cells in the respective cultures. The reason for the abrupt cessation of synthetase activity is not clear. Conceivably, cessation might occur because of initiation of differentiation with concomitant degradation of synthetases; however, cells that do not differentiate likewise abruptly terminate the phase of secondary metabolism. Perhaps continued formation of synthetase molecules does not occur because of end-product repression. In any case, extended synthetase activity presumably is not necessary because the pools of primary metabolic substrates that had accumulated when growth ceased apparently soon become exhausted.


## II.  PROPOSED FUNCTIONS OF SECONDARY METABOLISM

The proposed functions of secondary metabolism fall into three groups: (a) general functions that apply to all secondary metabolic processes; (b) special functions in which some secondary metabolites might affect differentiation of the producer cells; and (c) special functions in which some secondary metabolites might affect growth, health, or behavior of non-producer cells [6].


### A.  General Functions That Apply to All Secondary Metabolic Processes

An early proposal in this category was the suggestion that secondary metabolism is an evolutionary relic. According to this idea, the many thousands of secondary products had specific functions in past eons but are now nearly all obsolete. However, secondary metabolism occupies a prominent part of the life cycle of most plant and microbial cells; thus it is unlikely that it would have been retained by present forms of life were it merely a relic.

A second proposed general function for all secondary metabolites is that they might serve as reserve food-storage materials. This suggestion, too, is improbable, inasmuch as secondary products are often excreted as they are formed and in many cases are not catabolized by cells of the producer species. In contrast, authentic food-storage materials such as polyglucoses and poly-$\beta$-hydroxybutyric acid are anhydrous polymers, have broad taxonomic distribution, and are retained within and digested by producer cells.

A third proposed general function is that secondary materials are products of catabolism of cellular macromolecules that are released during death and autolysis of producer cells. This suggestion has been disproved by observations from several laboratories that secondary substances are formed de novo by living rather than dying cells and that autolysis does not occur during the secondary process. Moreover, many secondary metabolites can be formed in cell-free systems provided that the appropriate low-molecular-weight precursors and the correct synthetases are supplied.

The fourth and most probable of the proposed general functions of secondary metabolism is that the process is one of detoxification. The most dangerous phase in the growth cycle in regard to accumulation of potentially poisonous pools of primary metabolites and intermediates would be during the time that cell growth is slowing as nutrients become limiting; this is precisely when secondary metabolism begins. Moreover, since animal tissues have more efficient ways of excreting potentially poisonous primary metabolites than do plant tissues, dependence on secondary metabolism in animals would be expected to be much less than in plants; and this is indeed the case. For example, plants produce more than 5,000 different alkaloids; animals only 30 to 50 [5].

If the detoxification proposal is correct, evidence must be obtained that unused pools of primary metabolites or intermediates are indeed toxic to the producer cells. Direct evidence is not yet available. However, it has been shown that cells that are prevented from engaging in secondary metabolism die quickly. In contrast, cells that are permitted to complete the secondary process remain viable for many weeks or months [4,7,8].

B.  Special Functions That Affect
    Producer Cells

Inasmuch as secondary metabolites are formed just prior to the onset of differentiation, some investigators have suggested that such secondary compounds as the polypeptide antibiotics of <u>Bacillus</u> might function to control the ordered sequence of events that occurs during differentiation [e.g., 9]. In this case, the cytotoxic properties of these metabolites are believed to be gratuitous. It is proposed that selection has eliminated "functionless" peptides and that the quantity of secondary metabolites required for differentiation is only a few molecules per cell. As a corollary, high yielding antibiotic producer strains are considered to be simply derepressed mutants.

The incompatibility between high yields of secondary metabolites and ability to differentiate is well established [e.g., 6,10]. However, differentiation does occur in mutant strains that fail to produce detectable levels of secondary products [11,12]. Also, secondary metabolism occurs in procaryotes that are unable to differentiate. Nevertheless, secondary metabolism and differentiation are controlled by identical environmental factors (discussed in Section III). Many of the structural components of differentiated cells are, in fact, polymers of secondary substances; thus, strains that can make structural use of their "waste" products might have an advantage over organisms unable to do so. Still, of the total number of secondary metabolites, the percentage of compounds structurally useful to producer cells is small and their utility has probably evolved at random.

Although many secondary metabolites are good metal-binding agents, there is little evidence that they might be required by producer cells for either the solubilization, transport, or storage of specific trace elements. Unlike such primary metabolites as microbial siderophores (see Chapters 5 and 6), which are formed in large quantity in iron-deficient environments, the synthesis of secondary metabolites requires concentrations of metals greater than those needed for cell growth. Nevertheless, such antibiotics as bacitracin enhance the toxicity of specific metals and thus, under special circumstances, may function as trace element ionophores [12a, 12b]. A discussion of possible ionophoric roles of antibiotics is contained in Chapters 2 and 3.

C.  Special Functions That
    Affect Nonproducer Cells

A minority of secondary metabolites can affect growth, health, or behavior of nonproducer cells. Included in this category are hormones, pheromones, allelochemics, toxins, and antibiotics [4, 5, 13, 14]. The great diversity of pharmacological activities possessed by various secondary products of microorganisms is illustrated in Table 7-2.

Secondary metabolites of plants and animals that have hormonal action are essential for the organized multicellular existence of those creatures. Likewise, pheromones are needed for communication between individuals of a species; moreover, many allelopathic compounds are known to be useful in interspecies warfare. However, some pharmacologically active secondary metabolites appear to confer no ecological benefit on their producer cells. For example, the formation of potent antimammalian toxins by strains of Aspergillus flavus, Bacillus anthracis, or Clostridium botulinum is unlikely to favor their growth in their natural soil habitat over that of nontoxigenic strains. Moreover, plants toxic to animals do not necessarily enjoy a selective ecological advantage over nontoxic plants. Fortunately for the existence of higher forms of life, the great majority of microbial and plant secondary metabolites are pharmacologically inert.

Nevertheless, more antimicrobial factors are obtained from soil microorganisms than from plants; more alkaloids toxic to herbivorous arthropods and animals are produced by plants than by soil microorganisms. Perhaps this is because the latter might use some secondary metabolites to inhibit other microbial species in the soil habitat, whereas it would be advantageous to plants to inhibit herbivorous creatures [15]. Thus, some

TABLE 7-2 Examples of Pharmacological Activities of Microbial Secondary Metabolites

| | | | |
|---|---|---|---|
| Anabolic | Dermonecrotic | Hypocholesterolemic | Parasympathomimetic |
| Anesthetic | Edematous | Hypoglycemic | Photosensitizing |
| Analeptic | Emetic | Hypotensive | Pyrogenic |
| Anorectic | Erythematous | Hypersensitizing | Sedative |
| Anticoagulant | Estrogenic | Inflammatory | Spasmolytic |
| Antiinfective | Hallucinogenic | Insecticidal | Teleocidal |
| Antiinflammatory | Hemolytic | Leukemogenic | Teratogenic |
| Antilipemic | Hemostatic | Nephrotoxic | Tremorigenic |
| Cardiotoxic | Hepatotoxic | Neurotoxic | Ulcerative |
| Convulsant | Herbicidal | Paralytic | Vasodilatory |

secondary metabolites may have evolved from simply being trivial end products of a vital process into substances that have important coordinating, communicating, or combative functions.

In the case of secondary metabolites that are cytotoxic, producer cells must prevent autotoxicity. In plants, the secondary substances are kept inactive as crystals, glycosides, polymers, or in a reduced state; or they may be sequestered into heartwood [5]. Some toxic substances are discharged from the plant by leaching from the leaf surface, exudation from the roots, or volatilization [5]. In microbial cells, the metabolites often are secreted into the extracellular milieu from which they are not reassimilable; still other secondary materials are inactivated by polymerization and retained within the cells. As indicated previously, some of the polymers may then be incorporated into the structure of such entities as spores or cysts.

III.  ENVIRONMENTAL CONTROL
      OF SECONDARY METABOLISM

Despite the existence of an exceedingly diverse array of both inert and pharmacologically active products of secondary metabolism, the following generalization underlying their formation has long been recognized. Secondary metabolism has considerably narrower tolerances for concentrations of specific trace elements and inorganic phosphate ($P_i$), as well as for ranges of temperature, pH, and redox potential than does growth of the producer cells [4,16]. These restrictions obtain irrespective of whether the synthesis is occurring in a laboratory or a natural system. Generally, the restrictions function during the transitional period between cessation of vegetative growth and the onset of synthetase function. This period includes sequentially (a) repression of that portion of the genome responsible for cell multiplication; (b) derepression of that portion required for transcription of mRNA molecules that code for secondary metabolism; and (c) translation of the messages. After the synthetases have been formed and begin to function, the environmental restrictions generally are no longer manifest. The precise molecular sites at which the restrictions operate are not yet known. Information concerning trace metals, $P_i$, and temperature is surveyed briefly in the following subsections.

A.  Trace Elements

Examples of suppression or stimulation of biosynthesis of secondary metabolites or of differentiation by concentrations of trace metals that have no effect on vegetative growth are contained in Figure 7-2 [1,17-29] and in Table 7-3 [17-22,30-68]. Of the nine biologically important trace metals (atomic numbers 23 through 30 and number 42), the quantities of manganese,

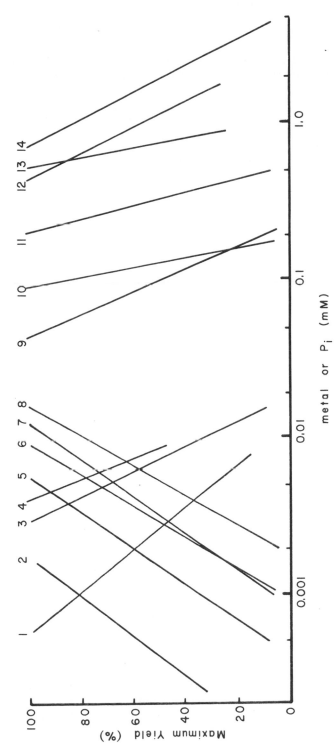

FIG. 7-2 Examples of the linear relationship between the concentration of metals or $P_i$ and the yield of secondary metabolite or process. Key: (1) Fe and fluorescein [17]; (2) Mn and bacitracin [18]; (3) Fe and diphtheria toxin [1]; (4) Fe and Shigella neurotoxin [19]; (5) Zn and penicillin [20]; (6) Fe and pyocyarine [21]; (7) Fe and cyanide [22]; (8) Zn and Fusarium phytotoxins [23]; (9) $P_i$ and prodigiosene [24]; (10) $P_i$ and pyocyanine [21]; (7) Fe and cyanide [25]; (11) $P_i$ and monamycin [26]; (12) $P_i$ and pyocyanine [27]; (13) $P_i$ and candicidin [28]; (14) $P_i$ and vanco-mycin [29].

TABLE 7-3  Secondary Metabolites or Structures Whose Yield is Affected by Concentrations of Trace Metals That Have No Effect on Vegetative Growth in the System Being Examined

| Product or structure | Metal concentration ($\mu$M) | |
|---|---|---|
| | Zinc [a] | Others |
| **I.  Fungi** | | |
| Aflatoxin [30,31] | R-5.0 | I-Cd-10 |
| | I-375 | I-Cu-200 |
| Malformin [32] | I-200 | R-Mn-1.0 |
| | | I-Mn-10 |
| Candida mycelia [33] | I-9.0 | |
| Lysergic acid [34] | R-5.0 | |
| Ergotamine [35] | R-10 | |
| Naphthazarin [23] | R-6.0 | |
| Fusaric acid [36] | R-3.0 | |
| Cynodontin [37] | I-12 | |
| Penicillin [20] | R-1.0 | R-Fe-20 |
| | I-30 | I-Cu-10 |
| Penicillium vesicles [39] | R-120 | |
| | I-300 | |
| Griseofulvin [40] | I-200 | |
| Gentisyl alcohol [41] | R-1.0 | I-Fe-15 |
| Patulin [42] | R-1.0 | R-Fe-15 |
| Pythium oogonia [43] | R-20 | R-Mn-0.3 |
| **II.  Bacteria (other than Bacillus sp.)** | Iron [a] | Others |
| Tetanus toxin [44] | R-30 | |
| Alkyl quinolinols [45] | I-20 | |
| Hydrogen cyanide [22] | R-3.0 | |
| Fluorescein [17] | I-3.0 | |
| Pyocyanine [21] | R-3.0 | |
| Prodigiosene [46] | R-3.0 | |
| | I-20 | |
| Shigella neurotoxin [19] | I-6.0 | |
| Staphylococcus enterotoxin [47] | R-100 | |
| Actinorubin[b] [48] | R-20 | |
| Actinomycin[b] [49] | R-100 | R-Zn-100 |
| Monensin[b] [50] | R-1000 | |
| Neomycin[b] [51] | R-10 | R-Zn-1.0 |
| | I-150 | I-Zn-10 |
| | | I-Mn-100 |

TABLE 7-3 (continued)

| Product or structure | Metal concentration (μM) | |
|---|---|---|
| | Iron[a] | Others |
| Candicidin[b] [52] | R-40 | R-Zn-40 |
| Grisein[b] [53] | R-40 | |
| Streptomycin[b] [54] | R-10 | R-Zn-3.0 |
| | | I-Zn-200 |
| Chloramphenicol[b] [55] | R-20 | R-Zn-20 |
| Mitomycin[b] [56] | R-400 | |
| Kanamycin [b] [57] | I-20 | R-Zn-9 |
| III.  Bacillus sp. | Manganese[a] | Others |
| Anthrax antigen [58] | R-5.0 | |
| | I-20 | |
| Bacitracin [18] | R-0.7 | |
| | I-40 | |
| Transformants [59,60] | R-200. | I-Cu-10 |
| Bacillin [61] | R-100 | |
| D-Glutamyl capsule [62] | R-1.5 | |
| Mycobacillin [63] | R-6.0 | R-Fe-5 |
| Subtilin [64] | R-5.0 | |
| Transfectants [65] | I 200 | |
| Bacteriophage [66] | R-100 | |
| Spores [67,68] | R-5.0 | R-Cu-0.3 |

[a] R, required; I, inhibitory.

[b] Produced by actinomycetes.

iron, and zinc are most critical in secondary metabolism.  Note that to obtain secondary metabolism and differentiation in Bacillus species, the metal whose concentration must be adjusted is manganese.  In contrast, for other bacteria, including actinomycetes, the quantity of iron must be controlled in order to obtain the production of secondary metabolites.  In the case of actinomycetes, as well as fungi, the concentration of zinc is critical for secondary metabolism and for differentiation.  These generalizations are summarized in Figure 7-3.

In many (if not all) cases, the maximum yield of secondary product occurs through a relatively small range of concentrations of the critical

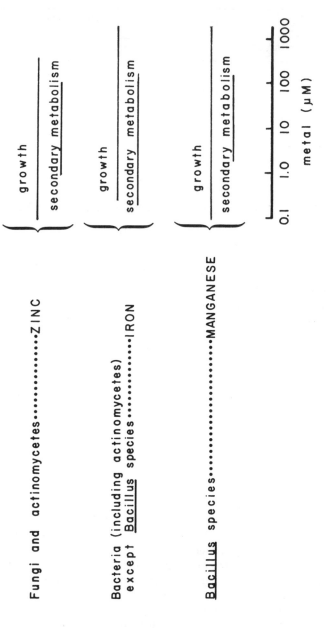

FIG. 7-3  Permissible range of trace metal concentrations that differ for growth and secondary metabolism (summation of Table 7-3).

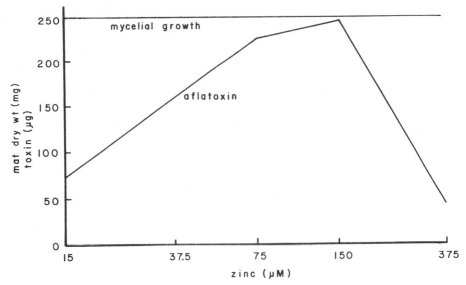

FIG. 7-4   Effect of zinc on growth and toxin production of <u>Aspergillus</u>
<u>parasiticus.</u>  [From Ref. 31.]

metal; an example is shown in Figure 7-4 [31].  The proportion of members of a family of secondary metabolites may be altered with the metal
gradient; in one study, for example, the ratio of aflatoxin $B_1$ to $G_1$ was
6:1, 4:1, and 1:1, respectively, at 15, 75, and 150 μM zinc [31].  In
<u>Penicillium griseofulvum,</u> fulvic acid was formed instead of griseofulvin
when Raulin-Thom rather than Czapek-Dox medium was used [40]; the
former medium contains ten times more zinc than the latter.

The action of the metals is associated with an event prior to the
activity of the synthetases.  For example, when inhibitors of either RNA
or protein synthesis were added to post-logarithmic-phase cultures of
<u>Bacillus licheniformis</u> within 2 hr subsequent to addition of manganese,
bacitracin formation did not occur [18].  In this system, the metal apparently functions in a molecular event that occurs prior to translation of
bacitracin synthetases.  Furthermore, the "correct" quantity of metal
essential for secondary metabolism must be present at a time in the history
of the culture in which the cells are competent to make use of it.  Note,
for example, in Figure 7-5 that iron required for cyanide synthesis was
maximally effective if added up to 4 hr after completion of exponential
growth.  If addition was delayed for 6 hr, secondary metabolism occurred
at a slower rate.  If the essential metal was withheld for 9 hr, very little
cyanide was produced; and for 11 hr, none at all [69].

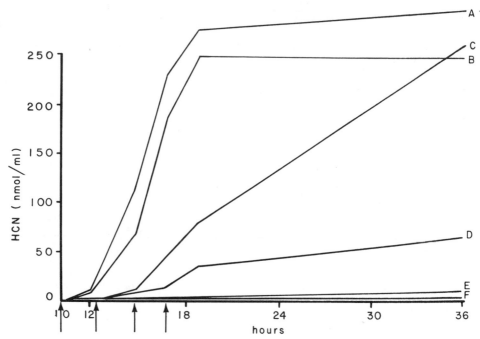

FIG. 7-5  Effect of time of addition of iron on its ability to induce secondary metabolism.  Culture medium with $1.0\,\mu M$ iron was inoculated at zero time with <u>Pseudomonas aeruginosa;</u> growth was completed by 10 hr. Twenty micromolar iron was added to medium at zero time (A), 10 hr, (B) 12.5 hr, (C), 15 hr, (D), 17 hr, (E), never (F).  [From Ref. 69.]

    The quantities of specific trace metals available to microorganisms that parasitize plants or animals or that contaminate food can be the critical determinant in the establishment of a disease condition.  For example, the amount of zinc in the soil of some areas of Madras, India, permits formation of fusarium wilt toxin; in other soil areas, the zinc content allows fungal growth but not toxigenesis [70].  Although the zinc content of grains is sufficient to permit a small amount of aflatoxin formation, the metal is generally complexed with phytic acid.  If the latter is inactivated by autoclaving, more toxin is produced [e.g., 71].

    Vertebrate hosts employ such proteins as transferrin and lactoferrin to withhold iron from microbial invaders (see Chapter 6).  To reinforce this withholding process, invaded hosts promptly become hypoferremic (e.g., Figure 7-6) [72] by halting intestinal assimilation of iron and by increasing liver storage of the metal.  Zinc is also removed from the plasma and stored in the liver during infection.  In contrast, copper levels

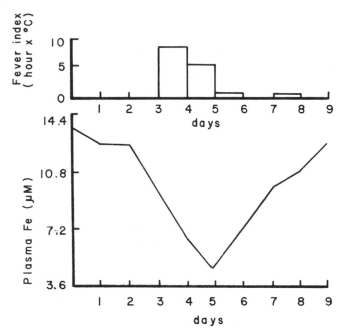

FIG. 7-6  Hypoferremic and hyperthermic response to infection.  Six humans were exposed to <u>Francisella tularensis</u> on day 0.  They developed a
mild illness, as indicated by their mean fever index on days 3 to 8.  Note
that their mean plasma iron level began to decline prior to the onset of
fever and started to return to normal as their clinical symptoms waned on
day 5.  (From Ref. 72.)

are elevated during the disease; as the patient recovers, the plasma quantities of iron, zinc, and copper promptly return to normal.  The extent to
which these shifts in mineral metabolism might suppress synthesis of secondary metabolic toxins in the host is not yet apparent.  Nor is it known if
comparable shifts occur in infected invertebrate hosts.

B.  Inorganic Phosphate

Examples of suppression of biosynthesis of secondary metabolites by concentrations of $P_i$ that permit vegetative growth are contained in Figure
7-7 [7,22,24,26,28,73-91].  The lengths of the arrows indicate the range
of $P_i$ through which increasing suppression of yield occurs; however, the
tips do not necessarily represent 100% suppression.  The very wide range
of $P_i$ concentrations that permit maximal vegetative growth of microbial,

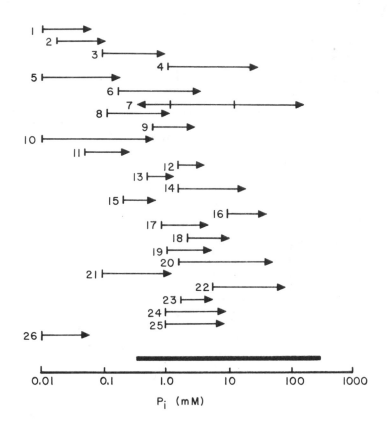

FIG. 7-7 Suppression of secondary metabolism, differentiation, and
culture viability by inorganic phosphate: vertical lines at base of arrows,
highest concentration of $P_i$ that permitted maximum yield of metabolite or
process; arrow tips, lowest concentration of $P_i$ at which minimum yield
was obtained; thick bar, range of concentrations throughout which maximum vegetative growth occurs. Quantities of $P_i$ were those in culture
media at time of inoculation except in No. 10, in which the amounts were
those contained in a resuspension medium. Key: (1) Anabaena sporulation [73]; (2) Bacillus microcycle sporulation [74]; (3) bacitracin [24];
(4) Bacillus sporulation [75]; (5) Caulobacter stalk elongation [76]; (6)
ristomycin [77]; (7) hydrogen cyanide [22,69]; (8) Pseudomonas culture
viability [7]; (9) pyocyanine [27]; (10) phenazine-1-carboxylic acid [78];
(11) prodiogiosene [24]; (12) chlortetracycline [79,80]; (13) candicidin
[28]; (14) streptomycin [81]; (15) monamycin [20]; (16) novobiocin [82];
(17) vancomycin [29]; (18) oxytetracycline [83]; (19) viomycin [84]; (20)
Acanthamoeba encystment [85]; (21) citric acid [86]; (22) aflatoxin [87];
(23) ergoline alkaloids [88]; (24) ergot [89]; (25) mycorrhizal infectivity
[90]; (26) Prymnesium toxin [91].

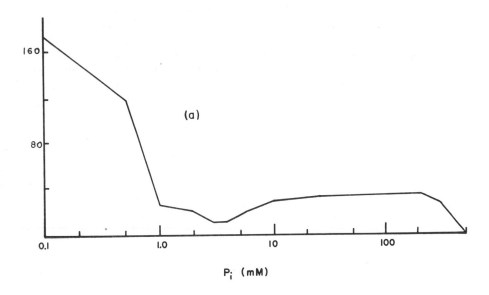

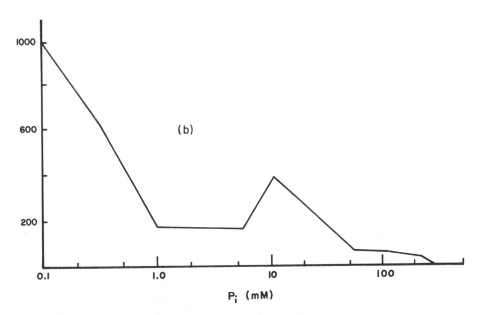

FIG. 7-8  Effect of phosphate concentration on yield of prodigiosene [92] and bacitracin [93].  Ordinate in (a):  micrograms of prodigiosene per milligram of cell protein.  Ordinate in (b):  micrograms of bacitracin per milliliter of supernatant.

plant, or animal cells is indicated by the thickened bar; smaller amounts of growth are obtained at lower and at higher quantities. In 11 of the 26 examples in Figure 7-7, best yield of the secondary metabolite or differentiated cells was obtained at $P_i$ concentrations that are suboptimal for vegetative growth. In one of the examples, that of cyanide, the yield was decreased at quantities of $P_i$ below 1.0 mM as well as above 10 mM [69].

In two systems in which a large number of concentrations of $P_i$ over a very wide range was examined, two peaks of secondary metabolite production were observed. With prodigiosene formed by <u>Serratia marcescens,</u> a high peak at 0.00-0.5 mM and a broader but smaller peak at 10-200 mM $P_i$ were obtained consistently (Fig. 7-8) [92]. With bacitracin formed by <u>Bacillus licheniformis</u>, the high peak occurred at 0.03-0.3 mM and the smaller peak at 10-20 mM $P_i$ [93]. A double peak of bacitracin formation was independently reported by a different laboratory [94]. Qualitative and quantitative differences in yield of members of the prodigiosene chemical family occurred at various concentrations of $P_i$ [92]. In cultures of <u>Streptomyces jamaicensis</u>, monamycin was formed maximally at 0.2 mM $P_i$, whereas an antibiotic of a different family was synthesized at 0.6 mM $P_i$ [26]. Dose response curves of six processes inhibited by $P_i$ are presented in Figure 7-2.

Numerous phosphatases are required in the formation of secondary metabolites, and the synthesis of alkaline phosphatases as well as some acid phosphatases is derepressed in low $P_i$ environments; thus, $P_i$ might function simply by preventing the production of phosphatases [16,95]. Nucleotides such as cyclic AMP, ATP, and 5'-AMP were found to be even more potent than 5.0 mM $P_i$ in inhibiting candicidin synthesis; these materials might be active by maintaining a high energy charge in the cells [96]. Moreover, in the prodigiosene fermentation described previously, alkaline phosphatase activity was depressed by concentrations of $P_i$ greater than 0.05 mM; nevertheless, a small peak of secondary metabolite did appear at 10-150 mM $P_i$ (Fig. 7-8). The ability of $P_i$ to suppress ergot production can be overcome by such inducers of the synthesis of this secondary metabolite as tryptophan and thiotryptophan but not 5-methyltryptophan [96a].

Possibly, high quantities of $P_i$ alter secondary metabolism either by preventing uptake or by modifying storage of trace elements that are critical for the secondary process. We have observed that although 50 mM $P_i$ could not suppress energy-dependent uptake of $^{59}Fe$ and $^{65}Zn$ by cells of <u>S. marcescens</u>, this concentration did alter (as compared with 0.3 mM $P_i$) distribution of iron among soluble intracytoplasmic components [92].

Complex media derived from plant and animal tissues contain a considerable amount of $P_i$. In fermentation media, one of the functions of calcium ions in improving yields of desired secondary metabolites may be that of precipitating excess $P_i$. On the other hand, it might be possible to suppress the formation of undesirable secondary products by enriching environments with additional $P_i$. For example, if toxigenic fungi or bacteria are present in small amounts in stored foods, the safety of the foods might be improved by raising slightly the concentration of $P_i$ so that no toxins could be formed following cessation of microbial growth.

## C.  Temperature

Examples of suppression of biosynthesis of secondary metabolites by temperatures that permit vegetative growth are contained in Figure 7-9 [22, 26, 50, 97-130]. As with $P_i$ in Figure 7-7, the lengths of the arrows indicate the range of temperatures through which increasing suppression of yield occurs; however, the tips of the arrows do not represent 100% suppression. In the polymyxin system, for instance, 90-95% inhibition was obtained at $37^\circ$ C. In the rubratoxin system, the arrow tip at $30^\circ$ C represents 50% inhibition. The amount of inhibition over an entire range of temperatures for cyanide formation is shown in Figure 7-10 [22].

In most of the 37 systems represented in Figure 7-9, only a single temperature was indicated by the authors as permitting the maximum yield of the particular secondary metabolite. In nine cases in this figure, narrow ranges of optimum temperature are shown on the arrows. Probably such narrow ranges occur for each of the other systems, but complete data are not yet available. Generally, excellent vegetative growth occurred at the temperatures represented by the arrow tips (and even beyond in some cases); as may be seen in Figure 7-10, the temperature range shown had no inhibitory effect on growth of the producer cells.

As is true for trace metals and $P_i$, the proportion of members of a family of secondary metabolites can be shifted by changes in temperature. In one study, for example, the ratio of aflatoxin $B_1$ to $G_1$ was 1:3, 2:1, and 1:1, respectively, at $20^\circ$ C, $25^\circ$ C, and $30^\circ$ C [117]. Also, the kinds of secondary metabolites produced may be temperature dependent. For example, in Aspergillus ochraceus, a temperature of $10^\circ$ C to $20^\circ$ C favored penicillic acid synthesis, whereas $28^\circ$ C stimulated ochratoxin A production [131].

As with trace metals and $P_i$, the influence of temperature on secondary metabolism occurs prior to the functioning of the synthetases in most cases. In some systems, three optimal temperatures are recognized: for growth, for synthetase formation, and for synthetase activity. In some fermentations, synthetase activity proceeds efficiently at both the temperature optimal for growth as well as that optimal for synthetase formation. As with trace metals, no unitary molecular mechanism is known that can explain the role of temperature in the production of secondary metabolic synthetases. A reasonable suggestion, however, is that synthesis of the corresponding repressors is the actual temperature-sensitive reaction [132].

Temperature may also control secondary metabolism indirectly by altering uptake of essential trace elements. Such microorganisms as

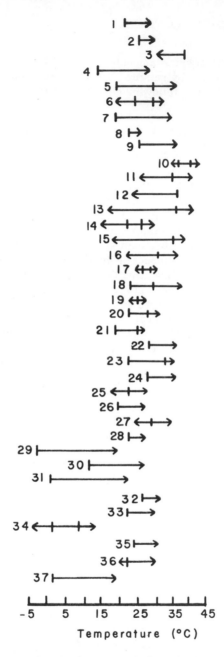

FIG. 7-9 Suppression of secondary metabolism by temperatures that permit vegetative growth: vertical bars, maximal yield of metabolite; arrow tips, minimal yield of metabolite. Key: (1) <u>Anabaena</u> toxin [97]; (2) <u>Aphanizomcron</u> toxin [97]; (3) <u>Bacillus</u> toxin [98]; (4) <u>Bacillus</u> pigment [99]; (5) polymyxin [100]; (6) <u>Clostridium botulinum</u> toxin [101,102]; (7) <u>Listeria</u> hemolysin [103]; (8) <u>Microcystis</u> toxin [97]; (9) plague toxin [104];

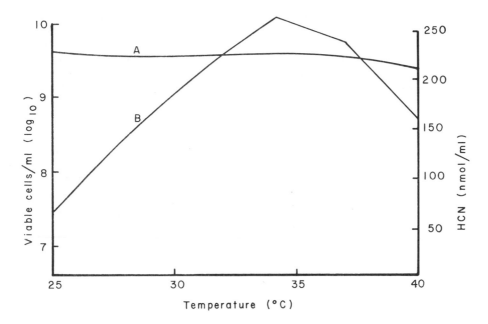

FIG. 7-10  Effect of temperature on growth and secondary metabolism of
Pseudomonas aeruginosa:  (A) viable cells; (B) HCN.  (From Ref. 22.)

Escherichia (cf. Chapter 6), Pseudomonas [133] , Salmonella [134] , and
Ochromonas, Euglena, Saccharomyces and Candida [135, 136] have, for
vegetative growth, elevated trace metal requirements at elevated temperatures.  For example, at 36.9 °C, Salmonella can produce iron-transporting compounds and consequently requires only 0.5 μM of the element.
However at 40.3 °C, detectable levels of the compounds are not formed, and
thus 50 μM iron is needed to obtain growth [134].  Inasmuch as the toler-

---

FIG. 7-9 (continued)

(10) plague virulence antigens [105]; (11) Pseudomonas HCN [22]; pyocyanine [106]; (13) Salmonella virulence antigen [107]; (14) prodigiosene
[108]; Staphylococcus enterotoxin [109]; (16) monensin [50]; (17) streptomycin [110]; (18) viomycin [111]; (19) monamycin [26]; (20) novobiocin
[112]; (21) rifamycin [113]; (22) nebramycin [114]; (23) choleratoxin [115];
(24) riboflavin [116]; (25) aflatoxin [117, 118]; citric acid [119]; (27) mal-
formin [120]; (28) ergoline alkaloids [121]; (29) Fusarium toxin [122];
(30) zearalenone [123]; (31) scirpene toxin [124]; (32) gibberellin [125];
(33) β-carotene [126]; (34) penicillic acid [127]; (35) penicillin [128];
(36) rubratoxin B [129]; (37) tremortin [130].

ance for iron concentration for bacterial secondary metabolism is far narrower than is the tolerance for vegetative growth, the consequences of temperature restriction of synthesis of compounds that accumulate (and possibly of compounds that export) iron should have profound effects on yields
of bacterial secondary substances.

   Potential applications of temperature control of secondary metabolism
remain underdeveloped. In most industrial fermentations, a single temperature somewhere between the optimum for vegetative growth and that
for production of the metabolite is employed. Although many kinds of
warm-blooded hosts respond to a variety of infectious agents by raising
their body temperature, it is not known if the fever can actually depress the
formation of secondary metabolites by the pathogens. The ability of high
temperature to interfere with the microbial assimilation of trace metals
may be useful to potential hosts by depressing both primary and secondary
metabolism of the pathogens. At the same time, the invaded host is also
defending itself from microbial growth and toxigenesis by halting the intestinal assimilation of iron and by removing iron and zinc from the plasma
[137].


IV.  CONCLUSIONS

Secondary metabolites consist of an enormous diversity of natural products, narrowly restricted in taxonomic distribution and without known
function in growth of the producer cells; they are formed for a short period
of time by cells that have recently stopped dividing. Of the many thousands of microbial secondary metabolites that have been characterized
chemically, those of most interest are compounds toxic either to various
microbial species (i.e., antibiotics) or to cells of plants or animals (i.e.,
toxins). The range of concentrations of trace metals and inorganic phosphate as well as of temperature, pH values, and oxygen tension is considerably narrower for efficient production of secondary metabolites and of
differentiation than is the range tolerated for growth of the producer microorganisms. Especially important in both natural and laboratory systems
for secondary metabolism and differentiation are the concentrations of
manganese for species of Bacillus, of iron for species of all other bacterial
groups including actinomycetes, and of zinc for species of actinomycetes
and fungi. The various environmental restrictions generally operate at
the level of formation rather than of activity of the secondary metabolic
synthetases. By manipulation of the restrictions, industrial microbiologists have often been able to enhance yields of desirable secondary metabolites. In contrast, environmental and medical microbiologists have had
less success in attempting to use the restrictions to suppress the yield of
undesirable secondary products in foods, waters, or hosts.

RECENT DEVELOPMENTS

An extensive review of ecological roles of plant secondary metabolites [138], a useful study of environmental control of rubratoxin formation [139], and a function of gramicidin in bacterial sporulation [140] have been published recently.

ACKNOWLEDGMENT

This work was supported in part by National Science Foundation Research grant BMS 75-16753.

REFERENCES

1. J. H. Mueller, J. Immunol., 42, 343 (1941).
2. J. W. Foster, Chemical Activities of Fungi, Academic Press, New York, 1949.
3. J. D. Bu'Lock, Advan. Appl. Microbiol., 3, 293 (1961).
4. E. D. Weinberg, Advan. Microb. Physiol., 4, 1 (1970).
5. M. Luckner, Secondary Metabolism in Plants and Animals, Chapman & Hall, London, 1972.
6. E. D. Weinberg, Persp. Biol. Med., 14, 565 (1971).
7. M. J. Gentry, D. K. Smith, S. F. Schnute, S. L. Werber, and E. D. Weinberg, Microbios., 4, 205 (1971).
8. D. K. Smith, C. D. Benedict, and E. D. Weinberg, Appl. Microbiol., 27, 292 (1974).
9. H. L. Sadoff, in Spores V (H. O. Halverson, R. Hanson, and L. L. Campbell, eds.), American Society for Microbiology, Washington, D.C., 1972, p. 157.
10. G. Sermonti, Genetics of Antibiotic-Producing Microorganisms, Wiley (Interscience), New York, 1969.
11. C. H. Nash and F. M. Huber, Appl. Microbiol., 22, 6 (1971).
12. H. I. Haavik and S. Thomassen, J. Gen. Microbiol., 76, 451 (1973).
12a. E. D. Weinberg, Antibiotics Annual 1958-1959, p. 924 (1959).
12b. H. I. Haavik, J. Gen. Microbiol., 96, 393 (1976).
13. R. H. Whittaker and P. P. Feeny, Science, 171, 757 (1971).
14. R. H. Whittaker, in Chemical Ecology (E. Sondheimer and J. B. Simeone, eds.), Academic Press, New York, 1970.
15. H. G. Floss, J. E. Robbers, and R. F. Heinstein, Rec. Advan. Phytochem., 8, 141 (1974).
16. E. D. Weinberg, Devel. Ind. Microbiol., 15, 70 (1974).
17. J. R. Totter and F. T. Mosely, J. Bacteriol., 65, 45 (1953).
18. E. D. Weinberg and S. M. Tonnis, Appl. Microbiol., 14, 850 (1966).

19. W. E. van Heyningen, Brit. J. Exp. Path., 36, 373 (1955).
20. J. W. Foster, H. B. Woodruff, and L. E. McDaniel, J. Bacteriol., 46, 421 (1943).
21. M. Kurachi, Bull. Inst. Chem. Res. Kyoto Univ., 36, 188 (1958).
22. P. A. Castric, Can. J. Microbiol., 21, 613 (1975).
23. H. Kern, S. Naef-Roth, and F. Ruffner, Phytopathol. Z., 74, 272 (1972).
24. E. D. Weinberg, M. A. Beattie, B. New, C. Peterson, and R. Raelson, Abstr. Ann. Mtg., Amer. Soc. Microbiol., p. 13 (1973).
25. D. K. Smith and E. D. Weinberg, unpublished data.
26. M. J. Hall and C. H. Hassall, Appl. Microbiol., 19, 109 (1970).
27. L. H. Frank and R. D. DeMoss, J. Bacteriol., 77, 776 (1959).
28. C.-M. Liu, L. E. McDaniel, and C. P. Schaffner, Antimicrob. Agents Chemother., 7, 196 (1975).
29. E. P. Mertz and L. E. Doolin, Can. J. Microbiol., 19, 263 (1973).
30. R. I. Mateles and J. C. Adye, Appl. Microbiol., 13, 208 (1965).
31. P. B. Marsh, M. E. Simpson, and M. W. Trucksess, Appl. Microbiol., 30, 52 (1975).
32. S. M. Steenbergen and E. D. Weinberg, Growth, 32, 125 (1968).
33. H. Yamaguchi, J. Gen. Microbiol., 86, 370 (1975).
34. J. P. Rosazza, W. J. Kelleher, and A. E. Schwarting, Appl. Microbiol., 15, 1270 (1967).
35. A. Stoll, A. Brack, A. Hofmann, and H. Kobel, U.S. Patent 2,809,920 (1957).
36. R. Kalyanasundaram and L. Saraswathi-Devi, Nature, 175, 945 (1955).
37. J. P. White and G. T. Johnson, Mycologia, 63, 548 (1971).
38. H. Koffler, S. G. Knight, and W. C. Frazier, J. Bacteriol., 53, 115 (1947).
39. E. L. Sharp and F. G. Smith, Phytopathology, 42, 581 (1952).
40. J. F. Grove, in Antibiotics (D. Gottlieb and P. D. Shaw, eds.), Vol. 2. Springer-Verlag, Berlin, 1967, p. 123.
41. G. Ehrensvärd, Exp. Cell Res. Suppl., 3, 102 (1955).
42. A. Brack, Helv. Chim. Acta, 30, 1 (1947).
43. J. F. Lenny and H. W. Klemner, Nature, 209, 1365 (1966).
44. W. C. Latham, D. F. Bent, and L. Levine, Appl. Microbiol., 10, 146 (1962).
45. F. Wensinck, A. van Dalen, and M. Wedema, Antonie van Leeuwenhoek J. Microbiol. Serol., 33, 73 (1967).
46. W. S. Waring and C. H. Werkman, Arch. Biochem., 1, 425 (1943).
47. E. P. Casman, Publ. Health Rep., Wash., 73, 599 (1958).
48. A. Kelner and H. E. Morton, J. Bacteriol., 53, 695 (1947).
49. E. Katz, P. Pienta, and A. Sivak, Appl. Microbiol., 6, 236 (1958).
50. W. M. Stark, N. C. Knox, and J. E. Westhead, Antimicrob. Agents Chemother., 1967, p. 353 (1968).

51. M. K. Majumdar and S. K. Majumdar, Appl. Microbiol. , 13, 190 (1965).

52. R. F. Acker and H. Lechevalier, Appl. Microbiol. , 2, 152 (1954).

53. D. M. Reynolds and S. A. Waksman, J. Bacteriol. , 55, 739 (1948).

54. C. G. C. Chesters and G. N. Rolinson, J. Gen. Microbiol. , 5, 559 (1951).

55. V. Gallichio, D. Gottlieb, and H. E. Carter, Mycologia, 50, 490 (1958).

56. E. J. Kirsch, in Antibiotics (D. Gottlieb and P. D. Shaw, eds.), Vol. 2, Springer-Verlag, Berlin, 1967, p. 66.

57. K. Basak and S. K. Majumdar, Antimicrob. Agents Chemother. , 8, 391 (1975).

58. G. G. Wright, M. A. Hedberg, and J. B. Slein, J. Immunol. , 72, 263 (1954).

59. C. B. Thorne and H. B. Stull, J. Bacteriol. , 91, 1012 (1966).

60. C. Anagnostopoulus and J. Spizien, J. Bacteriol. , 81, 741 (1961).

61. J. W. Foster and H. B. Woodruff, J. Bacteriol. , 51, 363 (1946).

62. C. G. Leonard, R. D. Housewright, and C. B. Thorne, J. Bacteriol. , 76, 499 (1958).

63. S. K. Majumdar and S. K. Bose, J. Bacteriol. , 79, 564 (1960).

64. E. F. Jansen and D. J. Hirschmann, Arch. Biochem. , 4, 297 (1944).

65. K. F. Bott and G. A. Wilson, Bacteriol. Rev. , 32, 370 (1968).

66. K. Huybers, Ann. Inst. Pasteur, 84, 242 (1953).

67. E. D. Weinberg, Appl. Microbiol. , 12, 436 (1964).

68. B. J. Kolodziej and R. A. Slepecky, Nature, 194, 504 (1962).

69. J. A. Castric, personal communication (1976).

70. T. S. Sadasivan, in Ecology of Soil-Borne Plant Pathogens (K. F. Baker and W. C. Snyder, eds.), University of California Press, Berkeley, 1965, p. 460.

71. S. K. Gupta, and T. A. Venkitasubramanian, Appl. Microbiol. , 29, 834 (1975).

72. R. S. Pekarek, R. W. Wannemacher, Jr. , and W. R. Beisel, Amer. J. Med. Sci. , 258, 14 (1969).

73. C. P. Wolk, Devel. Biol. , 12, 15 (1965).

74. I. MacKechnie and R. S. Hanson, J. Bacteriol. , 95, 355 (1968).

75. K. W. Hutchison and R. S. Hanson, Abstr. Ann. Mtg. Amer. Soc. Microbiol. , p. 50 (1973).

76. J. M. Schmidt and R. Y. Stanier, J. Cell Biol. , 28, 423 (1966).

77. N. S. Egorov, E. G. Toropova, and L. A. Suchkova, Mikrobiologiya, 40, 475 (1971).

78. M. E. Levitch and E. R. Stadtman, Arch. Biochem. Biophys. , 106, 194 (1964).

79. A. Prokofieva-Belgovskaya and L. Popova, J. Gen. Microbiol. , 20, 462 (1959).

80.  J. Doskocil, Z. Hostalek, J. Kasparova, J. Zajicek, and M. Herold,
     Biotechnol. Bioeng., 1, 261 (1959).
81.  H. B. Woodruff and M. Ruger, J. Bacteriol., 56, 315 (1948).
82.  H. Hoeksema and C. G. Smith, Progr. Ind. Microbiol., 3, 91 (1961).
83.  W. A. Zygmunt, Appl. Microbiol., 12, 195 (1964).
84.  L. Pass and K. Raczynska-Bojanowska, Acta Biochim. Polon., 15,
     355 (1968).
85.  R. J. Neff, S. A. Ray, W. F. Benton, and M. Wilbern, in Methods
     in Cell Physiology (D. M. Prescott, ed.), Vol. 1, Academic Press,
     New York, 1964, p. 455.
86.  R. C. Steel, C. P. Lentz, and S. M. Martin, Can. J. Microbiol., 1,
     299 (1954).
87.  R. V. Reddy, L. Viswanathan, and T. A. Venkitasubramanian,
     Appl. Microbiol., 22, 393 (1971).
88.  W. A. Taber and L. C. Vining, Can. J. Microbiol., 9, 1 (1963).
89.  J. E. Robbers, L. W. Robertson, K. M. Hornemann, A. Jindra,
     and H. G. Floss, J. Bacteriol., 112, 791 (1972).
90.  B. Mosse and J. M. Phillips, J. Gen. Microbiol., 69, 157 (1971).
91.  Z. Dafni, S. Ulitzur, and M. Shilo, J. Gen. Microbiol., 70, 199
     (1972).
92.  F. R. Witney, M. L. Failla, and E. D. Weinberg, Appl. Environ.
     Microbiol., 33, 1042 (1977).
93.  M. A. Beattie, In vivo effects of inorganic phosphate on the biosyn-
     thesis of bacitracin in Bacillus licheniformis. MA thesis, Indiana
     University, Bloomington, Indiana, 1973.
94.  H. I. Haavik, J. Gen. Microbiol., 84, 226 (1974).
95.  A. L. Demaine and E. Inamine, Bacteriol. Rev., 34, 1 (1970).
96.  J. F. Martin, L. E. McDaniel, and A. L. Demain, Abstr. Ann.
     Mtg., Amer. Soc. Microbiol., p. 194 (1975).
96a. V. M. Krupinski, J. E. Robbers, and H. G. Floss, J. Bacteriol.,
     125, 158 (1976).
97.  J. H. Gentile, in Microbial Toxins (S. Kadis, A. Ciegler, and
     S. J. Ajl, eds.), Vol. 7, Academic Press, New York, 1971, p. 27.
98.  P. F. Bonventre and C. E. Johnson, in Microbial Toxins (T. C.
     Montie, S. Kadis, and S. J. Ajl, eds.), Vol. 3, Academic Press,
     New York, 1970, p. 145.
99.  R. L. Uffen and E. Canale-Parola, Can. J. Microbiol., 12, 590
     (1965).
100. H. Paulus, in Antibiotics (D. Gottlieb and P. D. Shaw, eds.), Vol. 2,
     Springer-Verlag, Berlin, 1967, p. 254.
101. K. Abrahamsson, B. Gullivar, and N. Molin, Can. J. Microbiol.,
     12, 385 (1966).
102. A. C. Baird-Parker, in The Bacterial Spore (G. W. Gould and A.
     Hurst, eds.), Academic Press, New York, 1969, p. 517.

103. K. F. Girard, A. J. Sbarra, and W. A. Bardawil, J. Bacteriol., 85, 349 (1963).
104. T. C. Montie and S. J. Ajl, in Microbial Toxins (T. C. Montie, S. Kadis, and S. J. Ajl, eds.), Vol. 3, Academic Press, New York, 1970, p. 1.
105. H. B. Naylor, G. M. Fukui, and C. R. McDuff, J. Bacteriol., 81, 649 (1961).
106. M. Kurachi, Bull. Inst. Chem. Res. Kyoto Univ., 36, 163 (1958).
107. A. Jude and P. Nicolle, C. R. Acad. Sci., Paris, 234, 1718 (1952).
108. R. P. Williams, C. L. Gott, S. M. H. Qadri, and R. H. Scott, J. Bacteriol., 106, 438 (1972).
109. G. G. Dietrich, R. J. Watson, and G. J. Silverman, Appl. Microbiol., 24, 561 (1972).
110. D. J. D. Hockenhull, Progr. Ind. Microbiol., 2, 131 (1960).
111. A. H-K. Tam and D. C. Jordan, J. Antibiot., 25, 524 (1972).
112. C. G. Smith, Appl. Microbiol., 4, 232 (1956).
113. P. Sensi and J. E. Thiemann, Progr. Ind. Microbiol., 6, 21 (1967).
114. W. M. Stark, M. M. Hoehn, and N. G. Knox, Antimicrob. Agents Chemother., 1967, p. 314 (1968).
115. S. H. Richardson, J. Bacteriol., 100, 27 (1969).
116. L. Kaplan and A. L. Demaine, in Recent Trends in Yeast Research (D. S. Ahearn, ed.), Georgia State University Press, Atlanta, 1970, p. 137.
117. A. Ciegler, R. E. Peterson, A. A. Lagoda, and H. H. Hall, Appl. Microbiol., 14, 826 (1966).
118. A. Z. Joffe and N. Lisker, Appl. Microbiol., 18, 517 (1969).
119. K. Yamada and H. Hidaka, Agr. Biol. Chem., 28, 876 (1964).
120. M. Yukioka and T. Winnick, Biochim. Biophys. Acta, 119, 614 (1966).
121. M. Abe and S. Yamatodani, Progr. Ind. Microbiol., 5, 204 (1964).
122. A. Z. Joffe, in Microbial Toxins (S. Kadis, A. Ciegler, and S. J. Ajl, eds.), Vol. 7, Academic Press, New York, 1971, p. 139.
123. C. J. Mirocha, C. M. Christensen, and G. H. Nelson, in Microbial Toxins (S. Kadis, A. Ciegler, and S. J. Ajl, eds.), Vol. 7, Academic Press, New York, 1971, p. 107.
124. J. R. Banburg, W. F. Marasas, N. V. Riggs, E. B. Smalley, and F. M. Strong, Biotechnol. Bioeng., 10, 445 (1968).
125. E. G. Jeffreys, Advan. Appl. Microbiol., 13, 283 (1970).
126. M. M. Attwood, Antonie van Leeuwenhoek J. Microbial. Serol., 37, 369 (1971).
127. C. P. Kurtzman and A. Ciegler, Appl. Microbiol., 20, 204 (1970).
128. A. Constantinides, J. L. Spencer, and E. L. Gaden, Jr., Biotechnol. Bioeng., 12, 1081 (1970).
129. A. W. Hayes, E. P. Wyatt, and P. A. King, Appl. Microbiol., 20, 469 (1970).

130. C. T. Hou, A. Ciegler, and C. W. Hesseltine, Appl. Microbiol.,
     21, 1101 (1971).
131. A. Ciegler, Can. J. Microbiol., 18, 631 (1972).
132. A. Demain, Ann. Rev. Microbiol., 26, 369 (1972).
133. J. A. Garibaldi, J. Bacteriol., 105, 1036 (1971).
134. J. A. Garibaldi, J. Bacteriol., 110, 262 (1972).
135. S. H. Hutner, S. Aaronson, H. A. Nathan, H. Baker, S. Scher, and
     A. Cury, in Trace Elements (C. A. Lamb, O. G. Bentley, and J. M.
     Beattie, eds.), Academic Press, New York, 1958, p. 47.
136. I. Roitman, L. R. Travassos, H. P. Azenedo, and A. Cury,
     Sabouraudia, 7, 15 (1969).
137. E. D. Weinberg, Science, 184, 952 (1974).
138. D. A. Levin, Ann. Rev. Ecol. Syst., 7, 121 (1976).
139. C. O. Emeh and E. H. Marth, J. Milk Food Technol., 39, 184
     (1976).
140. P. K. Mukherjee and H. Paulus, Proc. Nat. Acad. Sci. U.S., 74,
     780 (1977).

Chapter 8

MICROBIOLOGY OF METAL TRANSFORMATIONS

Walter A. Konetzka

Department of Microbiology
Indiana University
Bloomington, Indiana

I. INTRODUCTION

Our understanding of the cycling of the major elements which constitute
living matter is fairly well understood on the global level, and even elementary textbooks in microbiology present the so-called nitrogen cycle,
carbon (actually carbon, oxygen, and hydrogen) cycle, the sulfur cycle,
and less frequently a phosphorus cycle, which is actually not a cycle in the
classical sense. All these cycles possess two significant characteristics:
(1) in at least one stage of the transformations the element exists as a gas
under normal atmospheric conditions (e.g., carbon dioxide, methane,
nitrogen, hydrogen sulfide), and (2) the element undergoes a change in its
oxidation state (nitrogen, 0; nitrate, +5; nitrite, +3, $NH_3$, -3).

The apparent exception to these generalizations is phosphorus. This element is metabolized through all living organisms at the same oxidation level, the +5 of $PO_4^{3-}$. There are, however, some hints that phosphorus may undergo changes in oxidation state and a gaseous phase of phosphorus does exist, namely, phosphine ($PH_3$). A number of investigators have shown that a variety of microorganisms are able to utilize phosphite as a sole source of phosphorus [1-3]. One of the bacteria, _Pseudomonas_, possesses a specific inducible NAD-phosphite oxidoreductase [4]. There is also a report that an extreme thermophilic _Bacillus_ is capable of oxidizing hypophosphite [5]. Although there are reports of organisms reducing phosphate to phosphite, hypophosphite, and $PH_3$ [6-8], they have not been confirmed.

The biogeochemical cycling of the so-called minor elements or trace metals [9], namely, those not represented as major components of the major macromolecules, has received much less attention. However, with the recognition that many microbial transformations of these elements can result in the formation of serious environmental pollutants, a renewed interest in the microbial metabolism of these metals has occurred [10]. The interaction of metals with organisms has been viewed by Heinen [11] as occurring on four distinct levels. The first level of interaction refers to the transport of an element in or out of a cell where it functions as a "charge carrier." The second level is that of a "charge transfer interaction," characterized by charge alteration of the element. At the next higher levels, the element interacts with a low-molecular-weight or a macromolecular compound to become a functional part of its reaction partner. This chapter concerns itself with the second level of interaction, the level where the metal functions as an electron donor or acceptor in an oxidation-reduction process and does not become an integral part of the organism; in such a process, large quantities of a particular form of the metal may accumulate in the microbe's environment. If, indeed, changes in oxidation state and the existence of a volatile phase are conditions necessary for the cycling of elements, we should search out such conditions in the cycling of the minor elements to determine where they may or may not exist. The existence of a gaseous state becomes significant when it becomes necessary to rid an environment of a toxic form of a metal. Obviously, not all metals can be examined in a review of moderate length, and therefore our examples must be limited to those where substantial research is available.

## II.  MERCURY

As is so often the case in the history of science, the occurrence of major disasters unfortunately must precede important research. Such is the case with our understanding of the microbial metabolism of mercury that developed only after a series of catastrophes resulted from the ingestion of excessive quantities of methylmercury. These disasters have occurred in widely separated geographical areas, such as Japan, Sweden, Guatemala, Pakistan, and Iraq. As recently as 1972, 459 deaths out of 6,530 reported cases in Iraq resulted from the ingestion of flour and wheat seed treated with methyl- and ethylmercury fungicides [12]. In fact, modern research into the microbial transformations of mercury began in Japan and Sweden during the 1960s soon after the identification of methylmercury as an environmental pollutant in those countries. Progress in our understanding of the role of microorganisms was then extraordinarily rapid, and in 1974 a rather elaborate geochemical cycle was proposed by Wood [13].

The evident involvement of methylmercury with these various catastrophes obviously directed the attention of investigators to the possible role of microorganisms in the methylation of mercury. Numerous studies revealed that mono- and dimethylmercury could be produced in aquarium and lake sediments [14,15], in river sediments [16], in soils [17], and even in human feces [18], and it was surmised that this production was brought about by microorganisms. Not surprisingly, in the complex microbial environments, the demethylation of organomercurials was also reported, but in these early studies little attention was paid to the consequences of this demethylation.

The actual production of dimethylmercury by an extract of Methano-bacterium strain M.o.H. was described by Wood et al. in 1968 [19] and served as the impetus for additional studies implicating microorganisms in the production of methylmercury. The anaerobic sporeformer Clostridium cochlearium was shown to produce considerable quantities of methylmercury from a variety of mercury compounds ($HgCl_2$, $HgI_2$, $HgO$, $HgNO_3$, $Hg(CN)_2$, $Hg(SCN)_2$, and $Hg(CH_3COO)_2$, and the production of methylmercury was stimulated by the addition of vitamin $B_{12}$ [20]. Interestingly, no methylmercury was produced from HgS, a form of mercury certainly present in anaerobic environments.

Because of the importance of methylmercury as the neurotoxic agent in these disasters, perhaps overemphasis was placed on the anaerobic production of the compound. The mechanism of the methylation of mercury both in the test tube and in the environment still is not fully understood [10]. Methylation can occur abiotically [21,22], and the transmethylation by methylcobalamin is inhibited in vitro by cellular proteins and sulfhydryl groups [21], which lends support to the observation by Yamada and Tonomura [20] that methylmercury is not produced from HgS. Moreover, methylation of mercury is brought about by Neurospora crassa, an aerobic organism, in which vitamin $B_{12}$ is not known to be involved [23]. The aerobic production of methylmercury from mercuric chloride has been observed in a large variety of microorganisms: Enterobacter aerogenes, Pseudomonas fluorescens, Mycobacterium phlei, Escherichia coli, Aspergillus niger, Scopulariopsis brevicaulis, and Saccharomyces cerevisiae [24,25]. These determinations were made long after growth had

ceased, approximately 7 days.  It would be much more informative to have
followed the production during the active growth of the organisms in order
to determine whether methylmercury was being synthesized during vigorous
growth.  Nevertheless, it is quite evident that substantial amounts of methylmercury can be produced in aerobic environments.

It would appear that much more widespread is the reduction of mercuric
compounds to elemental mercury.  From the identification of metallic mercury as a decomposition product of phenylmercuric acetate, ethylmercuric
phosphate, and methylmercuric chloride by a mercury-resistant pseudomonad [26], the list of microorganisms capable of this reduction continues
to grow [27-31].

In 1968, Tonomura and his coworkers [32] reported on the isolation of
a mercury-resistant pseudomonad which had been isolated from enrichment
cultures containing phenylmercuric acetate as the selective agent.  This
bacterium, designated Pseudomonas sp. K62, served as the stimulus for
research which resulted in the identification of metallic mercury as the
volatile compound that accounted for the loss of radioactive mercury from
cultures [33-38b].

Probably the most significant discovery was made in 1967 by Smith
[39], who examined for resistance to each of ten metals some fifty-five
clinical isolates and laboratory stocks of E. coli and Salmonella typhimurium which carried multiple-drug-resistant factors to various antibiotics.
Eleven clinical isolates were resistant to mercury, and this resistance
was carried by an extrachromosomal factor, namely, a plasmid.  Plasmids are small DNA molecules which are identical in physical structure
to the bacterial chromosome but which are separate autonomously
replicating structures.  They may contain the genes for resistance to a
variety of agents, for enzymes required for the dissimilation of certain
compounds, and for other specific functions.  Plasmids can be transferred
from cell to cell [40].  At the same time, Novick [41] reported that the
penicillinase plasmid of Staphylococcus aureus also conferred resistance
to a variety of metals, including mercury.  These observations naturally
led to an examination of the physiological basis for resistance to mercury,
because of the interest, no doubt, in the mercury disasters which had occurred just previously.  This suggestion is supported by the fact that the
organisms carrying these plasmids are also resistant to such metal ions
as aluminum, cadmium, chromium, copper, lead, silver, thallium,
arsenate, arsenite, and zinc.  However, with the exception of cadmium
resistance [42], little has been done to elucidate the mechanism of
resistance to these ions.

The physiological basis for mercury resistance in organisms containing the plasmid was attributed to the cell's ability to convert the mercuric
compounds to a volatile form.  Komura and coworkers [43-45] transferred
multiple-drug-resistant plasmids which also had mercury resistance to
E. coli and Aerobacter aerogenes and demonstrated that the resistant

recipients were capable of volatilizing $^{203}Hg^{2+}$. Crude cell-free extracts of E. coli possessed volatilization activity which was NADPH dependent. It was Summers and Silver [46a] , however, who presented convincing evidence for elementary mercury as the volatile product produced by a multiple-resistance plasmid containing E. coli. The identification was made by coupled gas chromatography and mass spectroscopy on toluene and chloroform extracts of the bacterial cultures containing $^{203}HgCl_2$. The rate of conversion of $10^{-5}$ M $Hg^{2+}$ to $Hg^{\circ}$ in resistant cells was calculated to be 4-5 nanomol of $Hg^{2+}$ per minute per $10^8$ cells. This rate would account for the rapid elimination of mercury from a culture medium. A further study of the properties of the enzymes involved was made by Summers and Sugarman [46b] .

A comprehensive study on resistance to mercuric chloride and phenylmercuric acetate (PMA) was performed by Schottel et al. [47] on over 30 plasmid-bearing strains of E. coli K12 and clinical isolates containing multiple-drug-resistant plasmids. The plasmids were transferred to E. coli K12 from a variety of bacteria. The resistance was determined by disk assay [47,48] and volatilization of mercury by measuring the loss of $^{203}Hg^{2+}$ or $^{203}HgPMA$ from an assay system containing a cell suspension which had been uninduced or induced with $Hg^{2+}$ or PMA. The results were unequivocal in that sensitive cells failed to volatilize either $Hg^{2+}$ or PMA. However, strains containing a plasmid that conferred resistance to $Hg^{2+}$ volatilized only $Hg^{2+}$, whereas strains containing a plasmid that conferred resistance to both $Hg^{2+}$ and PMA volatilized both compounds. The mercury volatilizing activity in all strains was inducible. In addition, some unusual patterns of induction-volatilization were observed in strains containing certain plasmids. For example, some showed low volatilization, and in one strain exposure to PMA inhibited $Hg^{2+}$ volatilizations. Thus, metabolic or physiological differences must exist and certainly bear investigation, but there can be no doubt that the majority of resistant strains possess the capacity to produce elemental mercury. There must exist in nature a high frequency of this capacity, especially since the resistant plasmids were obtained from such diverse organisms as Serratia marcescens, Ps. aeruginosa, Proteus vulgaris, Providencia, Shigella dysenteriae, and Salmonella paratyphi B. Moreover, the ability to convert $Hg^{2+}$ to $Hg^0$ was thought to occur in several resistant strains of E. coli, Staphylococcus aureus, and Ps. aeruginosa when it was found that these organisms produced a volatile form of mercury which is soluble in organic solvents [29]. By employing a closed system incorporating a vapor phase chromatograph and a flameless atomic absorption spectrophotometer, Nelson et al. [27] were able to show that $Hg^0$ and benzene were products of PMA degradation by selected resistant bacteria isolated from the Chesapeake Bay.

The reduction of mercury to this elemental state is not limited to procaryotes. The conversion of mercuric chloride to a volatile form, presumably elemental mercury, has been reported in the eucaryotic alga Chlamydomonas [28] and in the yeast Cryptococcus [31].

Despite the extensive research performed on the reduction of mercury, little has been done on the oxidation of elemental mercury. Holm and Cox devised a simple method for introducing and examining the fate of elementary mercury into a closed growth system [49] and then used the procedure to study conversions of the element accomplished by a number of bacteria [50]. Bacillus subtilis and B. megaterium oxidized within 48 hr essentially all of the elemental mercury present in a basal salts medium plus yeast extract, whereas Ps. aeruginosa, Ps. fluorescens, Citrobacter sp., and E. coli oxidized very little. The results, however, are difficult to interpret, because oxidation of the mercury occurred in the basal salts medium plus yeast extract before growth of the organisms and even greater oxidation occurred with sterile filtrates of the culture of B. megaterium. Obviously, further studies will be required to ascertain the significance of the microbial oxidations of elemental mercury.

The microbial conversion to elemental mercury is certainly no laboratory curiosity, for the production of elementary mercury has been shown to occur in lake sediments. Spangler et al. [15] performed long-term studies on sediments taken from the delta area of the St. Clair River in Michigan. Methylmercury was produced from the added $^{203}HgCl_2$, but upon continued incubation the methylmercury concentration decreased with a concomitant increase in elemental mercury. In a continuation of these studies, it was demonstrated that some 207 bacterial isolates were capable of degrading aerobically methylmercury to methane and $Hg^0$. Twenty-one of the isolates were also able to accomplish the conversion under anaerobic conditions [51].

The considerable evidence for the various transformations of mercury might be summarized by Scheme 8-1, a simple diagram of a mercury cycle [13, 18, 52].

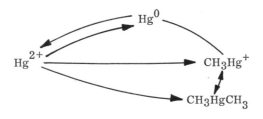

SCHEME 8-1

However, perhaps because of this evidence, numerous questions can be posed about the reactions that occur in vitro and how this information can be applied to a natural situation. It is not clear which transformations are most important in the cycling of mercury in the environment [53], and the application of laboratory data to studies of soil and bodies of water is no simple matter. A vast number of organisms may well carry out these reactions, and the plasmid-mediated resistances to other elements may be more widespread than hitherto expected. Perhaps these questions can be investigated and resolved before some new catastrophe occurs.


III.  SELENIUM AND TELLURIUM

Microbial interactions with selenium and tellurium have been known since the late 1800s, when numerous scientists studied the effects of additions of these metals to bacteriological media and the action of microorganisms on the reduction of the oxidized compounds [54]. In spite of these early studies, our understanding of the microbial metabolism of selenium is rather minimal compared to that of animal and plant metabolism and our knowledge of the microbial metabolism of tellurium is very sketchy, although the reduction of tellurite to elemental tellurium is readily observed in a variety of microorganisms [55-59]. The fact that considerable information has accumulated on the metabolism of selenium in plants and animals can be attributed to the fact that it is an essential trace element for these organisms [60-62] and that it manifests toxic properties at levels not much higher than the normal ingestion levels. Copeland [63] reported that since the normal levels are 0.2 mg/day whereas the toxic levels are 5 mg/day, selenium may represent a potentially serious environmental pollutant when the element is accumulated in the food chain as a result of industrial pollution.

The microbial metabolism of selenium has been reviewed extensively by Shrift [64,65] and by Ehrlich [9]. In addition to the interest in selenite as a selective agent in bacteriological media [66], there seems to have been four main lines of research on the microbial metabolism of selenium: (1) the substitution of selenium for sulfur and its incorporation into cellular components; (2) the requirement for selenium as an essential element in certain enzymes [67]; (3) the reduction of selenium compounds to amorphous red elemental selenium; and (4) the production of dimethyl and trimethyl selenide.

The competitive antagonism between sulfur and selenium was investigated by Fels and Cheldelin [68,69], and they were able to show that such compounds as methionine, cysteine, and, to a lesser extent, even thiamine could overcome selenate toxicity in E. coli and Saccharomyces cerevisiae. It soon became apparent that selenium was substituting for sulfur in a variety of compounds: adenosine phosphoselenate, adenosylselenomethionine, selenocystine, selenocystathionine, selenomethionine, and 4-selenouridine [70-74]. Selenium can also replace the acid-labile sulfur in putidaredoxin with no loss of activity [75]. Thus, the toxicity of selenite for microorganisms can be attributed, in part, to the incorporation of seleno analogs of sulfur-containing compounds into cellular components. Apparently, a considerable amount of selenium may be incorporated into cellular materials and still have no deleterious effects on the metabolism of the organism [76-78].

Although the essentiality of selenium in animal and plant nutrition is well documented, the requirement for selenium by microorganisms has only recently become apparent, in spite of the fact that as early as 1954 Pinsent [79] demonstrated a requirement for selenium and molybdenum in the production of formic dehydrogenase in E. coli. At least three microbial enzymes have been shown to be selenoproteins, the formic dehydrogenase of Escherichia coli [80], a formate dehydrogenase of Clostridium thermoaceticum [81], and the glycine reductase system of Clostridium stricklandii [67,82]. Perhaps these studies will provide the impetus to search for other selenoproteins in microorganisms, yielding data which in turn would contribute to our understanding of the significance of selenium as an essential nutrient for microorganisms.

Probably the most striking microbial transformation of selenium observed is the reduction of selenate or selenite to red amorphous selenium, which accumulates in cells and the culture medium to such an extent that the culture takes on a brick-red appearance. The reduction occurs in a variety of microorganisms: numerous bacteria, blue-green bacteria, yeasts, and Neurospora [68,83-87]. The biochemical mechanism for the reduction of selenate and selenite to selenium in microorganisms is not entirely clear. Woolfolk and Whiteley [88a], using $H_2$ uptake in conventional Warburg vessels as a measure of selenite and selenium reduction, concluded that enzyme preparations of Micrococcus lactolyticus reduced selenite and selenate to elemental selenium, which was further reduced to selenide. The reduction of selenium to selenide bears further investigation because essentially all other studies report no further reduction of elemental selenium by bacteria (with the possible exception of Methanobacterium [88b]). Falcone and Nickerson [83,87] were able to show that activity of dialyzed enzyme extracts of Candida albicans could be restored by the addition of glucose-6-phosphate, NADP, glutathione, and menadione. On the other hand, Tilton et al. [89,90] in a study of the cell-free activity of Streptococcus faecium and Streptococcus faecalis were able to show a stimulation by flavin adenine dinucleotide (FAD), whereas Ahluwalia et al. [91] were unable to obtain a cell-free preparation from E. coli. Unfortunately, these studies have not clarified the mechanism, and perhaps it would be appropriate to reexamine these preparations in the light of the rather detailed biochemical studies that are available from animal systems [92]. The mechanisms may differ considerably in microorganisms, because their resistance to selenium compounds varies markedly. This resistance serves as the basis for a variety of selective procedures for the isolation of certain groups of organisms [93,94].

The accumulation of the red granules of elemental selenium results in damage to the cell [91,95] and thus further confuses the toxicity picture, because not only is there an incorporation of selenium in place of sulfur in a variety of cellular components but there is a disruption of cellular integrity. The observations of McCready et al. [95] are interesting in a number of respects. They presented evidence that selenium with a 2+ valence state may have been an intermediate in the reduction to elemental selenium. This would be a reasonable intermediate and bears further investigation. They observed that filamentous cells were produced containing red granules and then amorphous red precipitated material accumulated in the background, suggesting that cells were lysing or that the selenium was being produced outside the cells. In addition, granule-containing ghosts were observed. The results suggested that the accumulation was deleterious to the cells. With one bacterium, Salmonella heidelberg, no volatile form of selenide was detected in the cultures, and the authors concluded that elemental selenium was the sole end product. This lack of a volatile product has been noted by others [83, 84].

A number of studies have focused on the cellular localization of the selenium granules. Gerrard et al. [96] present some convincing electron micrographs which show the granules accumulating in the cell membrane and cell wall but not in the cytoplasm. Silverberg et al. [97a], on the other hand, found that isolates from lake sediments which were classified as E. coli, Ps. aeruginosa, Aeromonas, and Flavobacterium accumulated selenium granules in the cytoplasm and not the membrane or wall. The reasons for these diametrically opposed observations are not readily apparent.

In spite of the dramatic appearance of cultures containing elemental selenium, there have not been any critical studies on how an organism can rid itself of the accumulated metal. There are some obvious possibilities. If presented with an aerobic environment, the cells may be capable of reoxidizing the element. Unfortunately, our knowledge of the oxidation of selenium compounds is nil [64]. Under anaerobic conditions, if the cells find themselves in a low-selenium environment, they may simply dilute out the elemental selenium until it is no longer a serious problem to the organism. The most reasonable alternative would be to convert the element to a volatile form. The particular organisms mentioned here may have been unable to carry out such a conversion under the cultural conditions, or the analytical procedures may not have been appropriate for detection of a gas, but we do know that organisms, macro and micro, are capable of producing dimethyl and trimethyl selenide from selenate and selenite.

In spite of the fact that the production of methylated selenides by microorganisms was known for some time [97b], the elucidation of biochemical reactions resulting in the formation of dimethyl and trimethyl selenide has been accomplished not through the study of microbial systems, but rather through animal [92] and plant systems [98]. Ganther and Hsieh [92] have studied the production of methyl selenides in rat liver and kidney and have identified several important characteristics of the biochemical reactions. They are "(1) anaerobic conditions were necessary for optimal activity; (2) there was an absolute requirement for glutathione [GSH] that could not be met by other thiols; (3) TPNH [i.e., NADPH[1]] stimulated the system even in the presence of high levels of glutathione; (4) S-adenosylmethionine [S-AM] was the methyl donor; (5) the system was strongly inhibited by arsenite...." They proposed the accompanying pathway (Scheme 8-2) for the dimethylselenide synthesis.

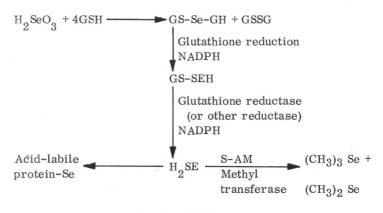

SCHEME 8-2

The production of $H_2Se$ as a formal intermediate prior to methylation was not meant to imply an obligatory step, for it is possible the methylation could occur in the intermediate GS-SeH to form GS-SeCH$_3$ and then methylation of HSeCH$_3$ to the dimethyl form. Lewis et al. [98] presented evidence for the formation of dimethyl selenide in cabbage leaves by the cleavage of Se-methyl selenomethionine selenonium salt, which they argued was analogous to the production of dimethyl sulfide from the cleavage of S-methyl methionine sulfonium salt. The equivalent level of biochemical reactions in microorganisms are not available.

The recent interest in the possible environmental effects of selenium has stimulated a renewed examination of the production of volatile selenium products by microorganisms. Fleming and Alexander [99] reported on a strain of Penicillium, isolated from sewage, which under certain conditions could convert 13 to 24% of the added selenite (10 μg/ml) to dimethyl selenide. In a more comprehensive screening procedure, Barkes and Fleming [100] isolated eleven distinct fungal strains which were shown to produce dimethyl selenide. The strains were initially selected because the colonies

---

[1] TPNH (reduced triphosphopyridine nucleotide) is called reduced nicotinamide adenine dinucleotide phosphate (NADPH) elsewhere in this volume, in accordance with recent usage.

exhibited the reddish hue suggestive of the production of elemental selenium.
The fungi were identified as belonging to the genera <u>Penicillium</u>, <u>Fusarium</u>,
Cephalasporium, and <u>Scopulariopsis</u>.  The production of dimethyl selenide
in some of the strains was dependent on the presence of selenite, for no
volatile product was detected when grown on selenate.  The authors also reported that a large number of bacterial isolates that had been selected because they reduced the selenite to elemental selenium were unable to produce dimethyl selenide.  This interesting result may explain why tremendous
quantities of elemental selenium accumulate ,in bacterial cells and may
indicate that bacteria are unable to dispose of the accumulated selenium.
This would imply that elemental selenium may accumulate as a result of
bacterial metabolism, and in turn would suggest that microorganisms
might also be able to oxidize or reduce extracellular elemental selenium.
With the exception of a report of an elemental selenium-oxidizing bacterium
[101], a finding which apparently has never been repeated, there are no
reports on the microbial attack on elemental selenium added to a microbial
environment.

Biogeochemical cycles for selenium have been suggested [64,102],
and it might be appropriate here to propose a revised cycle as a consequence of recent developments.  The accompanying scheme (8-3) is a mod-
ification of the one proposed by Shrift [102].

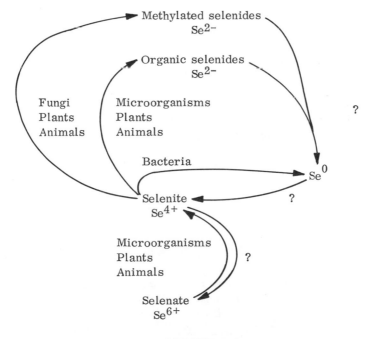

SCHEME 8-3

The evidence for such a cycle is tentative at best.  There is substantial evidence for the reduction of selenium compounds.  The oxidation of selenides, inorganic and organic, and of elemental selenium is still an unexplored area and presents some interesting microbiological problems. The oxidation of elemental selenium certainly bears investigation in light of the fact that bacterial metabolism results in the accumulation of selenium.

A. Shrift, the individual most responsible for drawing our attention to the microbial metabolism of selenium, has stated that "unlike investigations conducted with plants and animals, there is little sustained effort with bacteria, fungi, and algae" [64].  With the exception of Shrift's own studies, that same statement can be made some 10 years later.

## IV.  IRON

In spite of the vital role iron plays in the metabolism of all microorganisms, the transformation of macroquantities of the element is limited to a very small number of bacteria.  Such lesser-known bacteria as the Sphaerotilus-Leptothrix group and Gallionella have been referred to as the "iron bacteria" [103], but there is no evidence that the deposition of ferric oxide in the sheath of the Sphaerotilus-Leptothrix group is the result of direct metabolism of iron [104,105].  The suggestion has been made that Gallionella may be able to oxidize ferrous ion, because the organism will grow in a mineral medium with ferrous sulfate as a source of reduced iron. Unfortunately, these observations were made with impure cultures [106, 107].  Consequently, until more definitive studies have been performed, the best that can be said about these organisms is that they can cause the localization of insoluble iron oxides in certain environments.  However, a filamentous bacterium similar to Gallionella and Metallogenium [108] has been reported to require ferrous ions for growth and to oxidize the ions under "meso-acidic" conditions [109].

An unequivocal "iron bacterium" is Thiobacillus ferrooxidans [110, 111], which grows under strictly autotrophic conditions and obtains all the energy required for growth from the oxidation of ferrous ions or inorganic sulfur compounds.  The oxidation of ferrous ions can be accounted for by the following reaction [112] :

$$4Fe^{2+} + O_2 + 4H^+ \rightarrow 4Fe^{3+} + 2H_2O$$

This remarkable bacterium has been rather thoroughly studied because of its unique metabolism [113], its association with mine drainage [114-116], and its application to the leaching of sulfide ores [117].

The conditions for growth of T. ferrooxidans certainly represent an extreme environment [118].  The optimum pH for growth is between 2.0

and 3.5, and as the organism grows it produces sufficient sulfuric acid to lower the pH. At this low pH, carbon dioxide, its sole carbon source, is liberated from solution. Consequently, unless vigorous agitation or mixing occurs, the growth environment can become limiting for $CO_2$. The fact that more metal can be released when a mixed culture of Thiobacillus and Beijerinckia [119] was employed might be explained by the production of $CO_2$ by the heterotroph. The production of $CO_2$ from organic compounds could explain also the observation that after a brief exposure to glucose a new species, T. acidophilus, could be isolated from a culture of T. ferrooxidans [120]; T. acidophilus may increase the $CO_2$ content of the growth medium by its ability to oxidize organic compounds and thus form a "microbial consortium" with T. ferrooxidans.

Thiobaccilus ferrooxidans can metabolize a variety of ferrous compounds and in laboratory cultures substantial amounts of $Fe(OH)_3$ accumulate [121], although other salts (e.g., ferric oxide and phosphate) may precipitate, since most of them are insoluble. The low pH of the environment in which this organism grows prohibits the rapid chemical oxidation of ferrous ion and, therefore, it is well established that the organism is capable of enzymatically oxidizing the ferrous ion. The exact mechanism for the oxidation has not been elucidated, but it has been demonstrated quite conclusively that the oxidation occurs in the cell envelope [122,123].

No significant ultrastructural differences were observed in the cell envelopes of T. ferrooxidans when compared to the typical Gram-negative envelope. However, marked structural and functional differences must occur at the molecular level. The membrane and complex cell wall must be unreactive to the ferric hydroxide which accumulates or the cell would become encrusted with the compound. The membrane must obviously exclude and be resistant to the high concentration of $H^+$ when growing at such a low pH, because the oxidation of iron by isolated cell envelopes occurs at pH 3.0-3.5, and in whole cells the pH optimum is between 2.0 and 2.5 [124]. Since the oxidation takes place on the membrane and the organism can utilize such insoluble substrates as $FeS_2$ and elemental sulfur, the outer layers must indeed possess some unique molecules. Although specific binding proteins for iron have not been identified in these bacteria, a lipopolysaccharide-phospholipid-iron complex was isolated and partially characterized [125], but no physiological function was attributed to the complex. Ferrous ions bind rapidly to whole cells [122], and the oxidation of iron is incredibly resistant to high concentrations of other ions [117]. These observations would suggest that either a potent iron-binding site is present or that the site is well protected. In either event, the iron-binding protein should possess some rather interesting properties, and it certainly should be sought out in this bacterium.

The application of T. ferrooxidans to the leaching of sulfide metals has already provided considerable information on the physiological properties of this organism [117]. The kinds of ferrous and nonferrous minerals exposed to the action of the bacterium extend from orpiment ($As_2S_3$) to sphalerite (ZnS) [126-128]. Most of these studies are directed toward determining whether a particular metal can be leached from a complex or what conditions will enhance the leaching process [119,129]. However, many of these investigations describe some interesting or unusual interactions of metals with microorganisms.

Weinberg [130] has summarized the range of metal concentrations which influence growth and the release of secondary metabolites. For example, for iron, a concentration of $2 \times 10^{-7}$ M is required for bacterial growth whereas $1.0 \times 10^{-3}$ M is inhibitory. In the case of zinc, $1 \times 10^{-7}$ M is required for fungal growth and $5 \times 10^{-3}$ M is inhibitory. These values are exceeded enormously in cultures of T. ferrooxidans. In studies designed to determine the limiting factors in the leaching of zinc sulfide, concentrations of approximately 70 g of zinc per liter ($\sim 1.0$ M) were routinely obtained [129,131]. This remarkable resistance was also observed in the oxidation of iron by suspensions of T. ferrooxidans [117]. At concentrations of 0.1 M in buffer, $Co^{2+}$, $Zn^{2+}$, $Ni^{2+}$, and $Cu^{2+}$ were not inhibitory or very slightly so. At concentrations of 1.0 M, the inhibition of oxidation varied from a high of 75% for nickel to a low of 30% for copper. Another study on ore leaching suggests that, in general, microorganisms capable of growth at a low pH may be resistant to high concentrations of cations. In an interesting approach to the extraction of ores, Tsuchuja et al. [119] cultivated the autotrophic T. ferrooxidans in the presence of the acid-tolerant dinitrogen-fixing heterotroph, Beijerinckia lacticogenes. In a mixed culture, the autotroph would supply the carbon compounds for the heterotroph, which in turn would supply fixed nitrogen to the autotroph. Surprisingly, these investigators were able to show that growth of both organisms was enhanced in mixed cultures. The mixture also leached much larger amounts of copper and nickel from a number of minerals, and the final concentration of copper reached 8.44 g/liter. The data do not allow one to conclude that this concentration is the upper limit, because other factors may have been limiting the extraction. However, the results suggest that the heterotroph is also much more resistant to these cations than is generally the case. The versatility of T. ferrooxidans is further expanded by the fact that this bacterium may be capable of oxidizing a variety of metals other than iron. Nielsen and Beck [132] in examining the reaction of cells of T. ferrooxidans on chalcocite ($Cu_2S$) were able to detect only cupric ions and the more oxidized forms of the sulfide mineral, namely, digenite ($Cu_9S_5$) and covellite (CuS). They suggested that the organism was capable of fixing carbon dioxide by utilizing the energy provided by the oxidation of the cuprous ion. There is also a report that T. ferrooxidans can oxidize copper selenide, resulting in copper going into solution and elemental selenium being deposited [133]. Silver and Torma [134] in a study of the oxidation of various metal sulfides presented evidence that $Cu_2S$ may have been converted to CuS and metallic copper. There are enough reports on the leaching of mineral sulfide that hint at the oxidation of the metal that a thorough and systemic study of T. ferrooxidans' oxidative capacities would seem warranted.

The microbial reduction of ferric compounds, on the other hand, can be accomplished by practically every facultative or anaerobic heterotroph examined [135]. It is not clear in some of the reports whether the microorganisms are involved directly in the reduction or whether the reduction is brought about by organic compounds produced by the organisms. Takai and Kamura [136a] in a study of the mechanisms of reduction in soil present some convincing evidence that the reduction may be brought about by direct microbial action and that microorganisms can utilize ferric ions as electron acceptors under anaerobic conditions. They were able to show that isolates from rice paddy soils were capable of reducing nitrates, manganates, and ferric hydroxide, that in the presence of ferric hydroxide the bacteria would grow five times faster than in its absence and growth was proportional to the amount of ferrous ion produced, and that cell-free extracts were capable of reducing potassium ferricyanide or ferric hydroxide in the presence of organic acids. There is, nonetheless, a need for a more thorough biochemical analysis of the reductive process in microorganisms. An interesting observation on the reduction of ferric iron was made by Brock and Gustafson [136b]. They demonstrated that T. thiooxidans, Sulfolobus acidocaldarius, and T. ferrooxidans were able to reduce ferric ion using elemental sulfur as the electron donor. These observations have important implications in the generation of acid mine drainage and in the microbial leaching of ores.

A very exciting aspect of microbial iron metabolism has been reported recently by Blakemore [137], who described the discovery of magnetotactic bacteria. These bacteria respond to a magnetic field and thus can be collected and studied in spite of the fact that they have not been obtained in pure culture. Iron-rich particles can be observed in electron micrographs of the organisms, and this represents the first description of a microorganism which accumulates excessive quantities of iron intracellularly. The significant point is that the iron accumulates intracellularly, for the production of magnetic iron sulfide by microorganisms has been observed. Freke and Tate [138a] reported that Desulfovibrio can, under certain conditions, produce magnetic iron sulfide, and they employed this property to remove iron from solution. In a recent report, Jones et al. [138b] showed that Desulfovibrio and Desulfotomaculum can quite effectively extract trace amounts of metals from the culture medium. In media with excess iron, electron-dense particles, apparently FeS, are found within the bacteria. Further studies on these interesting phenomena should certainly contribute important links to our understanding of the biogeochemistry of iron.

## V.  MANGANESE

The microbial metabolism of manganese bears some striking resemblances
and contrasts to that of iron.  Surprisingly, a considerable amount of information has accumulated on the action of microorganisms on this metal
[9,139], because of the action of soil microorganisms on the availability
of manganese to plants [140] and the possible role of microorganisms in
manganese nodules found on the ocean floor [141,142].

Since the chemical oxidation of $Mn^{2+}$ occurs above pH 9, it poses no
serious problems to the study of the microbial oxidation of the ion.  However, there was some question as a result of an early observation by
Söhngen [143] that $Mn^{2+}$ may be chemically oxidized by certain hydroxy
acids at a pH of approximately 8.  These observations were shown by
Mulder [139] to play no significant role in the oxidation in natural environments and represent a laboratory artifact.  The oxidation of $Mn^{2+}$ can be
brought about by a substantial variety of microorganisms and has been
thoroughly studied [139,143-148].  The ease of detection of manganese-
oxidizing organisms probably accounts for the large number of papers
dealing with this process.  The insoluble brown oxide ($MnO_2$) precipitates
in or around colonies of oxidizing organisms which can be readily isolated
in pure culture [149].

In addition to the usual soil bacteria and fungi which have been shown
to oxidize manganese ions, three special groups deserve special mention.
The Sphaerotilus-Leptothrix group, dismissed as possible oxidizers of
iron, probably play a significant role in the oxidation of manganese.  These
bacteria not only oxidize $Mn^{2+}$ to $Mn^{4+}$ but also accumulate $MnO_2$ in their
sheaths and filaments.  Mulder, who has studied this group extensively,
reports that Leptothrix secretes a protein that catalyzes the oxidation of
manganese [139].  Addition of pronase to culture filtrates eliminates this
activity.  The presence of the protein in the culture filtrates was expected,
for the oxidation of manganese takes place at a considerable distance from
colonies on solid media.

There is still controversy on the possibility of an organism obtaining
energy from the oxidation of manganese ion, and there is no evidence that
the usual heterotroph is able to utilize the energy.  However, the heterotrophic growth of Sphaerotilus discophorus was shown by Ali and Stokes
[150] to be markedly enhanced by the addition of $MnSO_4$ to the culture
medium.  They also reported that growth could occur autotrophically with
$Mn^{2+}$ as the sole available energy source.  Mulder [139] questions these
results.  He was able to show a stimulation of heterotrophic growth by
$MnCO_3$ as measured by cellular nitrogen, but he suggests that the oxidized
manganate may coat the cell and prevent the loss of soluble nitrogenous
compounds.  He reports that there was a difference in the amount of soluble
nitrogen released from the cells.  It is not clear how these results could
account for the increased protein and DNA observed by Ali and Stokes in

their study. Mulder also questions the ability of Sphaerotilus to grow auto-trophically. In a study of the oxidation of manganese in soil, Mulder [139] was able to show that addition of manganous ion did not increase the num-ber of oxidizers, and he concluded that an increase would be expected if the oxidizers were utilizing the energy released in the oxidation. This conclu-sion is easily rebutted if one assumes that the cultural conditions for sel-ection of the autotrophic manganese oxidizers were not employed or known. Obviously, additional studies must be performed before the controversy will be settled.

In a study of the deposition of manganese in freshwater pipelines, Tyler and Marshall [151] demonstrated quite conclusively that the accumu-lation of manganese was brought about by microorganisms. They found that the most predominant organisms were the budding bacteria (hypho-microbia), although other bacteria and fungi were observed in the encrust-ations. Because of the similarity of the manganese deposits in pipelines throughout the world, it was suggested that these unusual bacteria play a more important role than has been suspected. The ability of hyphomicrobia to oxidize $Mn^{2+}$ was confirmed by Mulder [139].

A variety of fungi have been shown to oxidize manganese, but a most un-usual microbial oxidation was reported by Zavarzin [152-155]. Metallo-genium, a bacterium widely distributed in soil and water, resembles the hyphomicrobia in that the vegetative portion of the organism consists of very thin filaments [10-20 μm in diameter) encrusted with manganese oxide. The organism also produces round motile cells which are formed by budding off of the filaments. The organism only develops in symbiosis with a fungus. This association has also been observed by Schweisfurth [144] and Mulder [139]. There are reports of other microbial associa-tions which result in oxidation of manganese: Corynebacterium and Chromobacterium [156], and two species of pseudomonads [153].

The microbial reduction of manganese was first demonstrated by Mann and Quastel [157] who found that $Mn^{2+}$ could be detected in the per-fusate of soil perfusion columns supplemented with manganese oxide. Germon [158] in an examination of different forms of manganese concluded that $Mn^{2+}$ in soil resulted from microbial reducing activity and confirmed the conclusion by laboratory model experiments.

There seems to be little doubt that $Mn^{4+}$ is capable of serving as a final electron acceptor for microorganisms [158] and being reduced to $Mn^{2+}$ under anaerobic conditions [159]. In the classical study by Woolfolk and Whiteley [88a], a cell-free extract of Micrococcus lactilyticus was able to accomplish the reduction in the presence of $H_2$. Ehrlich ([160]; and with various coworkers [161-165])has most extensively studied the microbial reduction of $Mn^{4+}$. In an examination of ferromanganese nodules from the Atlantic Ocean, these investigators reported on the isolation of bacteria capable of reducing manganese oxides. The $MnO_2$-reductase was inducible with $Mn^{4+}$ serving as the inducer and, oddly, oxygen was necessary for the induction although it neither stimulated nor inhibited the activity once formed.

There can be little doubt of the role played in the geochemical cycling of manganese. Many organisms are capable of oxidizing $Mn^{2+}$, and there is no reason to doubt the fact that many other organisms are capable of utilizing $Mn^{4+}$ as an electron acceptor. There is evidence for the accumulation of manganese oxide both in the fouling of water pipes and the production of ferromanganese nodules. Microorganisms certainly play an important role in the uptake of manganese by plants [167-169].

## VI.   ARSENIC

The role of arsenic as an abnormal trace element in man was carefully assessed by Schroeder and Balassa [170], and they concluded that considerable confusion exists concerning the actual form of arsenic which may be responsible for an observed effect. They called for better analytical techniques for distinguishing the various oxidation states. Some 10 years later, the renewed interest in arsenic as a possible environmental pollutant emphasizes the validity of this insight (see the review by Penrose [171]).

Microbial metabolism of arsenic has been studied rather thoroughly and a geochemical cycle has been proposed [172,173]. In another one of his classic studies with soil perfusates, Quastel (with Scholefield [174]) demonstrated that arsenite was biologically oxidized to arsenate by soil organisms. In several papers [175-179], Turner and Legge reported on the isolation from an arsenical cattle dipping of a variety of heterotrophic bacteria, such as Pseudomonas, Xanthomonas, and Achromobacter, which could oxidize arsenite to arsenate. The enzyme responsible for the oxidation was inducible, since only cells grown in the presence of arsenite were able to oxidize it. The enzyme was a soluble arsenite dehydrogenase. The oxidation of arsenite resembles the oxidation of phosphite in its microbiology and biochemistry [3,4]. Ehrlich [180,181] was able to show that T. ferrooxidans was capable of oxidizing the arsenic in such minerals as arsenopyrite ($FeS_2 \cdot FeAs_2$), enargite ($3Cu_2S \cdot As_2S_5$), and orpiment ($As_2S_3$). Arsenite and arsenate were detected in the fluid.

The microbiology of the reduction of arsenate to arsenite is less clear. Woolfolk and Whiteley [88a] reported that arsenate could be reduced to arsenite by a cell-free extract of Micrococcus lactilyticus, but arsenite was not reduced further. (They also reported that bismuthate was probably reduced to elemental bismuth. This observation certainly bears further study.) No arsine ($AsH_3$) was detected, and no mention was made of the presence of methylated arsine. Additional studies on the reduction of the oxidized forms of arsenic by bacteria are needed. However, the reduction of arsenate to the most reduced state, arsine, is known to occur readily in a variety of fungi and has been shown to occur by cell extracts and whole cells of Methanobacterium [172].

The early studies on the production of methylated arsenic have been reviewed by Challenger [182], and he and his coworkers thoroughly studied the biochemistry of the process, identifying the methylated form as trimethylarsine and proposing a mechanism for its production [182-184]. The ability of the aerobic fungi to accomplish the reduction of arsenate to the methylated forms appears to be widespread [182,184,185], but there appears to be only one report of the production by a bacterium.

The reduction of arsenate resembles the reduction of selenate. Fungi are primarily responsible for the reduction and methylation, whereas perhaps only the methanogenic bacteria may be capable of this biochemical conversion. McBride and Wolfe [172] reported that not only were these bacteria capable of methylating arsenate, but they could also methylate selenate and tellurate.

Wood [173] has proposed a cycle for arsenic which is presented in Scheme 8-4 in a modified form. It is evident from this cycle that our knowledge of the reduction of arsenic is fairly well elucidated. The oxidation of the reduced compounds needs some serious investigation.

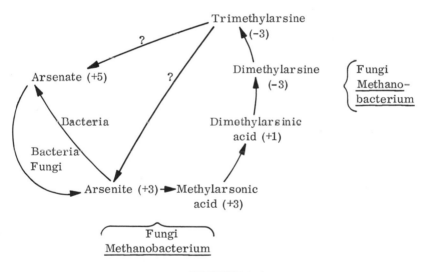

SCHEME 8-4

## VII.  CONCLUDING REMARKS

The resurgence in research into the microbiology of metal transformations can be attributed to the recent concern on the accumulation of toxic chemicals in our environment, especially in limited geographical areas as a result of industrial effluents.  Our understanding of the physiological and

biochemical aspects of the microbe-metal interactions may aid us in avoiding an accumulation of toxic compounds or, as is more likely, in ridding a
local environment of a toxic material after it has accumulated.  In the case
of mercury, selenium, and arsenic, the existence of significant volatile
forms of the element in the microbial cycling of the elements allow for a
ready means of detoxifying a limited environment.  Wood [13] has already
pointed out that such studies allow prediction to be made, and he has predicted that tin, palladium, platinum, gold, and thallium can be methylated
but that lead, cadmium, and zinc cannot, since the methylated compounds
are unstable in aqueous systems.  However, in some instances the excessive accumulation of the volatile forms in a confined environment might
cause serious problems because they are extremely toxic to higher forms
of life.  It would certainly be wise to examine the microbiology of other
industrially important metals in order to have available the information
before another environmental catastrophe occurs.  The ability of organisms
to reduce even some unlikely metals was elegantly demonstrated by
Woolfolk and Whiteley [88a].  An extract of Micrococcus lactolyticus was
able to accomplish the following reductions: arsenate to arsenite, penta-
valent and trivalent bismuth to the free element, selenite to elemental
selenium and possibly to selenide, tellurate to tellurium, lead and manganese dioxide to the divalent state, ferric to ferrous ions, osmium tetroxide
to osmate ion, osmium dioxide and trivalent osmium to the metal, uranyl
uranium to the tetravalent state, vanadate to the level of vanadyl, and
polymolybdate ions to molybdenum with an average valence of 5+.  This
incredible list of reductions certainly suggest that there may be microorganisms, heterotrophs or autotrophs, which can oxidize the reduced
ions, if from our current concept of the microbiology of metal transformation we can infer a general scheme.

   Another aspect of the environmental problems associated with metals
is the rapidity with which we are using up our easily attainable natural
resources of these elements.  An examination of the cycling of these elements offers a number of microbiological solutions to these problems.
The importance of T. ferrooxidans in leaching low-grade ores is already
well known.  However, it would appear that we could also make use of
those microorganisms which accumulate metals in the elemental state.  It
is now possible to attach bacteria to solid supports [186] and have them
transform substrates into useful products.  Surely, it should not be an
impossible task to use bacteria to accumulate metals from dilute solutions.
As a matter of fact, such a procedure is occurring constantly in water
pipes being fouled by manganese-oxidizing bacteria [151].

   The wealth of fundamental information already available on the microbiology of some metal transformations can certainly point the way to research on the ones less studied and to new approaches to the solution of
our environmental problems.

REFERENCES

1. F. Adams and J. P. Conrad, Soil Sci., 75, 361 (1953).
2. L. E. Casida, Jr., J. Bacteriol., 80, 237 (1960).
3. G. Malacinski and W. A. Konetzka, J. Bacteriol., 91, 578 (1966).
4. G. M. Malacinski and W. A. Konetzka, J. Bacteriol., 98, 1906 (1967).
5. W. Heinen and A. M. Lauwers, Arch. Microbiol., 95, 267 (1974).
6. K. J. Rudakow, Zentr. Bakteriol. Parasitenk. [II], 79, 229 (1929).
7. G. Toubota, Soil Plant Food, 5, 10 (1959).
8. W. P. Iverson, Nature, 217, 1265 (1968).
9. H. L. Ehrlich, in Soil Biochemistry (A. D. McLaren and J. Skujins, eds.), Vol. 2, Marcel Dekker, New York, 1971, p. 361.
10. A. Jernelov and A. L. Martin, Ann. Rev. Microbiol., 29, 61 (1975).
11. W. Heinen, Biosystems, 6, 133 (1974).
12. F. Bakir, S. F. Damluji, L. Amin-Zaki, M. Murtadha, A. Khalidi, N. Y. Al-Rawi, S. Tukriti, H. I. Dhahir, T. W. Clarkson, J. C. Smith, and R. A. Doherty, Science, 181, 230 (1973).
13. J. M. Wood, Science, 183, 1049 (1974).
14. S. Jensen and A. Jernelov, Nature, 223, 753 (1969).
15. W. J. Spangler, J. L. Spigarelli, J. M. Rose, and H. M. Mitter, Science, 180, 192 (1973).
16. G. Billen, C. Joiris, and R. Wollast, Water Res., 8(4), 219 (1974).
17. W. F. Beckert, A. A. Moghissi, F. H. F. Au, E. W. Bretthauer, and J. C. McFarland, Nature, 249, 674 (1974).
18. T. Edward and B. C. McBride, Nature, 253, 462 (1975).
19. J. M. Wood, F. S. Kennedy, and C. G. Rosen, Nature, 220, 173 (1968).
20. M. Yamada and K. Tonomura, J. Ferment. Tech., 50, 159 (1972).
21. L. Bertilsson and H. Y. Neujahr, Biochemistry, 10, 2805 (1971).
22. N. Imura, E. Sukejawa, S. Pan, K. Nagao, J. Kim, T. Kwan, and T. Ukita, Science, 172, 1248 (1971).
23. L. Landner, Nature, 230, 462 (1971).
24. M. K. Hamdy and O. R. Noyes, Appl. Microbiol., 30, 424 (1973).
25. J. W. Vonk and A. Kaars Sijpesteijn, Antonie van Leeuwenhoek J. Microbiol. Serol., 39, 505 (1973).
26. K. Furukawa, T. Suzuki, and K. Tonomura, Agr. Biol. Chem., 33, 128 (1969).
27. J. D. Nelson, W. Blair, F. E. Brinckman, R. R. Colwell, and W. P. Iverson, Appl. Microbiol., 26, 321 (1973).
28. D. Ben-Bassat, E. Shelef, N. Graner, and H. I. Shuval, Nature, 240, 43 (1972).
29. A. O. Summers and E. Lewis, J. Bacteriol., 113, 1070 (1973).
30. J. Schottel, A. Mandal, D. Clark, and S. Silver, Nature, 251, 335 (1974).

31.  R. L. Brunker and T. L. Bott, Appl. Microbiol., 27, 870 (1974).
32.  K. Tonomura, T. Nakagami, F. Futai, and K. Maeda, J. Ferment. Technol. (Osaka), 46, 506 (1968).
33.  K. Tonomura, K. Maeda, and F. Futai, J. Ferment. Technol. (Osaka), 46, 685 (1968).
34.  K. Furukama and K. Tonomura, Agr. Biol. Chem., 35, 604 (1971).
35.  K. Tonomura and F. Kanzaki, Biochim. Biophys. Acta, 184, 227 (1967).
36.  K. Furukawa and K. Tonomura, Agr. Biol. Chem., 36, 2441 (1972).
37.  K. Furukawa and K. Tonomura, Agr. Biol. Chem., 36, 217 (1972).
38a. K. Furukawa and K. Tonomura, Biochim. Biophys. Acta, 325, 413 (1973).
38b. T. Tezuka and K. Tonomura, J. Biochem. (Tokyo), 80, 79 (1976).
39.  D. H. Smith, Science, 156, 1114 (1967).
40.  D. Schlessinger (ed.), Microbiology — 1974, American Society for Microbiology, Washington, D.C., 1975.
41.  R. Novick, Fed. Proc., 26, 29 (1967).
42.  I. Kondo, T. Ishikawa, and H. Nakahara, J. Bacteriol., 117, 1 (1974).
43.  I. Komura, K. Izaki, and H. Takahashi, Agr. Biol. Chem., 34, 480 (1970).
44.  I. Komura and K. Izaki, J. Biochem. (Tokyo), 70, 885 (1971).
45.  I. Komura, T. Funaba, and K. Izaki, J. Biochem. (Tokyo), 70, 895 (1971).
46a. A. O. Summers and S. Silver, J. Bacteriol., 112, 1228 (1972).
46b. A. O. Summers and L. I. Sugarman, J. Bacteriol., 119, 242 (1974).
47.  J. Schottel, A. Mandal, D. Clark, and S. Silver, Nature, 251, 335 (1974).
48.  J. Schottel, A. Mandal, K. Toth, D. Clark, and S. Silver, Proc. Int. Conf. Transport and Persistence of Chemicals in Aquatic Ecosystems, Ottawa, Canada, Vol. 2, p. 65 (1974).
49.  H. W. Holm and M. F. Cox, Appl. Microbiol., 27, 622 (1974).
50.  H. Holm and M. F. Cox, Appl. Microbiol., 29, 491 (1975).
51.  W. J. Spangler, J. L. Spigarelli, J. M. Rose, R. S. Flippin, and H. H. Miller, Appl. Microbiol., 25, 488 (1973).
52.  S. Silver, J. Schottel, and A. Weiss, in Proc. 3rd Int. Biodegradation Symp. (J. M. Sharpley and A. Kaplan, eds.), Applied Science Publishers Ltd., London, 1975, p. 899.
53.  A. Jernelov, in The 15th Symp. of the Brit. Ecological Soc. (M. J. Chadwick and G. T. Goodman, ed.), Blackwell, Oxford, 1975, p. 49.
54.  V. E. Levine, J. Bacteriol., 10, 217 (1925).
55.  D. G. Smith, J. Gen. Microbiol., 83, 389 (1974).
56.  F. L. Tucker, J. F. Walper, M. D. Appleman, and J. Donohue, J. Bacteriol., 83, 1313 (1962).
57.  S. Nagai, J. Bacteriol., 90, 220 (1965).
58.  W. Van Iterson, J. Cell. Biol., 20, 377 (1964).

59. W. Van Iterson and W. Leene, J. Cell. Biol., 20, 361 (1964).
60. H. A. Schroeder, D. V. Frost, and J. J. Balassa, J. Chron. Dis., 23, 227 (1970).
61. I. Rosenfeld and O. A. Beath, Selenium, Academic Press, New York, 1964.
62. O. H. Muth (ed.), International Symposium on Selenium in Biomedicine, Avi Publ., Westport, Connecticut, 1967.
63. R. Copeland, Limnos, 3, 7 (1970).
64. A. Shrift, in International Symposium on Selenium in Biomedicine (O. H. Muth, ed.), Avi Publ., Westport, Connecticut, 1967, p. 241.
65. A. Shrift, in Organic Selenium Compounds: Their Chemistry and Biology (D. L. Klayman, ed.), Wiley, New York, 1973, p. 760.
66. A. Shrift, and R. F. Boulette, Appl. Microbiol., 27, 814 (1974).
67. T. C. Stadtman, Science, 183, 915 (1974).
68. I. G. Fels and V. H. Cheldelin, J. Biol. Chem., 185, 803 (1950).
69. I. G. Fels and V. H. Cheldelin, Arch. Biochem., 22, 323 (1949).
70. J. L. Hoffman and K. P. McConnell, Biochim. Biophys. Acta, 366, 109 (1974).
71. M. Blau, Biochim. Biophys. Acta, 49, 389 (1961).
72. T. Tuve and H. H. Williams, J. Biol. Chem., 236, 597 (1961).
73. K. F. Weiss, J. C. Ayres, and A. A. Kraft, J. Bacteriol., 90, 857 (1965).
74. S. H. Mudd and G. L. Cantoni, Nature, 180, 1052 (1957).
75. J. C. M. Tsibris, M. J. Namtvedt, and I. C. Gunsalus, Biochem. Biophys. Res. Comm., 30, 323 (1968).
76. D. Cowie and G. Cohen, Biochim. Biophys. Acta, 26, 252 (1957).
77. W. M. and J. T. Wachsman, J. Bacteriol., 104, 1393 (1971).
78. W. M. and J. T. Wachsman, J. Bacteriol., 105, 1222 (1971).
79. J. Pinsent, Biochem. J., 57, 10 (1954).
80. R. L. Lester and J. A. Demoss, J. Bacteriol., 105, 1006 (1971).
81. J. R. Andreesen and L. Ljungdahl, J. Bacteriol., 116, 867 (1973).
82. D. C. Turner and T. C. Stadtman, Arch. Biochem. Biophys., 154, 366 (1973).
83. G. Falcone and W. J. Nickerson, J. Bacteriol., 85, 754 (1963).
84. M. Zalokar, Arch. Biochem. Biophys., 44, 330 (1953).
85. V. Koval'skii, V. V. Ermakov, and S. V. Letunova, Microbiology (USSR), 37, 103 (1968).
86. H. D. Kumar and G. Prakash, Ann. Bot. [N.S.], 35, 697 (1971).
87. W. J. Nickerson and G. Falcone, J. Bacteriol., 85, 763 (1963).
88a. C. A. Woolfolk and H. R. Whiteley, J. Bacteriol., 84, 647 (1962).
88b. B. C. McBride and R. S. Wolfe, Biochemistry, 10, 4312 (1971).
89. R. C. Tilton, H. B. Gunner, and W. Litsky, Can. J. Microbiol., 13, 1175 (1967).
90. R. C. Tilton, H. B. Gunner, and W. Litsky, Can. J. Microbiol., 13, 1183 (1967).

340 W. A. KONETZKA

91. G. S. Ahluwalia, Y. R. Saxena, and H. H. Williams, Arch. Biochem. Biophys., 124, 79 (1968).
92. H. E. Ganther and H. S. Hsieh, in Trace Element Metabolism in Animals: Proc. 2nd Int. Symp. (W. G. Hoekstra, ed.), University Park Press, Baltimore, Maryland, 1974, p. 339.
93. F. Guth, Zentr. Bakteriol. Parasitenk. [I], 77, 487 (1916).
94. E. Leifson, Am. J. Hyg., 24, 423 (1936).
95. R. G. L. McCready, J. N. Campbell, and J. I. Payne, Can. J. Microbiol., 12, 703 (1966).
96. T. L. Gerrard, J. N. Telford, and H. H. Williams, J. Bacteriol., 119, 1057 (1974).
97a. B. A. Silverberg, P. T. S. Wong, and Y. K. Chan, Arch. Microbiol., 107, 1 (1976).
97b. F. Challenger, Advan. Enzymol., 12, 429 (1951).
98. B. G. Lewis, C. M. Johnson, and T. C. Broyer, Plant Soil, 40, 107 (1974).
99. R. W. Fleming and M. Alexander, Appl. Microbiol., 24, 424 (1972).
100. L. Barkes and R. W. Fleming, Bull. Environ. Contam. Toxicol., 12, 308 (1974).
101. J. G. Lipman and S. A. Waksman, Science, 57, 60 (1923).
102. A. Shrift, Nature, 201, 1304 (1964).
103. E. G. Pringsheim, Biol. Rev., 24, 200 (1949).
104. E. G. Mulder, J. Appl. Bacteriol., 27, 151 (1964).
105. N. C. Dondero, Ann. Rev. Microbiol., 29, 407 (1975).
106. S. Kucera and R. S. Wolfe, J. Bacteriol., 74, 344 (1957).
107. R. S. Wolfe, J. Amer. Water Wks Assoc., 50, 1241 (1958).
108. G. A. Zavarzin, Z. Allg. Mikrobiol., 4, 390 (1964).
109. F. Walsh and R. Mitchell, J. Gen. Microbiol., 72, 369 (1972).
110. A. R. Colmer and M. E. Hinkle, Science, 106, 253 (1947).
111. K. L. Temple and A. R. Colmer, J. Bacteriol., 63, 605 (1951).
112. M. P. Silverman and D. G. Lundgren, J. Bacteriol., 78, 326 (1959).
113. D. P. Kelly, Ann. Rev. Microbiol., 25, 177 (1971).
114. D. G. Lundgren, J. R. Vestal, and F. R. Tabita, in Water Pollution Microbiology (R. Mitchell, ed.), Wiley (Interscience), New York, 1972, p. 69.
115. R. T. Belly and T. D. Brock, J. Bacteriol., 117, 726 (1974).
116. P. R. Dugan, Ohio J. Sci., 75, 266 (1975).
117. O. H. Touvinen and D. P. Kelly, Z. Allg. Mikrobiol., 12, 311 (1972).
118. T. D. Brock, in Microbial Growth: Symposium of the Society for General Microbiology, Vol. 19 (P. M. Meadow and S. J. Pirt, eds.), Cambridge University Press, New York, 1969, p. 15.
119. H. M. Tsuchuja, N. C. Trivedi, and M. L. Shuler, Biotech. Bioeng., 16, 991 (1974).
120. R. Guay and M. Silver, Can. J. Microbiol., 21, 281 (1975).

121. H. Lees, S. C. Kwok, and I. Suzuki, Can. J. Microbiol., 15, 43 (1969).
122. P. R. Dugan and D. G. Lundgren, J. Bacteriol., 89, 825 (1965).
123. G. A. Din and I. Suzuki, Can. J. Biochem., 45, 1547 (1967).
124. C. A. Bodo, Jr., and D. G. Lundgren, Can. J. Microbiol., 20, 1647 (1974).
125. A. D. Agate, M. S. Korzynski, and D. G. Lundgren, Can. J. Microbiol., 15, 259 (1969).
126. M. P. Silverman and H. L. Ehrlich, Advan. Appl. Microbiol., 6, 153 (1964).
127. M. Silver and A. Torma, Can. J. Microbiol., 20, 141 (1974).
128. A. E. Torma, Rev. Can. Biol., 30, 209 (1971).
129. A. E. Torma, C. C. Walden, D. W. Duncan, and R. M. R. Branion, Biotechnol. Bioeng., 14, 777 (1972).
130. E. D. Weinberg, Trace Subst. Environ. Health, 4, 233 (1971).
131. A. E. Torma, C. C. Walden, and R. M. R. Branion, Biotechnol. Bioeng., 12, 501 (1970).
132. A. M. Nielsen and J. V. Beck, Science, 175, 1124 (1972).
133. A. E. Torma and F. Habashi, Microbiol. Abstr. 8A, 12113 (1973).
134. M. Silver and A. Torma, Can. J. Microbiol., 20, 141 (1974).
135. T. V. Aristovskaya and G. A. Zavarzin, in Soil Biochemistry (A. D. McLaren and J. Skujins, ed.), Vol. 2, Marcel Dekker, New York, 1971, p. 385.
136a. Y. Takai and T. Kamura, Folia Microbiol., 11, 304 (1966).
136b. T. D. Brock and J. Gustafson, Appl. Environ. Microbiol., 32, 567 (1976).
137. R. Blakemore, Science, 190, 377 (1975).
138a. A. M. Freke and D. Tate, J. Biochem. Microbiol. Technol. Eng., 3, 29 (1961).
138b. H. E. Jones, P. A. Trudinger, L. A. Chambers, and N. Pyliotis, Z. Allg. Microbiol., 16, 425 (1976).
139. E. G. Mulder, Rev. Ecol. Biol. Sci., 9, 321 (1972).
140. G. W. Leeper, Ann. Rev. Plant Physiol., 3, 1 (1952).
141. A. L. Hammond, Science, 183, 502 (1974).
142. J. Greendate, Nature, 249, 181 (1974).
143. N. L. Söhngen, Zentr. Bakteriol. Parasitenk. [II], 40, 545 (1914).
144. R. Schweisfurth, Z. Allg. Mikrobiol., 11, 415 (1971).
145. R. Schweisfurth, Z. Allg. Mikrobiol., 12, 667 (1972).
146. R. Schweisfurth, Z. Allg. Mikrobiol., 13, 341 (1973).
147. R. Schweisfurth, Zentr. Bakteriol. Parasitenk. [I], 233, 257 (1973).
148. R. Schweisfurth, Z. Allg. Mikrobiol., 16, 133 (1976).
149. S. M. Bromfield, Soil Biol. Biochem., 6(c), 383 (1974).
150. S. H. Ali and J. L. Stokes, Antonie van Leeuwenhoek J. Microbiol. Serol., 37, 519 (1971).

151. P. A. Tyler and K. C. Marshall, Antonie van Leeuwenhoek J. Micro-
     biol. Serol., 33, 171 (1967).
152. G. A. Zavarzin, Mikrobiologiya, 30, 774 (1961).
153. G. A. Zavarzin, Mikrobiologiya, 31, 481 (1962).
154. G. A. Zavarzin, Z. Allg. Mikrobiol., 4, 390 (1964).
155. G. A. Zavarzin, in The Ecology of Soil Bacteria (T. R. E. Gray and
     D. Parkinson, eds.), University of Toronto Press, Toronto, Canada,
     1968, p. 612.
156. S. M. Bromfield, Austral. J. Biol. Sci., 9, 238 (1956).
157. P. J. G. Mann and J. H. Quastel, Nature, 158, 154 (1946).
158. J. C. Germon, Rev. Ecol. Biol. Sol., 9, 451 (1972).
159. W. H. Patrick and F. T. Turner, Nature, 220, 476 (1968).
160. H. L. Ehrlich, Appl. Microbiol., 11, 15 (1963).
161. R. B. Trumble and H. L. Ehrlich, Appl. Microbiol., 16, 695 (1968).
162. R. B. Trumble and H. L. Ehrlich, Appl. Microbiol., 19, 966 (1970).
163. W. C. Ghiorse and H. L. Ehrlich, Appl. Microbiol., 28, 785 (1974).
164. H. L. Ehrlich, S. H. Yang, and J. D. Mainwaring, Z. Allg. Mikro-
     biol., 13, 39 (1973).
165. H. L. Ehrlich, Soil Sci., 119, 36 (1975).
166. S. M. Bromfield and V. B. D. Sherman, Soil Sci., 69, 337 (1950).
167. W. L. VanVeen, Antonie van Leeuwenhoek J. Microbiol. Serol.,
     39, 657 (1973).
168. D. A. Barber and R. B. Lee, New Phytol., 73, 97 (1974).
169. C. Sonneveld and S. J. Voogt, Plant Soil, 42, 49 (1975).
170. H. A. Schroeder and J. J. Balassa, J. Chron. Dis., 19, 85 (1966).
171. W. R. Penrose, CRC Crit. Rev. Environ. Control, 4, 465 (1974).
172. B. C. McBride and R. S. Wolfe, Biochemistry, 10, 4312 (1971).
173. J. M. Wood, Chem. Eng. News, p. 22 (1971).
174. J. H. Quastel and D. G. Scholefield, Soil Sci., 75, 279 (1953).
175. A. W. Turner, Nature, 164, 76 (1949).
176. A. W. Turner, Austral. J. Biol. Sci., 7, 452 (1954).
177. A. W. Turner and J. W. Legge, Austral. J. Biol. Sci., 7, 479
     (1954).
178. J. W. Legge, Austral. J. Biol. Sci., 7, 504 (1954).
179. J. W. Legge and A. W. Turner, Austral. J. Biol. Sci., 7, 496
     (1954).
180. H. L. Ehrlich, Econ. Geol., 58, 991 (1963).
181. H. L. Ehrlich, Econ. Geol., 59, 1306 (1964).
182. F. Challenger, Advan. Enzymol., 12, 429 (1951).
183. F. Challenger, D. Lisle, and P. Dransfield, J. Chem. Soc.,
     p. 1760 (1954).
184. R. Zussman, E. Vicher, and I. Lyon, J. Bacteriol., 81, 157 (1961).
185. D. P. Cox and M. Alexander, Bull. Contam. Toxicol., 9, 84 (1973).
186. J. F. Kennedy, S. A. Barker, and J. D. Humphreys, Nature, 261,
     242 (1976).

Chapter 9

MICROBIAL FACILITATION OF PLANT MINERAL NUTRITION

Michael R. Tansey

Department of Plant Sciences
Indiana University
Bloomington, Indiana

I. INTRODUCTION

Microorganisms facilitate plant mineral nutrition by changing the amounts, concentrations, and properties of minerals available to plants. These changes lead to changes in the growth, development, and chemical composition of plants that are common and substantial enough to encourage the exploitation of plant-microbe interactions for improvement of crop productivity. Possible approaches include both introduction of foreign microorganisms and capitalization on the indigenous microflora. This chapter reviews examples of plant-microbe interactions in which mineral nutrition of plants is aided. It will be demonstrated that these interactions are varied, widespread, common, and economically and ecologically significant for plant mineral nutrition. Three different degrees of interaction of plants and microorganisms are considered in this review: the symbiotic mycorrhizal association, the less well developed associations of microorganisms with leaves (the phyllosphere), and the fortuitous facilitation of plant mineral nutrition by lichens, especially by blue-green algae associated with lichens. Other examples of facilitation of plant mineral nutrition which are not discussed in this review are the following: Rhizobium and other root nodule associations [1,2], nonsymbiotic soil and rhizosphere interactions [3-6], symbiotic cyanophytes other than those associated with lichens [7-10], and the role of decay microorganisms in mineral cycles.

Microbial facilitation of plant mineral nutrition is increasingly recognized as a major factor in forestry and agriculture. If the recent rates of progress in several areas of research and application are sustained as expected, we may confidently predict that we are on the threshold of a new era of greatly increased application of sound knowledge of crop plant-microbe associations, and we may anticipate application of this knowledge to increase crop production and to establish economic plants in areas which have seemingly unfavorable soil and climatic conditions.

## II.   MYCORRHIZAE

### A.   Classification and Description

Most individuals of most species of angiosperms and gymnosperms are mycorrhizal. The term mycorrhiza (plural: mycorrhizae or mycorrhizas) is applied to associations of fungi with the absorbing organs of plants; more strictly, it is used only where the organ concerned is a root. The scope of the term and the systems of classification and nomenclature of mycorrhizae are controversial, due in part to diversity in nutritional interrelationships, in kinds of fungi and plants involved, and in morphological and anatomical details of different mycorrhizae. Traditionally, most mycorrhizae are classified as ectotrophic or endotrophic. Ectotrophic mycorrhizae (or ectomycorrhizae) are those in which the fungus is predominantly exogenous, forming a sheath around the root, and with penetration of hyphae between the cells of outer tissues of the root. Many familiar temperate forest trees are ectomycorrhizal, including beech, birch, and pine; their fungal partners are Basidiomycetes or Ascomycetes, and include many species of mushrooms well known for their conspicuous fruiting bodies.

Endotrophic mycorrhizae (or endomycorrhizae) have in common the absence (usually) of an external fungal sheath and the presence of hyphae within the cells of the root. Endotrophic mycorrhizae traditionally include three distinct types: those of (1) the Ericales and of (2) the Orchidaceae, with septate fungi, and (3) those of diverse groups of plants with species of the phycomycetous fungal family Endogonaceae. These latter mycorrhizae are called vesicular-arbuscular (VA) mycorrhizae because the fungus forms vesicles and intracellular arbuscules (hyphae which branch dichotomously and repeatedly until they end in a mass of fine hyphae). Additional miscellaneous kinds of associations are also included within the endotrophic group.

The foregoing dichotomous (ectotrophic/endotrophic) classification and brief descriptions are prerequisite for examination of the literature on mineral nutrition of mycorrhizae, but should not be allowed to camouflage the structural and functional diversity of mycorrhizae. Lewis's review [11] is recommended for further discussion of the concept of mycorrhizae and for comparison of different classifications. The biology of mycorrhizae has been reviewed by Harley [12], Hacskaylo [13], Shemakhanova [14], Marks and Kozlowski [15], Gerdemann [16], and Sanders et al. [17]; a bibliography is also available [18]. An intriguing hypothesis, that terrestrial plants are the product of an ancient and continuing symbiosis of an alga and a fungus, has been presented [19], a product and symptom of the growing enthusiasm for mycorrhizae engendered by increasing realization of their antiquity, ubiquity, and great significance to most plants.

B. Plant Responses

Some plant species are obligately mycorrhizal; for other species the association is facultative. Mycorrhizal and nonmycorrhizal individuals of the same species may differ in several major features, including mineral uptake and use. Quantitative and qualitative manipulations of the mycorrhizal status of plants can significantly affect their mineral nutrition and improve productivity. Much study of mycorrhizal associations has therefore concerned facilitation of plant mineral nutrition and potential improvement of agricultural and forestry practice.

1. Ectotrophic mycorrhizae

Although only approximately 3% of the species of flowering plants are ectomycorrhizal [20], most of these species are trees, and most tree species of cool and temperate forests are ectomycorrhizal. Ectomycorrhizae are therefore of considerable interest to forest scientists and have been and continue to be the subjects of much research. Many trees are obligately ectomycorrhizal under natural conditions (e.g., species of Abies, Larix, Picea, Pinus, Carpinus, Fagus, and Quercus), whereas others are facultative (e.g., Cupressus, Juniperus, Salix, Betula, Corylus,

Alnus, Ulmus, Pyrus, Acer, and Eucalyptus) [20]. Some of these latter
genera contain both VA mycorrhizae and ectomycorrhizae. Occurrence of
ectomycorrhizae in different forest types is discussed in detail by Meyer
[20], and soil factors influencing formation of mycorrhizae have recently
been reviewed [21,22]. There are broad differences in the nature and significance of mycorrhizae in different biomes [23].

The significance of ectomycorrhizae for successful growth and development of particular tree species can be great. Populations of obligately
mycorrhizal pines, for example, stagnate and may die in the absence of
their fungus partner. There are many regions of the world in which the
fungi of ectomycorrhizae do not naturally occur. Introduction of suitable
mycorrhizal fungi and establishment of mycorrhizal infection reverses
stagnation and results in approximately normal growth. Programs for
introduction of exotic tree species and establishment of man-made forests
have led to practical application of knowledge of the mycorrhizal association. Principles, practices, and examples of this important subject are
reviewed in detail by Mikola [24,25], Marx [26], and Meyer [20]. For
many tree species, successful establishment of plantings from seedlings
is absolutely dependent upon establishment of mycorrhizal associations,
involving both inoculation and site management to promote mycorrhizal
development.

Facilitation of mineral nutrition is one but not at all the only effect of
ectomycorrhizae on trees. The many effects are reviewed in Harley [12]
and Marks and Kozlowski [15] and include changes in disease resistance,
hormonal relationships, carbohydrate nutrition, root morphology and
anatomy, and other aspects of tree biology. Discussion of mineral nutrition necessarily includes several of these interacting factors. Bowen [27]
has critically reviewed the topic and provided extensive discussion of
various processes involved.

Increased growth of ectomycorrhizal plants, in contrast to nonmycorrhizal individuals under otherwise identical conditions, is usually interpreted to be due primarily to increased mineral uptake by the mycorrhiza.
This phenomenon is reported especially often for plants grown in soils or
in nutrient solutions which are relatively low (for normal plant nutrition)
in one or more nutrients. Evidence for facilitation of mineral uptake by
ectomycorrhizae is especially convincing in the case of controlled inoculation studies. The report of Mejstřík [28] is representative of such studies
and shows some of the many effects which occur. In this study, seedlings
of Pinus sylvestris grown in sterilized forest soil were inoculated with the
fungi Boletus luteus, B. bovinus, both species (together), or neither (i.e.,
uninoculated controls); plants were analyzed after one year's growth.
Inoculated plants were heavily mycorrhizal (e.g., 96% of the short roots
of plants inoculated with both species together), whereas only 3% of the
short roots of uninoculated plants were mycorrhizal. Dry weights of roots
of seedlings inoculated with B. luteus were 7 times larger than those of

uninoculated controls, shoot weights were 11 times larger, and the shoot/
root ratio was 3.2 in contrast to 2.1 for the uninoculated plants.  Phosphorus content of shoots of B. luteus-inoculated plants was 4.03 mg per
seedling; uninoculated controls were only 0.40 mg per seedling.  The concentration of phosphorus, however, was not significantly different in mycorrhizal plants.  In contrast, in the same study Mejstřík found that infection
of Picea abies by either Boletus species increased both the absolute levels
of phosphorus and the concentration.  Absolute increases were much greater
than 1,000% over controls; concentrations were approximately twice those
of controls.  In subsequent experiments with seedlings inoculated under
nonsterile conditions, differences in uptake of phosphorus were generally
less than those reported here in sterilized soils.

    Responses of ectomycorrhizal plants to different concentrations of
soil phosphorus are complex, as exemplified by the study of Malajczuk et
al. [29] on phosphorus uptake and growth of mycorrhizal and uninfected
seedlings of Eucalyptus calophylla.  Eucalypt ectomycorrhizae are anatom-
ically and morphologically similar to those of other trees such as conifers
which have been studied more intensively.  Similarly, ectomycorrhizae
have an important role in phosphorus uptake by eucalypts, and this is especially significant because phosphorus levels are notoriously low in the
Australian soils in which eucalypts commonly grow.  Malajczuk et al. [29]
grew young seedlings for a time in the presence or absence of mycorrhizal
fungi, and then raised them in sand which contained different concentrations
of phosphate.  Among the different inocula tested was nonsterile soil.  Control soils, and soils to which inocula were to be added, were sterilized by
$\gamma$-irradiation.  The sand used for later growth of seedlings was partially
sterilized by steaming, and phosphate was added in the form of $KH_2PO_4$.
Seedlings exposed to nonsterile soil produced mycorrhizae on a high proportion of the root systems of most plants at all levels (3-27 kg/ha phosphorus) of applied phosphate.  Phosphate applications had little if any
effect on the percentage of mycorrhizal infection.  In contrast, few uninoculated plants were ectomycorrhizal (these presumably arose from
contamination).  Total plant phosphorus (absolute content) was increased
by inoculation with unsterilized soil, and when the phosphorus content was
expressed per unit root dry weight it was apparent that efficiency of uptake
(concentration) by roots had been increased relative to uninoculated controls.  At the highest phosphate treatment, there was no significant effect
on phosphorus concentrations or contents of plants; the highest level of
phosphate application apparently replaced the effect of mycorrhizae in
promoting phosphate absorption.  This is a common observation for ectomycorrhizal plant species [27]: mineral absorption is enhanced in soils
having suboptimal concentrations, but not in those having higher concentrations; maximal nutritional responses occur upon establishment of
mycorrhizae in plants grown in relatively low fertility soils devoid of
mycorrhizal fungi.  Optimum phosphate concentrations for mycorrhizal

plants are much lower than optimum concentrations for nonmycorrhizal plants.

Soil phosphorus levels affect rates of establishment of ectomycorrhizal infection and are important in programs in which inoculation is required for establishment of plantations.  Lamb and Richards [30,31] found that addition of phosphate at 40 kgP/ha greatly increased formation of mycorrhizae in two species of pine and that phosphorus could be partly replaced by increasing inoculum density and by avoiding delay in inoculation.  At low available phosphorus concentrations (such as are common in many regions where such plantation programs are desired), the normal pattern of mycorrhizal infection of new short roots of pine is interrupted, and reinfection of the roots from fungi in the soil is of greater importance.  Hence, survival of potentially mycorrhizal fungi in the soil in the immediate absence of the potential host root might be a critical factor in inoculation programs. Greatly increased seedling yield was the net effect of optimizing inoculum species and spore type, time and spore type, time and density of inoculation, and phosphorus amendment.

Although most studies of facilitation of mineral uptake by ectomycorrhizae have involved macronutrients, especially phosphate, there is sufficient evidence to conclude that uptake of micronutrients is also enhanced. Bowen [27] noted that increased uptake of various elements could be expected as a result of increased root growth following relief of a major element deficiency following mycorrhizal inoculation.  Metabolically mediated absorption of zinc was 1.4 to 4.5 times greater for ectomycorrhizae of Pinus radiata than for uninfected roots; when corrected for the larger size of infected roots, the uptake per square centimeter was 1.4 to 3.0 times greater [32].

Increased absorbing power of the mycorrhizae is only one component of increased uptake of minerals from soil.  Transfer of ions to the root may itself be the limiting factor for uptake, especially for ions of phosphate, zinc, iron, and manganese which diffuse slowly through the soil.  In soil, uptake of these ions will be greatly increased by growth of the fungus out from the mycorrhizae into the soil, thereby effectively increasing the volume of soil available for supply of nutrients [27].  Species of ectomycorrhizal fungi differ in their ability to produce mycelial strands (cable-like aggregates of hyphal filaments), and Bowen [27] has suggested that ability to produce strands readily in a wide range of conditions should be considered in selecting mycorrhizal fungi for tree inoculation.  Strand-forming fungi could penetrate larger areas of soil than those species which did not form strands, thereby increasing the potential for absorption of soil minerals.  Skinner and Bowen [33] demonstrated that mycelial strands of the fungus Rhizopogon luteolus do take up and rapidly translocate [$^{32}$P]orthophosphate to Pinus radiata ectomycorrhizae from soil in both greenhouse material and in the field, and at distances of up to 12 cm.  They concluded that uptake and translocation by mycelial strands (as distinguished from

individual hyphae, which have less ability to penetrate soil) provide an effective method for mycorrhizal exploitation of the large interroot spaces
which occur with tree species, and assist the plants in competition for
nutrients.  Large differences occurred between their different strains of
R. luteolus in mycelial strand growth in soil, further emphasizing that
selection for large-scale field inoculation can profit from optimization of
fungal strain characteristics.  A subsequent report by the same investigators
[34] showed that mycelial strand growth can differ greatly between soils
even of the same general soil type and that compaction of soil greatly reduced mycelial strand penetration.

One of the most exciting and promising applications of ectomycorrhizae
is that of Marx and his colleagues, who are applying ectomycorrhizal fungi
in revegetation of strip-mined lands and other wasteland areas [26].  An
important aspect of their program is development of the fungus Pisolithus
tinctorius as an inoculum for seedlings of trees which are to be established
in these extreme environments; P. tinctorius is notable for its extensivity
developed mycelial strands which grow great distances through the nutrient-
poor waste material that remains after strip mining.  This current, rational application of mycological and forestry principles to "tailor" seedlings in the nursery with Pisolithus ectomycorrhizae is a model for affore-
station practice.  Survival and growth of Pisolothus-inoculated seedlings
has been much better than that of uninoculated seedlings.

The hazards in rejecting soil test systems in favor of more controllable
agar cultures in selection programs have recently been demonstrated by
Read and Stribley [35] in a study of diffusion and translocation of nutrients
(including $^{32}$P-labeled orthophosphate) in an ericaceous endomycorrhizal
fungus and in an ectomycorrhizal fungus.  Examination of the effects of
various diffusion barriers on living and dead material showed that some
of the nutrient movement observed using agar cultures may be due to passive diffusion and capillarity, in contrast to metabolically mediated active
translocation.  Although diffusive and capillary movement might well be
important in nature, the amount of transport observed in agar cultures
(with their characteristically dense aggregates of fungal hyphae, which
usually grow from a point source and are more or less in parallel) might
be unrepresentative of that in nature.  It is necessary to discriminate between active and passive modes of nutrient movement, because the capacity
for active translocation is likely to be an inherent characteristic of the
fungus, whereas a significant portion of the passive movement may be an
artifact of the culture system used.

In addition to the functions of absorption and transport just discussed,
there is the important ability of the fungal sheath of ectomycorrhizae to
store phosphate (and perhaps other minerals) when plentiful, and subsequently to move the phosphate into the plant when phosphate supply is low.
As much as 90% of the phosphate of some ectomycorrhizal fungi grown in
nutrient solution is in a nonmetabolic phosphate pool, probably in the

vacuole of the fungus [36]. Granules containing polyphosphate have been detected in the hyphae of ectomycorrhizae and of endomycorrhizae [37].

The detailed quantitative data necessary for ecosystem-level analysis of the quantitative significance of mineral accumulation by predominantly ectomycorrhizal fungi are only now being obtained for temperate conifer and deciduous forests [38,39]. It is likely that it will eventually be possible to place fairly exact values on the roles of these fungi in nutrient cycling in these ecosystems. This capability should improve the ability of forest managers to place a monetary value on their managerial decisions concerning inoculation and plantation management for optimal mycorrhizal development. Most ecosystems have so many different and potentially interacting species that attempts to analyze and experimentally manipulate the interactions of the few species of particular interest in a given situation get so thoroughly enmeshed in webs of interactions with other species that analysis and experimental design become hopelessly complex. But some interesting systems are much simpler than others because of their extreme environmental conditions which severely restrict species diversity. Thus, the study of relatively simple microbial communities in hot springs [40-42] has allowed study of microbial interactions on a more tractable level, resulting in significant discoveries and worthwhile generalizations. The study of plant-microbial interactions of the mycorrhizal association in such extreme habitats as anthracite wastes [43], arctic regions [44], and analogous places might provide valuable opportunities to study principles of mycorrhizal associations under situations which are relatively uncomplicated by interfering species.

2.  Vesicular-Arbuscular Mycorrhizae

a.  Mycorrhizae with Endogonaceae    Vesicular-arbuscular mycorrhizae are the most common kind of mycorrhizae; most plants growing under natural conditions possess VA mycorrhizal infection. It has been customary to assign the infecting fungi to the genus Endogone of the family Endogonaceae. Gerdemann and Trappe [45], however, have divided Endogone sensu lato into four genera: Endogone, which is associated with ectomycorrhizal hosts; Glomus and Gigaspora, which are endomycorrhizal; and Modicella, which is of unknown mycorrhizal status. Species of two other genera of the Endogonaceae, Acaulospora and Sclerocystis, also form VA mycorrhizae. In view of the thorough and scholarly nature of the monograph by Gerdemann and Trappe [45], the assignment of VA mycorrhizal fungi to genera other than Endogone will probably soon become commonplace in the scientific literature. Almost all of the articles discussed in the present review, however, have used the name Endogone for VA mycorrhizal fungi, and this name will be retained here.

The distinctive spores of Endogone are a regular component of the soil microflora [46]. Distribution of Endogone spores in the rhizosphere and also VA infection of roots are affected by light intensity, soil moisture, type of plant communities, temperature, seasons, and plant age. Spores are rare in permanently waterlogged soils, and hydrophytic plants are not, in general, infected with VA mycorrhizae. Although some families of vascular plants lack VA mycorrhizae, most families are infected [46-48]. In general, intensity of VA infection of roots is reduced in soils which have high levels of nutrients, especially of phosphate or nitrogen [48-50].

A few of the important plants that have VA mycorrhizae are maize, soybean, cotton, tobacco, potato, sugar cane, tomato, forage legumes, tea, citrus, coconut, Brazilian rubber tree, and sugar maple [47]. It is obvious that improvement of yield of crop plants by manipulation of their VA mycorrhizae would be of immense economic importance. Inoculation with Endogone and consequent establishment of mycorrhizae usually results in increased plant growth relative to nonmycorrhizal individuals. This has been demonstrated with maize [51], French bean [52], several herbage legumes [53], fourwing saltbush (an important range plant) [54], cotton [55], citrus [56-58], papaya [59], wheat [60], acacia [61], and many other plants [12,36,62,63]. Increased yield attributed to formation of VA mycorrhizae has been reported for soybean [64,65], strawberries [66], flowers of petunia [66], and for wheat (up to threefold) [60].

Endogone species are obligate symbionts. They have not yet been grown in axenic culture,[1] although a species has been grown monoxenically with clover seedlings [67], and in clover root organ culture [68]. In contrast, the VA mycorrhizal association is not obligatory for plant growth, but in most soils growth is improved if the plant is mycorrhizal. Especially noteworthy is the improved growth of mycorrhizal plants relative to nonmycorrhizal individuals grown in phosphate-deficient soils. Because phosphate is commonly a growth-limiting nutrient for plants in agricultural soils, the potential exists for improvement of crop productivity by stimulating formation of VA mycorrhizae in appropriate crop plants. Examples in which improved growth in phosphate-deficient soils can be stimulated by inoculation with Endogone species and subsequent establishment of VA mycorrhizae include onions [69], wheat [60], maize [70], and soybeans [64]. Additional examples are included in reviews by Gerdemann [47] and Mosse [63]. Growth responses may be dramatic. In onions [69], a 12-fold increase in shoot dry weight was obtained by inoculating seedlings grown in irradiated (i.e., Endogone-free) phosphate-deficient soil. In soils rich in phosphate, there was no response to inoculation. Mycorrhizae formed readily, but plant responses to mycorrhizae differed among soils obtained from different locations. In one of the phosphate-deficient soils, mycorrhizae developed extensively but did not increase plant growth, or total or percentage of percentage of phosphorus content. A possible explanation for this anomaly was provided by the observation that the external mycelium

---

[1] See the section Recent Developments at the end of this chapter.

of Endogone grew less in the one soil than in the other soils tested, perhaps decreasing possible benefits to the plant of additional absorbing surface produced by the fungus. This example reenforces the caveat that
attempted mycorrhizal inoculation in agriculture and forestry must consider
the complex variables of the soil environment, as well as the plant and
fungal partners of the mycorrhizal association.

Field trials of the feasibility of inoculation with Endogone are being
conducted with several economic plants. Mycorrhizal inoculation of wheat
seed may be beneficial [60]. The feasibility of inoculation on a field scale
by digging out mycorrhizal wheat roots and using them in a phosphorus-
deficient field having a low indigenous Endogone population is under investigation. Similarly, soils rich in certain types of Endogone spores are being
used to inoculate phosphorus-deficient field plots having fewer indigenous
spores [60]. There is a general correlation between the number of
Endogone spores in a particular soil and the amount of VA mycorrhizae of
wheat growing in it [71].

Growth, yield, nodulation, and nitrogen fixation by legumes can be
stimulated by VA mycorrhizae, especially in phosphorus-deficient soils
[52,53,64,72,73]. Although several attempts have been made to demonstrate that mycorrhizae can fix atmospheric nitrogen, this possibility remains unproven, and any nitrogen fixation which might be found to occur in
suspect species would be minor and insufficient to satisfy the species' requirements for nitrogen [22,74]. (See Becking [75] for a detailed review
and analysis of putative nitrogen-fixing mycorrhizae.)

The basis for the improved growth of VA mycorrhizal plants in phosphate-deficient soils appears to be that mycorrhizal plants take up and
accumulate soluble forms of phosphorus better than nonmycorrhizal individuals of the same species. In most cases of plants grown in phosphate-
deficient soils, the phosphorus content (percent dry weight) of mycorrhizal
plants is much higher than that of nonmycorrhizal plants. Radioactive
phosphorus was used by Gray and Gerdemann [62] to show that the sites of
increased accumulation are the VA mycorrhizae, and by Hattingh et al.
[76] to show that the mycelial network of Endogone enables plants to remove phosphate from a soil volume which extends beyond the immediate
vicinity of the root. Gray and Gerdemann [62] found that treatment with a
fungitoxic chemical reduced phosphate accumulation by mycorrhizae but
not by nonmycorrhizal roots. In further studies, Rhodes and Gerdemann
[77] have shown that absorption of phosphate and its translocation to the
host by hyphae of Glomus (Endogone) can extend the phosphate uptake zone
of mycorrhizal onions to at least 7 cm from the root surface, which is
considerably beyond the 1-2 mm zone normally assumed to be the region of
phosphate depletion. These data, and similar data reviewed by Smith [22]
and Tinker [3], suggest that the fungus absorbs phosphate and translocates
it to the plant. Thus, increased phosphate uptake by endomycorrhizae appears to be due to the fungus directly, rather than to the fungus increasing
uptake by the plant cells [32,78].

A second basis for the facilitation of nutrition might be the ability of VA mycorrhizae to increase phosphate uptake from apatite and other poorly soluble sources [79-82]. The case for solubilization of relatively insoluble forms of phosphate by the fungi of VA mycorrhizae remains undecided. Gerdemann [16] argues that there is no evidence that VA mycorrhizal fungi can dissolve insoluble forms of soil phosphate. He points out that phosphate is relatively immobile in soil and is present in the soil solution in very low concentration. If the amount in solution is depleted by roots, however, it is quickly replenished by release from solid-phase forms. Mycorrhizal fungi would bring about a more rapid utilization of soluble phosphate, resulting in release of phosphate into the soil solution from solid-phase phosphate. Presence of VA mycorrhizae does not ensure utilization of apatite or other poorly soluble sources of phosphate which are unavailable to the nonmycorrhizal individual [3,22,78,83-85]. Additional aspects of the mechanism of the nutritional affect of VA mycorrhizae are thoroughly discussed by Tinker [3].

Some Endogone strains are better than others in enhancing plant growth in phosphate-deficient soils. Most strains of Endogone can be transferred from one host plant species to another, and studies of the effect of different strains on growth of their host are beginning to provide data of practical significance [48,50,86]. Endogone strains also differ in their ability to improve zinc uptake of peach seedlings [87].

In some soils, endomycorrhizal plants may grow better than nonmycorrhizal plants do over a wide range of amounts of applied phosphate. In other soils, where high concentrations of phosphorus rapidly accumulate within the plant, those plants with mycorrhizae may grow worse than nonmycorrhizal plants do when more than a moderate amount of phosphate is added. It is proposed [86] that this response occurs because the mycorrhizal roots take up enough additional phosphate to reach supraoptimal phosphorus concentrations. Mycorrhizae tended to die out when high concentrations of phosphate were added to the soil. The type and quantity of phosphate fertilizer added, and its method of application under field conditions, influence the number of Endogone spores and the amount of VA mycorrhizae in the field [88]. At phosphate concentrations normally encountered in properly fertilized agricultural soils, it is possible that changes will be more moderate than those observed at very high phosphate concentrations in experimental pots. Nevertheless, the possibility of yield reduction in crops due to VA mycorrhizal infection does exist and must be taken very seriously [3].

Uptake of minerals other than phosphate can be facilitated by VA mycorrhizae. Results are inconsistent and are often confused by effects due to differences in phosphate uptake [63]. Gilmore [87], however, has provided convincing experimental evidence that inoculating peach seedlings with Endogone can prevent appearance of zinc-deficiency symptoms and result in greater growth and higher zinc content than for nonmycorrhizal seedlings. Metabolically mediated absorption of $^{65}$Zn was 2.6 times greater for Endogone-infected plants of hoop pine (Araucaria cunninghamii) than for uninfected plants [32], and uptake of $^{90}$Sr from soil by soybeans was markedly increased by infection with Endogone [89]. Increase of sulfur has been reported [90], and Powell [91] has reported that yield and potassium nutrition can be improved. VA mycorrhizal fungi can reduce nitrate, presumably increasing their effectiveness in terms of nitrogen assimilation by their hosts [92], although evidence of improved nitrogen nutrition has so far been conspicuously absent in studies of VA mycorrhizae, as opposed to those of ectomycorrhizae.

In contrast to the predominantly ectotrophic mycorrhizal status of temperate forest trees, trees of the tropical rain forest are predominantly endomycorrhizal [20, 93, 94]. Detailed studies of mycorrhizae of trees of tropical rain forests are not yet available, and the fungi involved are poorly defined; nevertheless a major hypothesis of possibly great significance has been proposed concerning these mycorrhizae. Based upon their field studies in the Amazonian rain forest, Went and Stark [93, 94] conclude that most of the leaf and other litter on the floor of the rain forest is decomposed by fungi and that most of these fungi are mycorrhizal. The fungi thus would form a link between forest litter and the tree roots, and would transmit most of the nutrients which they absorb from the litter to the tree roots. Nutrients are thus cycled from the tree to the litter and back to the tree, the mycorrhizal fungi closing the cycle. This phenomenon is called direct mineral cycling because the soluble minerals are not released into the soil. Once the cycle is broken by cutting and burning of the forest, as is practiced in certain forms of rain forest agriculture, only a poor secondary forest can be reestablished, which may further degenerate into a sandy semidesert. It is further suggested that direct nutrient recycling through mycorrhizal fungi is not only the secret of lush rain forest growth on poor tropical soils, but may also explain the existence of forests on very poor sandy soils such as those in Georgia and Florida. Quantitative values of the flow of nutrients in the tropical rain forest are given by Fittkau and and Klinge [95], who also conclude that fungi play a decisive value in concentrating the otherwise limited nutrient resources. The observations of Edmisten [96, 97] in a tropical rain forest in Puerto Rico also support the idea that mycorrhizae act as a trap for minerals, especially anions such as nitrate, and recycle them to higher plants instead of allowing loss in runoff. Of 32 species of rain forest dominants which were surveyed for mycorrhizae, 12 were endotrophic, 6 were ectotrophic, 5 were intermediate, and 9 lacked mycorrhizae.

Important questions remain unanswered concerning the relative contribution of endo- and ectomycorrhizae in plant nutrition in the tropical rain forest, including the identity and significance of the abundant mycelial strands found in the litter layer and the ability of the predominant mycorrhizal fungi to degrade litter. The significance of the potential applications of the concept of direct mineral cycling is enormous, but much more detailed studies are required before applications are made.

Similar direct observations of the abundance of mycorrhizal fungal hyphae and of fungal rhizomorphs in soils of deserts and summer-dry forests have been made by Went and Stark [93] and Went [98]. Again, detailed descriptions of the functional nature of the mycorrhizal associations are not provided, but the argument is made that the amount of biomass of mycorrhizal fungi is significant in these soils and that the role of these fungi in plant nutrition and soil physics deserves far more study than it has received.

b. <u>Ericaceous mycorrhizae</u> Mycorrhizae of members of the plant order Ericales (including species of the family Ericaceae and others), and the fungi which form them, are heterogeneous in morphology and anatomy. Penetration of host plant cells and formation of intracellular hyphal masses is common, and formation of a sheath of hyphae about the root occurs in some types. Several different species of septate fungi are involved in different ericaceous mycorrhizae, and some have been grown in pure culture. Considerable controversy exists over the species of fungi to be included as capable of forming mycorrhizae in nature with the Ericales, and whether or not infection is systemic [22]. Among the major mycorrhizal plants of the Ericales are species of <u>Arbutus</u>, <u>Arctostaphylos</u>, <u>Vaccinium</u>, <u>Calluna</u>, <u>Monotropa</u>, and <u>Rhododendron</u>.

The early observation that many members of the Ericales grow on nutrient poor soils suggested that the mycorrhizae might facilitate mineral nutrition [12]. On the most nutrient-deficient soil, for example, non-mycorrhizal plants did not grow at all, whereas mycorrhizal plants did grow [99]. Mycorrhizal infection of inoculated plants of cranberry (<u>Vaccinium macrocarpon</u>) increased dry weight of plants grown on infertile soils [100, 101]; mineral nutrition was also improved. Mycorrhizal plants contained higher concentrations of nitrogen and had higher specific absorption rates for nitrogen than did uninfected plants. Mycorrhizae did not stimulate growth of plants which were grown on a fertile soil. Thus, mycorrhizae are probably of advantage to <u>Vaccinium</u> when it is growing in infertile soils (as it characteristically does), but not elsewhere, at least in regard to facilitation of mineral nutrition. Phosphorus uptake is also facilitated by ericaceous mycorrhizae [102,103]. Most of the phosphorus and nitrogen in the soils to which ericaceous plants are normally confined is organically complexed. Pearson and Read [104] have shown that the major fungal partner of mycorrhizae of <u>Calluna vulgaris</u> is able to utilize organic sources of nitrogen and phosphorus; this is likely to be of considerable significance in facilitating the supply of these nutrients to the host plant.

c. <u>Orchidaceous mycorrhizae</u> Orchid seeds are extremely small, weighing as little as 0.3 μg. The small amount of food material stored by the seed is soon utilized as the seed germinates and growth begins. Organic matter required for further growth is then absorbed from the substrate. The embryo gradually becomes autotrophic as growth and differentiation proceed and as chlorophyll formation occurs in photosynthetic organs (in autotrophic species). There is always, therefore, a period of saprophytic absorption; in the case of the many saprophytic species of orchids, such absorption may be permanent.

All orchids, at some phases of their lives, are infected with fungi under natural conditions, so that their absorbing organs contain fungal hyphae. These fungi are Basidiomycetes or have affinities to the Basidiomycetes, and characteristically have a vigorous carbon nutrition, including the ability to utilize complex carbohydrates. Much of the research on the orchid/fungus association deals with the role of the fungus as a trigger of metabolic processes leading to normal utilization of seed food reserves and the initiation of normal growth and differentiation of the seedling. Another subject which has received considerable attention is the role of the fungus in supplying carbohydrates (and in some cases growth factors) to the orchid in the early stages of growth before the plant can become self-supporting (if ever) by photosynthesis. Effects of the addition of different exogenous carbohydrates at different times (relative to germination) and concentrations have received special treatment, as have biochemical and anatomical mechanisms for transfer of carbohydrates between the fungus and orchid [12,22,105-107].

Few studies have yet been done on the role of mycorrhizal fungi in facilitating mineral nutrition of orchids. Fungus-mediated transfer of labeled phosphate, for example, from exogenous sources into orchid cells has not yet been demonstrated [22], although Smith [108] has shown that the mycelium of the mycorrhizal fungus Rhizoctonia repens could translocate $^{32}$P-labeled orthophosphate and that the label did appear in orchid seedlings as a result. In view of the importance of mycorrhizal fungi in facilitation of mineral nutrition of other plants, it is surprising that this aspect of orchid physiology has been virtually neglected.

III.   LICHENS

Lichens increase the supply of minerals for higher plants by weathering rock and by accumulating minerals from the substrate, from rainwater, and from the atmosphere. The exaggerated view of eighteenth and nineteenth century naturalists that lichens are of foremost importance as primary colonizers of bare rock and as pioneer organisms in plant succession and soil development has given way to an underestimation of their role in rock weathering and soil development [109].

A lichen is a symbiotic association of an alga and a fungus that produces a vegetative structure (lichen thallus) which is unlike either component alone. The algal cells are completely surrounded by fungal tissue,

and in a few species fungal haustoria penetrate the algal cells -- evidence of
the close nutritional relationship of the partners of the thallus.   Lichens
can grow on rock, including newly exposed surfaces of boulders and tomb-
stones, and on such other stressful substrates as glass windows and animals.   Their resistance to desiccation, as well as their ability to regain
water rapidly and to take up needed water from atmospheric water vapor,
account in part for their remarkable occurrence on exposed rock and similar seemingly inhospitable substrates.   An introduction to the biology of
lichens is available in the book by Hale [110]; more specialized reviews
are presented by Ahmadjian and Hale [111], Brown et al. [112], and
Smith [112a].

A.   Weathering: Physical Processes

Two physical processes by which rock-inhabiting lichens may weather
their substrates are rhizine penetration and thallus expansion and contraction.   Rhizines are compacted strands of fungal hyphae that originate
largely from the lower portion of the lichen thallus and penetrate the surface of the substrate, e.g., rock.   Rhizines of lichens growing on lime-
stone penetrate up to 16 mm [113].   Various authors have suggested that
penetration by rhizines contributes mechanically to the disintegration of
rock which occurs beneath several rock-inhabiting species [109].

Fry [114] presented detailed observational and experimental evidence
that lichen thallus expansion and contraction contribute to the physical
weathering of rock surfaces.   Crustose lichens are firmly attached to rock
substrates.   These lichens often contain gelatinous material which causes
the lichen to expand and contract with wetting and drying.   As contraction
(including arching of the lichen) occurs, fragments of rock are pulled loose
from near the growing margin of the thallus.   The effect of this small-
scale disintegration of the rock surface is important in increasing the surface area of the rock available to chemical action, thereby rendering it
more susceptible to chemical weathering.   Furthermore, long-term
studies [115] of rock lichen communities have shown that recycling of
lichen species occurs; that is, in certain situations foliose (leafy) lichens
grow for a while, then the thalli crumble and disintegrate, to be replaced
by crustose lichens, which are in turn replaced by foliose lichens, and so
on.   This recycling may occur over periods of time which are measured in
a few decades.   Presumably, changes in the cycle are accompanied by
mechanical transfer of substrate mineral material into nearby soil as the
thalli disintegrate.   Since lichens cover virtually all available rock surfaces
in some ecosystems, the amount of mineral material involved in this transfer may be appreciable.   Lichens may also leak a considerable amount of
minerals when a dried thallus is rewetted [116].

B.   Weathering:  Chemical Processes

Lichens produce chemical compounds which can contribute to chemical
weathering of rocks.   These compounds include the lichen compounds (also
called lichen substances, lichen acids, lichenic acids, etc.), $CO_2$, and
oxalic acid.   Lichen compounds, although variously defined, are a large
group of compounds, mainly weak phenolic acids, that are synthesized by
lichens.   Their function is unknown, although some have antibiotic activity.
Their structural chemistry and distribution have been intensively studied
[117-122].

    Lichen compounds contribute to rock weathering by forming metal
complexes with rock minerals.   The lichen compound squamatic acid, a
depside extracted from the lichen Cladonia squamosa, is capable of solubilizing iron from the sandstone on which the lichen grows, and also
chelates ferric ions in solution [123].   Algal and fungal isolates obtained
from this and from other lichens showed essentially no ability to chelate
iron, but many free-living fungi isolated from the same sandstone chelated
considerable amounts.   Citric acid was thought to be the chelator produced
by free-living fungi.   Iskandar and Syres [124] found that soluble complexes
formed when lichen compounds were shaken with suspensions of granite,
basalt, and biotite.   Release of cations (of Ca, Mg, Fe, and Al) from the
silicate materials resulted largely from metal complex formation rather
than from reactions directly involving hydrogen ions (in general, lichen
compounds are rather weak acids).   As in the study of Williams and
Rudolph [123], citric acid released considerably greater amounts of
cations.

    If the metal complexes are formed when lichen compounds react with
minerals and rocks in the laboratory, and also are formed under field
conditions, then their solubility should result in chemical weathering.
Although some authors [see Ref. 109] have questioned the solubility of
lichen compounds themselves in water and therefore doubt the importance
of lichens in chemical weathering, it has since been shown that representative lichen compounds are slightly soluble in water [123,125].   The solubility of the ten lichen compounds tested by Iskandar and Syers [125]
ranged from 5 to 57 mg per liter (in water).   Moreover, solid lichen compounds, shaken with water suspensions of biotite, formed soluble complexes [124].   The common lichen compound usnic acid is released from
lichen thalli sprayed with water in the field, suggesting that lichen compounds may be released by action of rainwater [126].   Lichen compounds
are extracellular and are often produced in great abundance, forming
crystals on the fungal hyphae of the thallus.   In crustose lichens, the
lichen-compound-rich layer of the thallus may be in direct contact or in
close proximity to the rock substrate.

    The blue-green alga Nostoc muscorum, which occurs in some lichen
thalli [127] and which is a common free living organism on soil, also

produces chelating agents in pure culture and in bacteria-associated cultures [128].

Respiratory $CO_2$, dissolved in water, would yield hydrogen ions which could promote decomposition of certain substrates, especially those composed of carbonates. The actual importance of biologically produced $CO_2$ in chemical weathering by lichens has not been determined. Similarly, several investigators have suggested that oxalic acid is excreted by lichens and may be important in rock decomposition because it removes calcium and deposits it as insoluble calcium oxalate. Some lichens contain appreciable amounts of calcium oxalate [129]. Syers and Iskandar [109] argue, however, that calcium oxalate formation by lichens would be of minor significance in chemical weathering.

Different species of rock-inhabiting lichens may be expected to have different effects on different substrates, just as the ability to form metal complexes differs with both the lichen compound used and with the type of mineral suspension tested. The effects of lichens on limestones is especially clear: the thallus may become immersed in the stone, with development of pits in the stone where fruiting bodies occur [113]. The weathering crust of lichen-covered basalt of recent Hawaiian lava flows is thicker than that which lacks lichens; this may indicate that chemical weathering has been enhanced by the presence of lichens. The lichen-covered crust is much richer in Fc and poorer in Si, Ti, and Ca than the lichen-free crust [130]. Additional examples of chemical and mineral changes caused by lichens in the substratum are reviewed by Syers and Iskandar [109] and include a variety of syntheses of new mineral phases (for example, the formation of colloids from the parent rock). The accumulation of primitive or lithomorphic soils beneath lichens is thus well documented [see Ref. 131].

The ability of lichens to accumulate mineral elements and radioactive nuclides is a well-known and remarkable property of these organisms that has received considerable attention. Lichens have the ability to accumulate high levels of heavy metals from industrial fallout and have been involved in accumulation of radioactive nuclides in human food chains [109,132, 133]. Absorption occurs both from the substrate of the lichen and, to a considerable extent, from the atmosphere (from dust and rainout). According to studies cited by Syers and Iskandar [109], several Russian workers have found that lichens accumulate phosphorus and transform apatite into forms of phosphorus which are available for plant growth. Approximately a 90-fold enrichment of phosphorus was found in the fine-earth fraction beneath lichen thalli. Other studies showed that other nutritionally important elements, including Mg, Ca, K, Fe, and various microelements, are accumulated by lichens and converted to forms available to higher plants [134].

Epiphytic (i.e., growing on other plants) lichens are a significant part of some plant communities. They may affect mineral nutrition of higher plants by altering the inorganic chemistry of bulk precipitation, through elution of captured aerosols, by surface exchange reactions with through-fall and stemflow solutions, and by ion uptake. Lang and Reiners [135] described enrichment in metallic cations (Ca, Mg, K) and decrease of ammonium in throughfall and stemflow, and found experimentally that lichens yielded hydrogen and metallic cations and took up ammonium. Hoffman [136] showed that epiphytic lichens were highly efficient in taking up $^{137}$Cs from stemflow. This study, in which $^{137}$Cs was introduced into tree stems of Liriodendron tulipifera through slits cut into the bark, suggests that mineral nutrients can be cycled through epiphytic lichens. The slow growth rates and long lives of these lichens may mean that they act as a sink for some elements in some ecosystems, holding elements out of more rapid circulation.

An interesting interaction between two of the groups of organisms considered in this review is described by Brown and Mikola [137]. Lichens were found to have an inhibitory effect on natural mycorrhizae and seedling growth of pines; removal of the abundant growth of soil lichens by grazing or other means removed this inhibition. Whether this inhibition is due to production of antibiotics, sequestration of minerals, interception of water, or other effects of the lichens is not known; as mentioned earlier, lichens do produce antibiotics [122].

## C.   Nitrogen Fixation

Large quantities of nitrogen may be converted into biologically usable forms by lichens under certain circumstances. This is suggested by observation that nitrogenase activity (and therefore nitrogen fixation) can occur at a high rate in lichens. The percentage of nitrogen content is relatively low, the rate of growth is slow [138], and the nitrogen-fixing blue-green algae of lichens liberate very large proportions of their assimilated nitrogenous substances into a medium. More than 60% of the assimilated nitrogen may be released, although the proportion is usually much less. Most of the extracellular material is in the form of polypeptides, with smaller amounts of free amino acids [139-142]. Millbank and Kershaw [143], using labeled nitrogen, showed that the nitrogen which was actively and rapidly fixed by intact cephalodia (wartlike outgrowths of the lichen thallus) containing the blue-green alga Nostoc was virtually all secreted. The cephalodia could be considered a nitrogen-fixing system, with the rate of fixation equal to the rate of secretion. The secreted nitrogen would then be taken up by the fungal component of the lichen. Release of nitrogen to the soil could occur by leaching, grazing, secretion, or decomposition of the thallus.

One measure of the possible contribution of minerals to their environment by lichens are quantitative changes in the mineral composition of the lichen biomass of an area; this would be a measure of mineral turnover.

In this regard, the total nitrogen content of non-nitrogen-fixing lichens on a sandy, dry, grassland soil in Hungary underwent major seasonal changes [144]. For example, the total amount of N per square meter of the dominant lichen species ranged in amount from 500-1,000 up to 2,400 mg/m$^2$ at different seasons. By implication, an appreciable amount of nitrogen, obtained from soil, rain, and dust, is released to the soil. Similar studies indicate that the role of lichens in calcium turnover in these plant communities is more important than that of the higher plants (grasses) associated with them [145].

Lichens which do not include blue-green algae do not fix nitrogen [138, 146], except at very low rates [147] which are attributable to the presence of nitrogen-fixing bacteria. It is probable that all lichens which contain filamentous blue-green algae are capable of fixing nitrogen [148]. In lichens in which the primary algal component is not a blue-green alga, nitrogen fixation occurs only if cephalodia containing blue-green algae are present [143,149]. Initial studies of the nitrogenase activity of the symbiotic blue-green alga Nostoc in the lichen Peltigera canina estimated that activity of the symbiont was two or three times that of the free-living blue-green alga [150]. In conjunction with the low proportion of heterocysts found, this led to the suggestion that nonheterocystous, vegetative cells of Nostoc fixed nitrogen in lichen thalli. (Heterocysts are cells with relatively undifferentiated interiors and thick refractive walls; they are thought to be the site of nitrogen fixation in species of blue-green algae which produce them [141,151].) Studies using improved techniques for estimating cell numbers, however, indicate that the rate of nitrogen fixation by the Nostoc in P. canina is of the same order as in free-living cultures of the alga [152]. This line of evidence cannot be used, then, for proof that nitrogen fixation occurs in vegetative cells of Nostoc.

Only 8% of the total lichen species of approximately 18,000 have blue-green algae as their algal component, and in general lichens containing green algae are more widely distributed than are those containing blue-green algae. The world distribution of the two groups does not suggest that nitrogen-fixing lichens have a gross advantage over non-nitrogen-fixing species [153]. However, lichen species of tropical regions are poorly studied. In a recent study it was found that virtually all of the canopy lichens of a Columbian rain forest contained only blue-green algae [154]. Our knowledge of the distribution of blue-green algae in lichens may have been biased by studies done in temperate, subarctic, and arctic ecosystems.

The annual increment to an ecosystem as a result of nitrogen fixation by lichens is the measurement of ultimate interest to us here. This is difficult to assess because it varies from area to area depending on numbers of nitrogen-fixing lichens and the abundance of each species, and may vary markedly throughout the year. We shall therefore consider studies of several different ecosystems.

Based on the mean acetylene reduction rate over a 9-month period, Hitch and Stewart [138] have calculated that the amount of nitrogen fixed by lichens at Scottish coastal habitats was comparable to, or considerably greater than, the average rates of nitrogen fixation due to free-living blue-green algae in natural ecosystems. Similar results have been obtained by Crittenden [155] for nitrogen fixation by lichens on glacial drift in Iceland. Tundra ecosystems characteristically have a low available nitrogen supply [156,157]. Although non-nitrogen-fixing species are usually the dominant lichens, nitrogen-fixing species are common and are locally abundant in less stable habitats such as some glacial deposits, fell-fields, and arctic barrens. Crittenden [155] found that some of the most abundant lichens on glacial drift are nitrogen fixing and that they include some of the first species to become established on recent drift. He concluded that lichens are probably the major contributors to fixation of nitrogen. Estimation of nitrogen inputs varied greatly among the four sites studied, ranging from 62 to 1,700 $\mu g/m^2$ per day. In view of the ready availability of moisture in the study area (precipitation was more than 2,000 mm per year, in addition to frequent sea mists) and comparatively mild temperature, he suggests there are extensive periods of the year when conditions are suitable for lichen nitrogen fixation.

Schell and Alexander [158] used acetylene reduction rates to estimate nitrogen-fixation rates in arctic coastal tundra near Barrow, Alaska. At their sites, the prevailing summer temperatures were near freezing, with brief exceptions. The highest values of nitrogen fixation were found in lichen communities and were attributed to the lichens Stereocaulon sp. and Peltigera sp. These nitrogen-fixing lichens, which formed a crust between 1 and 3 cm thick, were considered essential to the sustenance of the pioneer plant community there. The tundra found in the vicinity of Barrow conserves nitrogen present in the upper 15 cm is equivalent to the gross inputs over a period of several thousand years, despite losses due to denitrification. Inputs are due to precipitation and nitrogen fixation, with nitrogen fixation being the greater of the two. Marked variations in nitrogenase activity occur with small differences in terrain on the tundra and with year-to-year differences in precipitation [159,160]. The mean nitrogen input for all of the tested microhabitats was 69.3 $mg/m^2$ per year [160]. This amount is 75% of the annual input of nitrogen from all sources [161]. The principal organisms active in nitrogen fixation were blue-green algae; these were free-living, growing on or in mosses, and as components of lichens. Comparisons of nitrogen fixation regimes at different circumpolar tundra sites [see Refs. 160 and 162] show that, in all cases where lichens are involved, moisture is of primary importance. Given adequate moisture, the response to temperature or light becomes important. Kallio and Kallio [163] also found that moisture was a limiting factor for nitrogen fixation by lichens at sites in northern Finland.

Despite the demonstrated ability of tundra lichens and free-living bacteria and blue-green algae to fix nitrogen, nitrogen levels in tundra soil are suboptimal for most plant species, as already mentioned. When moisture levels are not limiting for nitrogen fixation, it is probably that the low temperatures which predominate in tundra regions even in summer seriously limit nitrogen fixation by lichens. Tundra lichens show some temperature adaptation of nitrogenase activity, in the sense that maximum acetylene reduction takes place at a temperature $5°C$ lower in subarctic material than in temperate material of the same species [164]. Optimal rates in tundra lichens, however, occur at temperatures which last for only brief periods in their natural habitat [156,164]. Nitrogen fixation does occur in appreciable amounts at $0°C$ and below in both free-living and lichen blue-green algae and would be expected to occur at low rates beneath moderate snow cover in the north temperate zone during winter [165].

In the antarctic habitats studied by Fogg and Stewart [149], several species of lichens were nitrogen fixing; one of these was recognized as one of the principal nitrogen-fixing organisms in the area [166].

One of the major questions in understanding desert plant nutrition is the source of usable nitrogen in desert ecosystems. Nodulated nitrogen-fixing plants seem to be too widely dispersed to provide sufficient nitrogen for the many nonnodulated species which occur, and soluble nitrogen from mineralization of the limited amount of organic matter available appears to be insufficient to sustain plant growth on sandy sites. Among the potential sources of usable nitrogen in the desert are nitrogen-fixing lichens. Deserts have a rich lichen flora [167]. Of the lichens and mosses collected in the high deserts of Utah and Idaho, Snyder and Wullstein [168] found that the lichens Peltigera rufescens and Dermatocarpon lachneum fixed significant amounts of nitrogen (measured by acetylene reduction). Herford and Edmisten [169] found that species of Peltigera, Cladonia, and Parnelia growing on a granite outcrop in Georgia fixed nitrogen. The amount of assimilable nitrogen produced by desert cryptogamic crusts (primarily lichens) would appear to be small relative to that of nodulated soybean roots, having approximately one-tenth the acetylene-reducing capacity of fresh nodulated soybean roots, but this amount may be essential to the survival of the slow-growing plants in local sites of the desert ecosystem. Shields et al. [170] studied the species of alga- and lichen-stabilized soil crusts from several contrasting semidesert substrates to provide a basis for evaluating these surface crusts as a source of nitrogen. Based on measurements of lichen and algal populations, presence or absence of nitrogen-fixing bacteria, and concentrations of amino nitrogen, nitrite, and nitrate in these crusts and in underlying strata, these investigators concluded that alga- and lichen-stabilized surface crusts, through death and decomposition of component cells, release amino and other nitrogen compounds. These compounds are mineralized when moisture becomes available. The surface growth of algae and lichens thus represents a continually renewable supply of soil nitrogen. Rogers et al. [147] demonstrated that nitrogen fixation occurred in the lichen Collema coccophorus, which is widely distributed in the soil lichen crust of arid regions of Australia and which was a member of the arid soil crust studied in California by Shields et al. [170]. Annual nitrogen fixation by blue-green algae-lichen crusts in the Great Basin Desert was estimated at 10 to 100 kg/ha per year, depending on microenvironmental conditions [171].

A fascinating study of the significance of epiphytic lichens in the nutrition of forest trees is described by Denison [172] and Pike et al. [173]. Treetop studies using mountaineering techniques allowed in situ measurement of the biomass of epiphytes in an old-growth Douglas-fir forest in Oregon. The lichen Lobaria oregana was by far the most abundant epiphyte in the treetops, and the forest floor was littered with fallen pieces of its thallus. This lichen fixes nitrogen; its primary algal component is a green alga, but it has cephalodia containing the blue-green alga Nostoc. Between 1.8 and 10.0 lb of nitrogen per acre (i.e., 2.0-11.2 kg/ha) per year are made available to the forest by L. oregana. Other epiphytes that do not fix nitrogen contribute approximately 5.4 lb/acre (6.1 kg/ha) per year; this nitrogen is obtained from rainwater flowing over the branches and leaves, and includes some nitrogen released by decay of nitrogen-fixing epiphytes. Denison [172] concludes that in an old-growth stand of Douglas fir, the nitrogen-fixing plant life in the canopy can serve as the main pathway for introduction of new nitrogen. If this is so, then the generally destructive effects of atmospheric pollution on lichens [174-177] becomes particularly alarming. Air pollution can decrease nitrogen fixation by lichens [178].

Another remarkable study of epiphytic lichens is the measurement of the lichen community in the canopy of a Columbian rain forest by Forman [154], who performed it by felling trees. These lichens fixed an estimated 1.5-8.0 kg nitrogen/ha per year, equivalent to the estimated range of total nitrogen input in precipitation. Moisture, light, and temperature levels in the canopy were probably suitable for metabolism and nitrogen fixation throughout the year. Forman's measurements of total nitrogen fixed were probably underestimates because only a portion of the potentially nitrogen-fixing lichen population was measured. Qualitative studies in similar tropical forests suggested that lichens are probably abundant in the canopy in regions where there is no pronounced dry season. The nitrogen fixed at the top of this ecosystem is presumably distributed by leaching and by decomposition of lichens; the steady precipitation in the rain forest would provide the transport mechanism for constant dispersal of this important input of nitrogen. J. Edmisten (personal communication, 1976) tested the tree-trunk lichens and mosses of a Puerto Rican rain forest for nitrogen fixing ability and found a wide range of abilities. More than one-half of the lichens tested (by acetylene reduction) were positive and had blue-green algal components. This source of nitrogen fixation was estimated to have annual inputs of between 4 and 10 lb/acre (4.4-11.1 kg/ha). Other studies of nitrogen fixation by lichens and their potential contribution of nitrogen to their environment are tabulated in Forman [154], Millbank [10], Kallio and Kallio [163], and Alexander [179].

## IV. THE PHYLLOSPHERE

Substantial populations of microorganisms live saprophytically on the surface of healthy leaves of most plants. The nature and degree of activity of this microflora is not yet clearly understood, but attention has focused on competitive effects against leaf pathogens, a possible role in hastening the process of leaf senescence, and facilitation of plant nitrogen nutrition by means of microbial nitrogen fixation [180-184]. The leaf (including its surface) may be conveniently referred to as the phyllosphere.

In considering the role of phyllosphere microorganisms in facilitation of plant mineral nutrition, we become interested in the passage of minerals into and out of the plant through its leaves. There are two main reasons for interest in the leaching of minerals from leaves. First, these leached minerals provide many of the nutrients needed for the growth of phyllosphere saprophytes. Secondly, phyllosphere microorganisms may act as a nutrient sink, trapping minerals and delaying cycling to higher plants; in many climates this decay would occur during the optimal growing season for plants. It has been proposed [185] that the phyllosphere and rhizosphere organisms of tropical plants are also a sink for naturally occurring organic volatiles in the air. Minerals and organic substances are leached from leaves by rain, dew, mist, fog, and agricultural sprays. All of the essential minerals, including both the micro- and macroelements are leached. Amounts leached are significant. For example, the leachate of 1 ha of sugar beet during 18-24 hr of rainfall was 62 kg of ash contents, 32 kg of phosphoric acid equivalents, and 5 kg of CaO [186]. Losses from apple foliage were 20-30 kg K, 10.5 kg Ca, and 9 kg Na per hectare per year [187]. During 30 days, 2-3 kg each of K, Na, and Ca were leached from trees of Picea abies and Pinus sylvestris [188]. According to Tukey [189] in his review of leaching from plants, all plants studied so far can be leached to some degree at least. Guttation and bleeding can also lead to loss of minerals from leaves [190].

Substances lost from leaves can have several possible fates. They can fall to the ground and be reabsorbed by mycorrhizae and roots. They can be intercepted and absorbed by stems, branches, and leaves of the same and adjacent plants [191,192]. Soil is in part a product of plant activity, and plant leachates have an effect on the nature of soil [189]. Deep roots extract minerals from lower layers in the soil and translocate them into stems and leaves, from which they are returned to the upper layers of the soil. This process is sufficient to influence the mineral content of soil beneath vegetation and is thus of broad ecological significance in development of plant communities.

Phyllosphere organisms help plants retain minerals which are deposited on leaves in leachates, dry fallout, or rainfall. Nuclide retention of tropical rain forest leaves was 1.7 to 20 times greater for leaf surfaces with phyllosphere organisms intact than for those which had their phyllosphere organisms wiped off with cotton swabs (Table 9-1 [193]).

TABLE 9-1   Retention of Nuclides by Tropical Rain Forest Leaves [a]

| Plant species | Leaf treatment | Nuclide (ppm) retained by each 2.5 cm$^2$ leaf disk | | | |
|---|---|---|---|---|---|
| | | $^{137}$Cs | $^{32}$P | $^{54}$Mn | $^{89}$Sr |
| Ormosia krugii | Unwiped | 1,054 | 336 | 878 | 6 |
| (tree) | Wiped | 116 | 18 | 186 | 3 |
| Euterpe globosa | Unwiped | 1,061 | 253 | 956 | 7 |
| (palm) | Wiped | 53 | 24 | 416 | 3 |
| Dacryodes excelsa | Unwiped | 1,726 | 218 | 386 | 7 |
| (tree) | Wiped | 148 | 13 | 155 | 4 |
| Manilkara bidentata | Unwiped | 1,193 | 408 | 632 | 10 |
| (tree) | Wiped | 182 | 86 | 370 | 4 |

[a] Adapted from Witkamp [193].

In this study, disks were floated upside down on solutions of nuclides; wiped and unwiped disks were cut from the same leaf. The leaves had hard waxy cuticles and smooth surfaces; it is assumed that mechanical damage caused by wiping was not significant. Differences in mineral retention by phyllosphere organisms may be even greater on intact leaves in the field, because the freshly cut edges of the disks used in these experiments might significantly facilitate uptake of isotopes by the tissue. Odum et al. [194] found that the amount of radioactivity (attributed to fallout radioisotopes) in leaves of tropical plants was in proportion to their phyllosphere growth of organisms.

Plants can absorb minerals through their leaves. Foliar application of nutrients has been a satisfactory way of dealing with a number of special problems which were not solved by other means [195]. There are special situations in which foliar applications of nitrogen and magnesium in particular give satisfactory plant responses. Urea spraying has found application in apple growing and in production of sugar cane as practical means for supplying nitrogen needed by the plant. Nitrogen from urea is readily absorbed by many and diverse plant species. It is doubtful that foliar application of nutrients will ever supplant soil application as a general practice; nevertheless, the fact that plants can absorb significant amounts of minerals through their leaves has been firmly established [192, 195-197].

Now we shall consider the production of fixed nitrogen, one of the most important (in the sense that it is often limiting for plant growth) nutrients in the phyllosphere. Nitrogen-fixing microorganisms are abundant in the phyllosphere. Reports are summarized in Table 9-2 [154, 182, 198-202].

TABLE 9-2   Nitrogen-Fixing Microorganisms in the Phyllosphere

| Microorganisms | Source | References |
|---|---|---|
| Azotobacter spirillum, Beijerinckia fluminense, B. lactigenes, Klebsiella, Pseudomonas ambigua | Sheath of rice plants | 182 |
| Azotobacter macrocytogenes, A. beijerincki, A. vinelandii | Nymphea, Elodea (Java) | 182 |
| Klebsiella, Beijerinckia | Grass sheath (Java) | 182 |
| Klebsiella, Beijerinckia, Anabaena, Nostoc, Scytonema, Microcystis | Moss and liverworts (Java) | 182 |
| Klebsiella, Achromobacter | Zea mays (Netherlands); Tripsacum laxum (Surinam) | 198 |
| Beijerinckia | Miscellaneous trees and shrubs (Java, Sumatra, Banka) | 182 |
| Nostoc-containing lichens | Rain forest trees (Columbia) | 154 |
| Azotobacter, Beijerinckia | Morus indica (India) | 199 |
| Cryptococcus albidus var. albidus, Torulopsis candida, Trichosporon cutaneum, Metchnikowia pulcherrima | Several plant species (Ukraine) | 200 |
| Pseudomonas ambigua, Spirillum, coliform bacteria | Tripsacum laxum (Ivory Coast) | 201 |
| Azotobacter chroococcum, A. vinelandii, A. agile | Miscellaneous plants (Iraq) | 202 |

Beijerinckia was observed in 192 out of 196 cases on tropical shrub and tree leaves studied [203]. A total of at least $1.25 \times 10^7$ cells/cm$^2$ of Azotobacter and Beijerinckia occurred on a tropical leaf surface; this was a conservative measure in view of the methods used [204]. Beijerinckia and Azotobacter comprised 5 to 10% of the bacterial population of mulberry leaves [199]. Populations of Beijerinckia estimated by leaf washings were $4.8 \times 10^6$ cells/cm$^2$, and Azotobacter populations were $2.5 \times 10^6$/cm$^2$, on leaves of garden-grown mulberry bushes. The apparent ubiquity of Beijerinckia in the phyllosphere of tropical plants has led Ruinen [182, 203]

TABLE 9-3　Nitrogen Fixation in the Phyllosphere

| Materials | Increase in nitrogen | Method | References |
|---|---|---|---|
| Lawn moss | $50\text{-}60\,\mu g/m^2$ per hour | Acetylene reduction | 182 |
| Lawn grass | $7.8\,\mu g/m^2$ per hour | Acetylene reduction | 182 |
| Imperata cylin-drica, leaf sheath plus washed stem | $4.27\text{-}47\,\mu g/m^2$ per hour | Acetylene reduction | 182 |
| Nymphea, leaf | $122\,\mu g/m^2$ per hour | Acetylene reduction | 182 |
| Oryza, leaf sheath plus stem | $1.5\,\mu g/culm$ per hour | Acetylene reduction | 182 |
| Sea grasses | $12.6\ mg/m^2$ per hour | Acetylene reduction | 206 |
| Dew from Theobroma | 49.5 ppm/2 days | Kjeldahl | 207 |
| Gossypium barbadense | 27.4% increase/ 7 days | Kjeldahl | 207 |
| Coffea liberica | 135.8% increase/ per 13 days | Kjeldahl | 207 |
| Manilkara bidentata | 5.65% of 23.4 mg total N, total leaf phyllosphere organisms | $^{15}N$ | 205 |
| Manilkara bidentata | 1.478% conversion of $C_2H_2$ to $C_2H_4$, whole leaves | Acetylene reduction | 205 |
| | 0.11% conversion of $C_2H_2$ to $C_2H_4$, scraped leaves | Acetylene reduction | 205 |
| Citrus paradisi | 5% conversion of $C_2H_2$ to $C_2H_4$ | Acetylene reduction | 205 |
| Pseudotsuga douglasii | $1.79\,\mu g$ N/shoot per day, last year's shoot | $^{15}N$ | 208 |

TABLE 9-3 (Cont.)

| Materials | Increase in nitrogen | Method | References |
|-----------|---------------------|--------|------------|
| Coprosma robusta | Enrichment 0.19 atom percent excess $^{15}$N in 12 days, plant tops | $^{15}$N | 209 |
| Prunus armeniaca | Enrichment of 0.058 atom percent excess $^{15}$N in 7 days, plant shoots | $^{15}$N | 209 |

to suggest that there is a relationship between this species' nitrogen-fixing ability and the luxuriant growth of tropical plants on soils which are poor in available nitrogen (as many tropical soils are). As described in Section III of this review, lichens often contain nitrogen-fixing Nostoc, especially in the tropical rain forest. Forman [154] found that, in the lowland Costa Rican rain forest, 96% of the ground-cover vascular plants had lichens on at least one upper leaf surface and frequently on all leaves. On saplings, lichens were prominent on 95% of the leaves. These data, and those of Harrelson [205] given in this section, suggest that nitrogen-fixing phyllo-sphere lichens are of major importance in the nitrogen budget of the tropical rain forest.

The large amount of nitrogen fixed in the phyllosphere is of significance in the nitrogen budgets of various ecosystems. Data are summarized in Table 9-3 [182, 205-209]; many data, however, were not presented in a form suitable for inclusion in this table and are discussed separately. Harrelson [205; see also Refs. 97 and 209a] surveyed the nitrogen-fixing ability of phyllosphere organisms on leaves in tropical forests of Central America. Leaves sealed in plastic bags on trees in the forest gave positive results using acetylene reduction, as did the phyllosphere organisms when scraped from leaves and tested in flasks. Bacteria, blue-green algae, lichens, and liverworts fixed nitrogen, the latter two presumably due to procaryotes occurring on and in them. Harrelson estimated that 25-75 kg/ha per year of fixed nitrogen could be added to the ecosystem by phyllo-sphere organisms. Edmisten [97] studied the role of phyllosphere organ-isms in the nitrogen budget of a tropical rain forest in Puerto Rico at a site also studied by Harrelson [205]. Based on measurement of the nitrogen content of various model compartments (leaves, mineral soil, etc.) and fluxes (stem-flow and leachates, runoff, rain, etc.), Edmisten [97] pro-poses a preliminary nitrogen budget which shows that nitrogen-fixing phyl-losphere organisms are quite important. This intensive study of the El

Verde forest includes measurement of actual nitrogen fixation by phyllosphere organisms, using acetylene reduction. An estimated 65 lb of nitrogen per acre (73 kg/ha) per year are added from leaf leachate and phyllosphere organisms; phyllosphere organisms account for approximately
55 lb/acre (61.8 kg/ha) per year.

A considerable proportion of the annual requirement for nitrogen by
Douglas fir in England is provided by nitrogen fixation on the leaves and in
the soil; a significant proportion of this occurs on the leaves. For example, for a particular Douglas-fir forest having a canopy nitrogen content
of 704 kg/ha, phyllosphere nitrogen fixation amounted to 64.5 g/ha per
day. Similar proportions held for other Douglas-fir forests. The methods
for measuring nitrogen fixation was $^{15}$N tracer [208].

Ruinen [207] measured the increase of total nitrogen of detached
leaves of Coffea, Gossypium, and Phaseolus which were floated on a
nitrogen-free medium. She also tested the effects of enrichment of the
phyllosphere microflora with nitrogen-fixing bacteria. Gains in total
nitrogen ranging from 20 to 105% occurred within 2 weeks. Bhat et al.
[210] found that inoculation of leaves of Dolichos lablab with Beijerinckia
resulted in increased rate of accumulation of dry weight by seedlings, relative to uninoculated plants. Roots of inoculated plants exuded more
amino acids, and these in greater amounts, than did those of uninoculated
plants. These experiments were carried out using axenically and semi-
axenically grown plants. In the latter case, only the roots were maintained
under sterile conditions.

The amount of nitrogen fixed by the microorganisms in the phyllosphere
of Guatemala grass, 260-400 g/ha per year, was estimated by Bessems
[198] by exposing entire plants to $^{15}$N in situ. Nitrogen fixation was attributed mainly to Klebsiella. Acetylene reduction was also used to demonstrate nitrogen fixation qualitatively.

Nitrogen fixation by fungi, especially by yeasts, has often been
claimed and has often been refuted [143,211,212]. Ruinen [182] found
that common phyllosphere yeast-like fungi did not fix nitrogen. It is possible that some fungi do fix nitrogen, but single reports to this effect
should be accepted cautiously until confirmed by detailed study employing
varied techniques. Kvasnikov et al. [200] have reported that a number of
species of yeasts (see Table 9-2) isolated from the phyllosphere were
capable of increasing the nitrogen content of culture media. Addition of
2.25-7.99 mg N per 100 ml medium (determined by micro-Kjeldahl analysis) was reported; this addition was interpreted as being the result of
fixation of atmospheric nitrogen.

Occasionally, surveys of phyllosphere microorganisms do not detect
nitrogen-fixing bacteria, despite use of techniques which would be expected to detect them. For example, no nitrogen-fixing bacteria were
isolated from leaves of Prunus persica, Citrus sinensis, Gossypium
barbadense, Punica granatum, or Saccharum officinarum in Egypt by

Wahab [213], although nitrogen fixers have been found on leaf surfaces of several species of plants growing in the same region [202].

If the role of phyllosphere microorganisms in nitrogen fixation and other processes is as significant as suggested here, then it will be important to know how they are affected by the fungicides, herbicides, insecticides, and other "unnatural" chemicals (including those of polluted air) to which human activity increasingly exposes them. Research in this area is limited but is increasing. Dickinson [214] found that a narrow-spectrum fungicide, ethirimol, had no obvious effect on saprophytic phyllosphere fungi of barley, whereas the broad spectrum fungicide zineb considerably reduced the phyllosphere fungal flora. Similar major reductions in viable populations of saprophytic phyllosphere fungi on plants treated with agricultural fungicides have been reported by several authors [215-220; see also Refs. 221 and 222]. Changes in fungal populations were often accompanied by development of an anomalous saprophytic microflora dominated by one or a few species which normally played a lesser role in the populations. The effects of these changes on the activities and species composition of the microbial flora, especially the nitrogen fixers, is presumably significant but is virtually unknown. Before considering the potential significance of human-induced changes in the phyllosphere microflora as they may relate to mineral nutrition of the plant, we need first to consider in more detail the composition and dynamics of these populations under normal conditions.

Although some of the microorganisms found on leaves are merely deposited there as dormant structures, there is a distinct population of microorganisms which is characteristic of phyllosphere microfloras, predominates there, and actively grows and metabolizes in that niche. Different techniques used for estimating the species composition of the phyllosphere microflora give different estimates of the relative abundance of yeasts, yeast-like fungi, and filamentous fungi [184, 223-227], but yeast-like fungi, together with bacteria, appear to predominate. Among the yeast-like fungi, Aureobasidium pullulans (a filamentous fungus which has a yeast-like phase), Candida, Rhodotorula, Tilletiopsis, and especially Sporobolomyces are common. Certain species of filamentous fungi are much more common than others; these include Aureobasidium pullulans, Cladosporium herbarum, Botrytis cinerea, Alternaria tenuis, and Epicoccum nigrum [215, 228-231]. Some of these can become pathogens of plants, and A. pullulans can grow as a nonpathogenic endophytic parasite as well as a superficial saprophyte on plants [23]. The species of fungi isolated from fronds of bracken fern are similar to those found on angiosperm leaves [229].

Most species of saprophytic bacteria and yeast-like fungi on leaves share several features [231]. The same relatively few species tend to occur on a wide range of host plant species. These species of microorganisms are widely distributed in temperate zones of both northern and southern hemispheres.  They rarely occur on soils as frequently as on leaves.  They are usually pigmented.  For example, 81% of the bacterial isolates from leaves of <u>Lolium perenne</u> (perennial rye grass) in England were chromogenic [225].  Actinomycetes are relative uncommon [225].

A major theme of many studies has been that a definite succession of species and groups of species occurs in the phyllosphere.  Hudson [230] has summarized reports of a succession of species of fungi.  Ruinen [204] described a succession of organisms on tropical plants, leading to formation of a thick layer.  The sequence which she found is bacteria and actinomycetes, fungi (including yeasts), algae, lichens, mosses and ferns, and occasionally seed plants or flowering plants.  The climax population was rapidly reached and would be maintained on the mature and fully active leaf.  The period of maturity was extended on many nondeciduous species for up to 2 years.  Studies of herbaceous plants by Dickinson [232] and Bainbridge and Dickinson [217] indicate that filamentous fungi are relatively inactive on undamaged, green leaves; in contrast, Pugh and Buckley [227, 233] found that filamentous fungi were active in the phyllosphere from the earliest stages of development of deciduous tree leaves.  Colonization of the leaf surface is also affected by the availability of pollen, which is occasionally abundant and can provide a concentrated nutrient input available for growth of saprophytic microorganisms, as well as potential for nutrient recycling in the phyllosphere [184,224,234,235].

With the foregoing details of phyllosphere composition and dynamics in mind, let us consider how disruption of populations of one microbial group by application of antimicrobial agents can lead to unwanted changes in the microbial flora.  For example, application of antibacterial antibiotics to leaves can improve germination of spores of the common phyllosphere fungus (and potential pathogen) <u>Botrytis cinerea.</u>  The amounts of nutrients present on plant leaves are sufficient to cause marked stimulation of germination and growth of germ tubes of <u>B. cinerea</u> spores (conidia).  Conidia of <u>B. cinerea</u>, however, do not normally germinate satisfactorily on leaf surfaces, apparently due in part to competition by phyllosphere bacteria for small quantities of certain nutrients.  Suppression of bacterial growth releases the fungus from nutrient insufficiency, and so growth occurs [236-240a].  Competition for nutrients on leaves is a major theme of life in the phyllosphere.  Numerous examples of competition (usually by unspecified means) between saprophytes and parasites have been reported by plant pathologists interested in using this principle in control of disease [181,183,184,241].

The adverse effects of air pollution on many species of plants and lichens are well known; the effects on the phyllosphere microflora are not.  One example [182] is partially pertinent because it suggests that the nitrogen-fixing ability of the phyllosphere microflora is threatened by air pollution.  The original suggestion by Ruinen [203] that the phyllosphere microflora played an important part in the nitrogen economy of tropical plants was based upon the abundance of <u>Beijerinckia</u> on leaves in the famous Botanic Garden in Bogor, Java. In recent years, air pollution in Bogor has considerably increased with the rapidly developing street traffic. Ruinen [182, p. 150] notes that the "exhaust fumes of the two-stroke combustion motors drift into the garden during the greater part of the day." (Press reports mention the incredible levels of air pollution caused by motor bicycle engines in cities of the area, and concomitant killing of street trees.) Ruinen attributes reduced phyllosphere bacterial numbers in part to the oxidizing effects of this air pollution. Sites with less air pollution had greater rates of acetylene reduction. Yeasts and other fungi seemed to be less affected. By using tetrazolium to demonstrate the redox potential in the phyllosphere, she found that reduction was strongest under the microbial layer and near the hyphae of the fungus network where nitrogen-fixing bacteria, including <u>Beijerinckia</u> and <u>Azotobacter</u>, had congregated. Where the air was highly polluted (near filling stations, for example), the reaction was virtually absent throughout the microbial layer. Similar observations were made in crusts of lichen and growths of blue-green algae on leaf surfaces. In the center of the crusts, where a composite growth of fungi, blue-green algae, and bacteria occurred, reduction was high, in contrast with the layers in contact with the air. The differences which are found in redox potential within the phyllosphere may be important in allowing nitrogen fixation. Oxygen inhibits nitrogen fixation, in part because nitrogenase is irreversibly inactivated by prolonged exposure to oxygen. It is clear that air pollution can adversely affect phyllosphere nitrogen fixation in several ways.

Treatment of seeds with antimicrobial substances prior to planting, a common agricultural practice, can alter the subsequent phyllosphere microflora. For example, Klincāre et al. [242] reported 2.4 X 10⁶ microorganisms per gram on leaves of lucerne grown from untreated seeds, but 1.25 X 10⁶ from Mercuran-treated seeds and 1.5 X 10⁶ from streptomycin treatment.

Indoleacetic acid is produced by the bacteria [243,244], yeasts [245], with filamentous fungi [246] that occur on leaf surfaces, and presence of phyllosphere auxin-producing microorganisms does increase the auxin content of the host plant. In view of the manifold effects of microbially produced, exogenously applied plant-growth substances on plant nutrition, including mineral nutrition, phyllosphere microorganisms should prove worthy of further study in this regard.

In summary, facilitation of plant mineral nutrition by activities of phyllosphere microorganisms is widespread and is significant in both temperate and tropical regions. There is evidence that these activities are affected by human activities. Further research is needed to determine whether manipulation of this microflora is economically feasible for agriculture and forest practice.

## V.  SUMMARY AND CONCLUSIONS

Microorganisms facilitate plant mineral nutrition in two main ways:  (1)
by producing metabolites which plants can use (e. g. , fixed nitrogen from
atmospheric nitrogen), and (2) by making soil and/or organically bound
minerals more readily available for absorption (e.g. , by solubilizing minerals and by increasing the volume of soil which is tapped for its minerals).
Microorganisms are not absolutely essential for the growth of most plants,
since healthy plants can be grown under sterile conditions if all required
minerals are provided in suitable forms.  But in practice, microorganisms
are always associated with plants and, as the examples given in this review
demonstrate, these microorganisms can greatly aid plant mineral nutrition.  The significance of these microorganisms is especially great in situations in which the supply of minerals is limiting for plant growth.  This
fact promises an important place for future research on microbial facilitation of plant mineral nutrition.  Past availability of ample supplies of
chemical fertilizers for agricultural use in developed nations, coupled with
the profits to be made by producing and using these fertilizers, has done
little to encourage efforts to develop alternatives to use of these chemical
fertilizers.  But the economics of production, transport, and use of chemical fertilizers is changing.  Production of ammonium (nitrogen) fertilizers
in particular requires large inputs of energy and of petroleum, both of
which are increasingly expensive and likely to remain so.  National goals
may also call for decreased dependence on imports of fertilizers or of
materials which are needed for production of chemical fertilizers.  Partly
in response to these changes, increased efforts are being made to manipulate
nitrogen-fixing microorganisms to increase the biological production of
nitrogenous fertilizers from atmospheric nitrogen [247,248].  Much of
this research is aimed at incorporating bacterial nitrogen-fixing genes
into non-nitrogen-fixing crop plants.  At the same time, there is increasing
interest in improving plant productivity, or in avoiding decreases in productivity, in tropical rain forests and savannahs, temperate forests, coal
wastes and other waste areas, and other habitats in which microbial
facilitation of mineral nutrition may be especially significant and use of
chemical fertilizers may be impractical.  An intriguing example which has
not been discussed in this review is given in Farnworth and Golley [249,
p. 156; see also Ref. 250] ; it is compelling in its appeal for application of
microbiological principles.  At moderate to high elevations within the
geographical tropics there are many regions which have cool nights and
warm days throughout the year.  The availability of inorganic forms of
nitrogen and sulfur for plant uptake is greatly diminished at low night temperatures because the release of these nutrients from organic matter is a
function of microbial activity.  Many areas of the high Andes and Central
American highlands are therefore quite deficient in nitrate and sulfate for
crop use, even where soil reserves of the two elements themselves are

high. Most plants take up nitrogen and sulfur most efficiently only in oxidized form. It is suggested that the mycorrhizal association might be managed in such a way that nutrient uptake by plants could occur directly from organic matter (including organic fertilizers), without the need for other biological transformations.

Wilde [251, p. 484], referring to the mycorrhizal association, has written that, "a tree removed from the soil is only part of the whole plant, a part surgically removed from its rhizosphere and digestive organ." Biologists, especially agricultural and forestry scientists, now realize that laboratory experiments on plant nutrition, and research and application in the field and forest, must consider the microbial contribution to mineral nutrition of their plants. Nevertheless, the proportion of research effort given to VA mycorrhizae, for example, is miniscule in comparison to that devoted to Rhizobium and other nitrogen-fixing root nodule and soil bacteria. Yet, as Gerdemann [47, p. 414] has observed, "The amount of plant tissue in the world infected by Endogone species must exceed that infected by any other group of fungi. A relationship as common and as biologically important as this deserves much greater recognition and study." Research on VA mycorrhizae has increased since that was written; this research, as shown in the present review, further demonstrates the desirability of manipulating VA mycorrhizal populations, but research remains insignificant to proportion to the importance of the subject.

This review has dealt with plant-microorganism relationships which are less familiar to most biologists than are some such as the Rhizobium-root nodule association but which are potentially of great significance in many important habitats. The microorganisms involved present challenges and opportunities that require the research skills of the microbiologist, whether it is to grow Endogone species in pure culture, determine the effects of air pollution on the nitrogen-fixing ability of blue-green algae or lichens, or manipulate phyllosphere microbial populations.

RECENT DEVELOPMENTS

Several reports which include new information about the significance of endomycorrhizae in plant mineral nutrition are included in Endomycorrhizas, edited by Sanders et al. [252]; this book includes a report by J. H. Warcup of growth of a species of Endogone in pure culture on agar media. Microbiology of Aerial Plant Surfaces, edited by Dickinson and Preece [253], includes new contributions concerning the roles of phyllosphere microorganisms in plant mineral nutrition.

ACKNOWLEDGMENTS

The author is indebted to W. Boss, W. C. Denison, J. A. Edmisten,
J. W. Gerdemann, P. Patterson, and L. Rhodes for reading and comment-
ing on portions of the manuscript of this review.

REFERENCES

1.  J. M. Vincent, in The Biology of Nitrogen Fixation (A. Quispel, ed.),
    North-Holland Publ., Amsterdam, 1974, pp. 265-341.
2.  G. Bond, in The Biology of Nitrogen Fixation (A. Quispel, ed.), North-
    Holland Publ., Amsterdam, 1974, pp. 342-378.
3.  P. B. H. Tinker, Symp. Soc. Exptl. Biol., 29, 325 (1975).
4.  D. A. Barber, Pestic. Sci., 4, 367 (1973).
5.  J. Döebereiner, in The Biology of Nitrogen Fixation (A. Quispel, ed.),
    North-Holland Publ., Amsterdam, 1974, pp. 86-120.
6.  W. D. P. Stewart (ed.), Nitrogen Fixation by Free-living Micro-
    organisms, Cambridge University Press, New York, 1975.
7.  P. J. Ashton and R. D. Walmsley, Endeavour, 124, 39 (1976).
8.  G. A. Peters, Arch. Microbiol., 103, 113 (1975).
9.  D. A. Dalton and A. W. Naylor, Amer. J. Bot., 62, 76 (1975).
10. J. W. Millbank, in The Biology of Nitrogen Fixation (A. Quispel, ed.),
    North-Holland Publ., Amsterdam, 1974, pp. 238-264.
11. D. H. Lewis, Biol. Rev., 48, 261 (1973).
12. J. L. Harley, The Biology of Mycorrhiza, 2nd ed., Leonard Hill,
    London, 1969.
13. E. Hacskaylo (ed.), Mycorrhizae, Superintendent of Documents, U.S.
    Government Printing Office, Washington, D.C., 1971.
14. N. M. Shemakhanova, Mycotrophy of Woody Plants, Israel Program
    for Scientific Translations, Jerusalem, 1967.
15. G. C. Marks and T. T. Kozlowski (eds.), Ectomycorrhizae,
    Academic Press, New York, 1973.
16. J. W. Gerdemann, in The Development and Function of Roots (J. G.
    Torrey and D. T. Clarkson, eds.), pp. 575-591, Academic Press,
    New York, 1975.
17. F. E. T. Sanders, B. Mosse, and P. B. H. Tinker (eds.), Endomycor-
    rhizas, Academic Press, New York, 1975.
18. E. Hacskaylo and C. M. Tompkins, World Literature on Mycorrhizae,
    Contribution of Reed Herbarium No. 22, Baltimore, Maryland, 1973.
19. K. A. Pirozynski and D. W. Malloch, Biosystems, 6, 153 (1975).
20. F. H. Meyer, in Ectomycorrhizae (G. C. Marks and T. T. Kozlowski,
    eds.), Academic Press, New York, 1973, pp. 79-105.
21. V. Slankis, Ann. Rev. Phytopathol., 12, 437 (1974).
22. S. E. Smith, CRC Crit. Rev. Microbiol., 3, 275 (1974).

23. I. A. Selivanov, Zh. Obshch. Biol., 36, 107 (1975).
24. P. Mikola, Unasylva, 23, 35 (1969).
25. P. Mikola, in Ectomycorrhizae (G. C. Marks and T. T. Kozlowski, eds.), Academic Press, New York, 1973, pp. 383-411.
26. D. H. Marx, Ohio J. Sci., 75, 288 (1975a).
27. G. D. Bowen, in Ectomycorrhizae (G. C. Marks and T. T. Kozlowski, eds.), Academic Press, New York, 1973, pp. 151-205.
28. V. K. Mejstřík, New Phytol., 74, 455 (1975).
29. N. Malajczuk, A. J. McComb, and J. F. Loneragan, Austral. J. Bot., 23, 231 (1975).
30. R. J. Lamb and B. N. Richards, Soil Biol. Biochem., 6, 167 (1974a).
31. R. J. Lamb and B. N. Richards, Soil Biol. Biochem., 6, 173 (1974b).
32. G. D. Bowen, M. F. Skinner, and D. I. Bevege, Soil Biol. Biochem., 6, 141 (1974).
33. M. F. Skinner and G. D. Bowen, Soil Biol. Biochem., 6, 53 (1974a).
34. M. F. Skinner and G. D. Bowen, Soil Biol. Biochem., 6, 57 (1974b).
35. D. J. Read and D. P. Stribley, Trans. Brit. Mycol. Soc., 64, 381 (1975).
36. G. D. Bowen and A. D. Rovira, in Root Growth (W. J. Whittington, ed.), Plenum, New York, 1969, pp. 170-201.
37. M. Ling-Lee, G. A. Chilvers, and A. E. Ashford, New Phytol., 75, 551 (1975).
38. K. Cromack, Jr., R. L. Todd, and C. D. Monk, Soil Biol. Biochem., 7, 265 (1975).
39. A. G. Wollum, II, and C. B. Davey, in Forest Soils and Forest Land Management (B. Bernier and C. H. Winget, eds.), Les Presses de l'Université Laval, Quebec, 1975, p. 67.
40. P. C. Fraleigh and R. G. Wiegert, Ecology, 56, 656 (1975).
41. T. D. Brock, Ann. Rev. Ecol. Systemat., 1, 191 (1970).
42. R. W. Castenholz, in The Biology of Blue-Green Algae (N. G. Carr and B. A. Whitton, eds.), Blackwell, Oxford, England, 1973, pp. 379-414.
43. J. R. Schramm, Trans. Amer. Phil. Soc. [N.S.], 56, 1 (1966).
44. E. Ohenoja, Rep. Kevo Subarctic Res. Sta., 8, 122 (1971).
45. J. W. Gerdemann and J. M. Trappe, The Endogonaceae in the Pacific Northwest, The New York Botanical Garden, New York, 1974.
46. A. G. Khan, J. Gen. Microbiol., 81, 7 (1974).
47. J. W. Gerdemann, Ann. Rev. Phytopathol., 6, 397 (1968).
48. H. W. Kruckelmann, in Endomycorrhizas (F. E. T. Sanders, B. Mosse, and P. B. H. Tinker, eds.), Academic Press, New York, 1975. Cited in Tinker [3].
49. D. S. Hayman, Trans. Brit. Mycol. Soc., 54, 53 (1970).
50. D. S. Hayman, in Endomycorrhizas (F. E. T. Sanders, B. Mosse and P. B. H. Tinker, eds.), Academic Press. Cited in Tinker [3].
51. J. W. Gerdemann, Mycologia, 57, 562 (1965).

52. M. J. Daft and A. A. El-Giahmi, New Phytol., 73, 1139 (1974).
53. J. R. Crush, New Phytol., 73, 743 (1974).
54. S. E. Williams, A. G. Wollum, II, and E. F. Aldon, Soil Sci. Soc. Amer. Proc., 38, 962 (1974).
55. J. R. Rich and G. W. Bird, Phytopathology, 64, 1421 (1974).
56. D. H. Marx, W. C. Bryan, and W. A. Campbell, Mycologia, 63, 1222 (1971).
57. G. D. Kleinschmidt and J. W. Gerdemann, Phytopathology, 62, 1447 (1972).
58. M. J. Hattingh and J. W. Gerdemann, Phytopathology, 65, 1013 (1975).
59. B. N. Ramirez, D. J. Mitchell, and N. C. Schenck, Mycologia, 57, 1039 (1975).
60. A. G. Khan, Ann. Appl. Biol., 80, 27 (1975).
61. C. R. Johnson and S. Michelini, Proc. Fla. St. Hortic. Soc., 87, 520 (1974).
62. L. E. Gray and J. W. Gerdemann, Plant Soil, 30, 415 (1969).
63. B. Mosse, Ann. Rev. Phytopathol., 11, 171 (1973a).
64. J. P. Ross, Phytopathology, 61, 1400 (1971).
65. N. C. Schenck and K. Hinson, Agron. J., 65, 849 (1973).
66. M. J. Daft and B. O. Okusanya, New Phytol., 72, 1333 (1973).
67. B. Mosse, J. Gen. Microbiol., 27, 509 (1962).
68. B. Mosse and C. Hepper, Physiol. Plant Pathol., 5, 215 (1975).
69. D. S. Hayman and B. Mosse, New Phytol., 70, 19 (1971).
70. A. G. Khan, Biologia, Spec. Suppl., 42 (April 1972).
71. S. R. Saif and A. G. Khan, Can. J. Microbiol., 21, 1020 (1975).
72. J. P. Ross and J. W. Gilliam, Soil Sci. Soc. Amer. Proc., 37, 237 (1973).
73. J. P. Ross and J. A. Harper, Phytopathology, 60, 1552 (1970).
74. M. K. Jain and K. Vlassak, Ann. Microbiol. (Inst. Pasteur), 126A, 119 (1975).
75. J. H. Becking, in The Biology of Nitrogen Fixation (A. Quispel, ed.), North-Holland Publ., Amsterdam, 1974, pp. 583-613.
76. M. J. Hattingh, J. E. Gray, and J. W. Gerdemann, Soil Sci., 116, 383 (1973).
77. L. H. Rhodes and J. W. Gerdemann, New Phytol., 75, 555 (1975).
78. F. E. Sanders and P. B. Tinker, Pestic. Sci., 4, 385 (1973).
79. A. K. Eglite, Trud. Inst. Mikrobiol. Akad. Nauk Latviisk. SSR, 7, 67 (1958).
80. I. R. Hall, N. Z. J. Bot., 13, 463 (1975).
81. M. J. Daft and T. H. Nicolson, New Phytol., 65, 343 (1966).
82. C. L. Murdoch, J. A. Jackobs, and J. W. Gerdemann, Plant Soil, 27, 329 (1967).
83. K. F. Abeyakoon and C. D. Pigott, New Phytol., 74, 147 (1975).
84. B. Mosse, D. S. Hayman, and D. J. Arnold, New Phytol., 72, 809 (1973).

85. C. Ll. Powell, New Phytol., 75, 563 (1975).
86. B. Mosse, New Phytol., 72, 127 (1973b).
87. A. E. Gilmore, J. Amer. Soc. Hort. Sci., 96, 35 (1971).
88. D. S. Hayman, A. M. Johnson, and I. Ruddlesdin, Plant Soil, 43, 489 (1975).
89. N. E. Jackson, R. H. Miller, and R. E. Franklin, Soil Biol. Biochem., 5, 205 (1973).
90. L. E. Gray and J. W. Gerdemann, Plant Soil, 39, 687 (1973).
91. C. Ll. Powell, in Endomycorrhizas (F. E. T. Sanders, B. Mosse, and P. B. H. Tinker, eds.), Academic Press, New York, 1975. Cited in Tinker [3].
92. I. Ho and J. M. Trappe, Mycologia, 67, 886 (1975).
93. F. W. Went and N. Stark, Proc. Nat. Acad. Sci. U.S., 60, 497 (1968).
94. F. W. Went and N. Stark, Bioscience, 18, 1035 (1968b).
95. E. J. Fittkau and H. Klinge, Biotropica, 5, 2 (1973).
96. J. Edmisten, in A Tropical Rain Forest (H. T. Odum and R. F. Pigeon, eds.), U.S. Atomic Energy Commission, Washington, D.C., 1970, pp. F15-F20.
97. J. Edmisten, in A Tropical Rain Forest (H. T. Odum and R. F. Pigeon, eds.), U.S. Atomic Energy Commission, Washington, D.C., 1970, pp. H211-H215.
98. F. W. Went, Amer. J. Bot., 60, 103 (1973).
99. T. M. Morrison, New Phytol., 56, 247 (1957).
100. D. P. Stribley, D. J. Read, and R. Hunt, New Phytol., 75, 119 (1975).
101. D. P. Stribley, D. J. Read, New Phytol., 73, 1149 (1974).
102. V. Pearson and D. J. Read, New Phytol., 72, 1325 (1973).
103. D. J. Read and D. P. Stribley, Nature New Biol., 244, 81 (1973).
104. V. Pearson and D. J. Read, Trans. Brit. Mycol. Soc., 64, 1 (1975).
105. G. Hadley and B. Williamson, New Phytol., 70, 445 (1971).
106. G. Harvais, Can. J. Bot., 50, 1223 (1972).
107. C. L. Withner, in The Orchids (C. L. Withner, ed.), Wiley, New York, 1974, pp. 129-168.
108. S. E. Smith, New Phytol., 65, 488 (1966).
109. J. K. Syers and I. K. Iskandar, in The Lichens (V. Ahmadjian and M. E. Hale, eds.), Academic Press, New York, 1973, pp. 225-248.
110. M. E. Hale, Jr., The Biology of Lichens, 2nd ed., American Elsevier, New York, 1974.
111. V. Ahmadjian and M. E. Hale (eds.), The Lichens, Academic Press, New York, 1973.
112. D. H. Brown, D. Hawksworth, and R. H. Bailey (eds.), Lichenology: Progress and Problems, Academic Press, New York, 1976.
112a. D. C. Smith, Symp. Soc. Exptl. Biol., 29, 373 (1975).

113. J. K. Syers, Ph.D. Thesis, University of Durham, England, 1964. Cited in Syers and Iskandar [109].

114. E. J. Fry, Ann. Bot., 41, 437 (1927).

115. M. E. Hale, Jr., Bull. Torrey Bot. Club, 86, 126 (1959).

116. J. F. Farrar, Thesis, University of Oxford, 1973. Cited in Smith [112a].

117. C. F. Culberson, Chemical and Botanical Guide to Lichen Products, University of North Carolina Press, Chapel Hill, North Carolina, 1969.

118. S. Huneck, in The Lichens (V. Ahmadjian and M. E. Hale, eds.), Academic Press, New York, 1973, pp. 495-522.

119. K. Mosbach, in The Lichens (V. Ahmadjian and M. E. Hale, eds.), Academic Press, New York, 1973, pp. 523-546.

120. J. Santesson, in The Lichens (V. Ahmadjian and M. E. Hale, eds.), Academic Press, New York, 1973, pp. 633-652.

121. Y. Asahina and S. Shibata, Chemistry of Lichen Substances, Japan Society for the Promotion of Science, Tokyo, 1954.

122. K. O. Vartia, in The Lichens (V. Ahmadjian and M. E. Hale, eds.), Academic Press, New York, 1973, p. 547-561.

123. M. E. Williams and E. D. Rudolph, Mycologia, 66, 648 (1974).

124. I. K. Iskandar and J. K. Syers, J. Soil Sci., 23, 255 (1972).

125. I. K. Iskandar and J. K. Syers, Lichenologist, 5, 45 (1971).

126. J. Malicki, Ann. Univ. Mariae Curie Skłodowska [Sect. C. Biol.], 20, 239 (1965).

127. V. Ahmadjian, Phycologia, 6, 127 (1967).

128. W. Lange, Can. J. Microbiol., 20, 1311 (1974).

129. J. K. Syers, A. C. Birnie, and B. D. Mitchell, Lichenologist, 3, 409 (1967).

130. T. A. Jackson and W. Keller, Amer. J. Sci., 269, 446 (1970).

131. G. V. Jacks, in Experimental Pedology (E. G. Hallsworth and D. V. Crawford, eds.), Butterworths, London, 1965, pp. 219-226.

132. Y. Tuominen and T. Jaakkola, in The Lichens (V. Ahmadjian and M. E. Hale, eds.), Academic Press, New York, 1973, p. 185-223.

133. S. J. Wainwright and P. J. Beckett, New Phytol., 75, 91 (1975).

134. V. B. Il'in, Soviet Soil Sci., 6, 283 (1974); Engl. transl. of Pochovovedeniye, 5, 89 (1974).

135. G. E. Lang and W. A. Reiners, Bull. Ecol. Soc. Amer., 56(2), 38 (1975).

136. G. R. Hoffman, Bot. Gaz., 133, 107 (1972).

137. R. T. Brown and P. Mikola, Acta Forest. Fenn., 141, 1 (1974).

138. C. J. B. Hitch and W. D. P. Stewart, New Phytol., 72, 509 (1973).

139. J. A. Hellebust, in Algal Physiology and Biochemistry (W. D. P. Stewart, ed.), Blackwell, Oxford, England, 1974, pp. 838-863.

140. J. W. Millbank, New Phytol., 73, 1171 (1974b).

141. G. E. Fogg, in Algal Physiology and Biochemistry (W. D. P. Stewart, ed.), Blackwell, Oxford, England, 1974, pp. 560-582.

142. J. W. Millbank and K. A. Kershaw, in The Lichens (V. Ahmadjian and M. E. Hale, eds.), Academic Press, New York, 1973, pp. 289-307.

143. J. W. Millbank and K. A. Kershaw, New Phytol., 68, 721 (1969).

144. K. Verseghy and E. Kovács-Láng, Acta Agron. Acad. Sci. Hungar., 24, 19 (1975).

145. E. Kovács-Láng and K. Verseghy, Acta Agron. Acad. Sci. Hungar., 23, 325 (1974).

146. G. D. Scott, New Phytol., 55, 111 (1956).

147. R. W. Rogers, R. T. Lange, and D. J. D. Nicholas, Nature, 209, 96 (1966).

148. C. J. B. Hitch and J. W. Millbank, New Phytol., 74, 473 (1975a).

149. G. E. Fogg and W. D. P. Stewart, Brit. Antarctic Surv. Bull., 15, 39 (1968).

150. J. W. Millbank, New Phytol., 71, 1 (1972).

151. V. V. S. Tyagi, Biol. Rev., 50, 247 (1975).

152. C. J. B. Hitch and J. W. Millbank, New Phytol., 75, 239 (1975b).

153. G. Bond, Advan. Sci., 15, 382 (1959).

154. R. T. T. Forman, Ecology, 56, 1176 (1975).

155. P. D. Crittenden, New Phytol., 74, 41 (1975).

156. P. Kallio, Oikos, 25, 194 (1974).

157. R. W. Haag, Can. J. Bot., 52, 103 (1974).

158. D. M. Schell and V. Alexander, Arctic, 26, 130 (1973).

159. V. Alexander and D. M. Schell, Arctic Alpine Res., 5, 77 (1973).

160. V. Alexander, M. Billington, and D. Schell, Rep. Kevo Subarctic Res. Stat., 11, 3 (1974).

161. R. J. Barsdate and V. Alexander, J. Environ. Qual., 4, 111 (1975).

162. V. Alexander, in Soil Organisms and Decomposition in Tundra (A. J. Holding, O. W. Heal, S. F. MacLean, and P. W. Flanagan, eds.), Swedish IBP Committee, Wenner-Gren Center, Stockholm, 1974, pp. 109-121.

163. S. Kallio and P. Kallio, in Fennoscandian Tundra Ecosystems (F. E. Wielgolaski, ed.), pt. 1, Springer-Verlag, New York, 1975, pp. 292-304.

164. E. Maikawa and K. A. Kershaw, Can. J. Bot., 53, 527 (1975).

165. B. Englund and H. Meyerson, Oikos, 25, 283 (1974).

166. A. J. Horne, Brit. Antarctic Surv. Bull., 27, 1 (1972).

167. E. I. Friedmann and M. Galun, in Desert Biology (G. W. Brown, Jr., ed.), Vol. 2, Academic Press, New York, 1974, pp. 165-212.

168. J. M. Snyder and L. H. Wullstein, Amer. Midland Naturalist, 90, 257 (1973).

169. J. R. Herford and J. A. Edmisten, A.S.B. Bull., 16, 54 (1969).

170. L. M. Shields, C. Mitchell, and F. Drouet, Amer. J. Bot., 44, 489 (1957).

171.  R. C. Rychert and J. Skujins, Soil Sci. Soc. Amer. Proc., 38, 768 (1974).

172.  W. C. Denison, Sci. Amer., 228 (6), 75 (1973).

173.  L. H. Pike, W. C. Denison, D. M. Tracy, M. A. Sherwood, and F. M. Rhoades, Bryologist, 78, 389 (1975).

174.  O. L. Gilbert, in The Lichens (V. Ahmadjian and M. E. Hale, eds.), Academic Press, New York, 1973, pp. 443-472.

175.  B. W. Ferry, M. S. Baddeley, and D. L. Hawksworth, eds., Air Pollution and Lichens, University of Toronto Press, Toronto, Canada, 1973.

176.  D. L. Hawksworth, Lichenologist, 7, 173 (1975).

177.  F. LeBlanc and D. N. Rao, Bryologist, 76, 1 (1973).

178.  S. Kallio and T. Varheenma, Rep. Kevo Subarctic Res. Stat., 11, 42 (1974).

179.  V. Alexander, in Nitrogen Fixation by Free-living Micro-organisms (W. D. P. Stewart, ed.), Cambridge University Press, New York, 1975.

180.  A. J. Skidmore and C. H. Dickinson, Trans. Brit. Mycol. Soc., 60, 107 (1973).

181.  T. F. Preece and C. H. Dickinson (eds.), Ecology of Leaf Surface Microorganisms, Academic Press, New York, 1971.

182.  J. Ruinen, in The Biology of Nitrogen Fixation (A. Quispel, ed.), North-Holland Publ., Amsterdam, 1974, pp. 121-167.

183.  C. Leben, Ann. Rev. Phytopathol., 3, 209 (1965).

184.  F. T. Last and R. C. Warren, Endeavour, 31, 143 (1972).

185.  R. A. Rasmussen and R. S. Hutton, Chemosphere, 1, 47 (1972).

186.  K. Arens, Jahrb. Wiss. Bot., 80, 248 (1934).

187.  S. Dalbro, Proc. 14th Int. Hort. Congr. (Paris), p. 770 (1956).

188.  C. O. Tamm, Physiol. Plant., 4, 184 (1951).

189.  H. B. Tukey, Jr., in Ecology of Leaf Surface Micro-organisms (T. F. Preece and C. H. Dickinson, eds.), Academic Press, New York, 1971, pp. 67-80.

190.  C. R. Stocking, Handb. PflPhysiol., 3, 489 (1956).

191.  H. B. Tukey, Jr., Bot. Rev., 35, 1 (1969).

192.  H. B. Tukey, Jr., in A Tropical Rain Forest (H. T. Odum and R. F. Pigeon, eds.), U.S. Atomic Energy Commission, Washington, D.C., 1970, pp. H155-H160.

193.  M. Witkamp, in A Tropical Rain Forest (H. T. Odum and R. F. Pigeon, eds.), U.S. Atomic Energy Commission, Washington, D.C., 1970, pp. H177-H179.

194.  H. T. Odum, G. A. Briscoe, and C. B. Briscoe, in A Tropical Rain Forest (H. T. Odum and R. F. Pigeon, eds.), U.S. Atomic Energy Commission, Washington, D.C., 1970, pp. H167-H176.

195.  D. Boynton, Ann. Rev. Plant Physiol., 5, 31 (1954).

196.  S. H. Wittwer and F. G. Teubner, Ann. Rev. Plant Physiol., 10, 13 (1959).

197. J. S. Pate, Soil Biol. Biochem., 5, 109 (1973).
198. E. P. M. Bessems, Agr. Res. Rept., No. 786, Centre for Agricultural Publishing and Documentation, Wageningen, Netherlands, 1973, pp. 1-68.
199. V. N. Vasantharajan and J. V. Bhat, Plant Soil, 28, 258 (1968).
200. E. I. Kvasnikov, T. M. Klyushnikova, and S. S. Nagornaya, Mikrobiol. Zh. (Kiev), 36, 790 (1974).
201. J. Ruinen, in Ecology of Leaf Surface Micro-organisms (T. F. Preece and C. H. Dickinson, eds.), Academic Press, New York, 1971, pp. 567-579.
202. Y. Abd-El-Malek, Plant Soil, Special Vol., p. 423 (1971).
203. J. Ruinen, Nature, 177, 220 (1956).
204. J. Ruinen, Plant Soil, 15, 81 (1961).
205. M. A. Harrelson, Ph.D. Thesis, University of Georgia, Athens, Georgia, 1969. Cited in Ruinen [182] and in Dissert. Abstr. Int., 30, 3513B (1970).
206. J. J. Goering and P. L. Parker, Limnol. Oceanogr., 17, 320 (1972).
207. J. Ruinen, Plant Soil, 22, 375 (1965).
208. K. Jones, Ann. Bot. [N.S.], 34, 239 (1970).
209. G. Stevenson, Ann. Bot. [N.S.], 23, 622 (1959).
209a. J. Ruinen, in Nitrogen Fixation by Free-living Micro-organisms (W. D. P. Stewart, ed.), Cambridge University Press, New York, pp. 85-100.
210. J. V. Bhat, K. S. Limaye, and V. N. Vasantharajan, in Ecology of Leaf Surface Micro-organisms (T. F. Preece and C. H. Dickinson, eds.), Academic Press, New York, 1971, pp. 581-595.
211. E. N. Mishustin and V. K. Shil'nikova, Biological Fixation of Atmospheric Nitrogen (transl. by A. Crozy), Macmillan, London, 1971.
212. H. H. Tabak and W. B. Cooke, Bot. Rev., 34, 126 (1968).
213. A. M. A. Wahab, Folia Microbiol., 20, 236 (1975).
214. C. H. Dickinson, Trans. Brit. Mycol. Soc., 60, 423 (1973).
215. M. A. Stott, in Ecology of Leaf Surface Micro-organisms (T. F. Preece and C. H. Dickinson, eds.), Academic Press, New York, 1971, pp. 203-210.
216. E. C. Hislop and T. W. Cox, Trans. Brit. Mycol. Soc., 52, 223 (1969).
217. A. Bainbridge and C. H. Dickinson, Trans. Brit. Mycol. Soc., 59, 31 (1972).
218. C. H. Dickinson, J. Watson, and B. Wallace, Trans. Brit. Mycol. Soc., 63, 616 (1974).
219. C. H. Dickinson, Pestic. Sci., 4, 563 (1973a).
220. R. C. Warren, Trans. Brit. Mycol. Soc., 62, 215 (1974).
221. R. R. Mishra and R. S. Kanaujia, Acta Soc. Bot. Polon., 43, 213 (1974).
222. R. R. Mishra and V. B. Srivastava, Acta Soc. Bot. Polon., 43, 203 (1974).

223.  K. R. Sharma, H. Behera, and K. G. Mukerji, Trans. Mycol. Soc. Japan, 15, 223 (1974).

224.  H. G. Diem, J. Gen. Microbiol., 80, 77 (1974).

225.  C. H. Dickinson, B. Austin, and M. Goodfellow, J. Gen. Microbiol., 91, 157 (1975).

226.  E. I. Kvasnikov, S. S. Nagornaya, and I. F. Shchelokova, Microbiology, 44, 299 (1975).

227.  G. J. F. Pugh and N. G. Buckley, in Ecology of Leaf Surface Micro-organisms (T. F. Preece and C. H. Dickinson, eds.), Academic Press, New York, 1971.

228.  C. H. Dickinson, Trans. Brit. Mycol. Soc., 60, 423 (1973).

229.  B. E. S. Godfrey, Trans. Brit. Mycol. Soc., 62, 305 (1974).

230.  H. J. Hudson, New Phytol., 67, 837 (1968).

231.  F. T. Last and F. C. Deighton, Trans. Brit. Mycol. Soc., 48, 83 (1965).

232.  C. H. Dickinson, Can. J. Bot., 45, 915 (1967).

233.  G. J. F. Pugh and N. G. Buckley, Trans. Brit. Mycol. Soc., 57, 227, (1971b).

234.  H. G. Diem, Can. J. Bot., 51, 1079 (1973).

235.  H. J. Fokkema, Neth. J. Plant Pathol., 77, Suppl. 1, 3 (1971).

236.  J. P. Blakeman, Trans. Brit. Mycol. Soc., 65, 239 (1975).

237.  J. P. Blakeman, Physiol. Plant Pathol., 2, 143 (1972).

238.  J. P. Blakeman and A. Sztejnberg, Trans. Brit. Mycol. Soc., 62, 537 (1974).

239.  J. P. Blakeman and A. K. Fraser, Physiol. Plant Pathol., 1, 45 (1971).

240.  A. Sztejnberg and J. P. Blakeman, J. Gen. Microbiol., 78, 15, (1973).

240a. I. D. S. Brodie and J. P. Blakeman, Physiol. Plant Pathol., 6, 125, (1975).

241.  J. E. Crosse, in Ecology of Leaf Surface Micro-organisms (T. F. Preece and C. H. Dickinson, eds.), Academic Press, New York, 1971, pp. 283-290.

242.  A. A. Klincāre, D. J. Krēslina, and I. V. Mishke, in Ecology of Leaf Surface Micro-organisms (T. F. Preece and C. H. Dickinson, eds.), Academic Press, New York, 1971, pp. 191-201.

243.  E. Libbert, W. Kaiser, and R. Kunert, Physiol. Plant., 22, 432 (1969).

244.  E. Libbert and R. Manteuffel, Physiol. Plant., 23, 93 (1970).

245.  H. G. Diem, C. R. Acad. Sci. (Paris), 272, 941 (1971).

246.  N. G. Buckley and G. J. F. Pugh, Nature, 231, 332 (1971).

247.  R. W. F. Hardy and U. D. Havelka, Science, 188, 633 (1975).

248.  K. T. Shanmugam and R. C. Valentine, Science, 187, 919 (1975).

249.  E. G. Farnworth and F. B. Golley, eds., Fragile Ecosystems, Springer-Verlag, New York, 1974.

250. D. H. Marx, Forest Sci., 21, 353 (1975b).
251. S. A. Wilde, Bioscience, 18, 482 (1968).
252. F. E. Sanders, B. Mosse, and P. B. Tinker (eds.), Endomycor-rhizas, Academic Press, New York, 1975.
253. C. H. Dickinson and T. F. Preece (eds.), Microbiology of Aerial Plant Surfaces, Academic Press, New York, 1976.

Chapter 10

ANTIMICROBIAL ACTIVITIES OF MINERAL ELEMENTS

William O. Foye

Department of Chemistry
Massachusetts College of Pharmacy
Boston, Massachusetts

I. ANTIMICROBIAL ACTIVITIES OF
COPPER, ZINC, MERCURY, AND SILVER IONS

A. Silver and Mercury

The use of silver and mercury for therapeutic purposes goes back to very
early times, but the employment of organic compounds of silver and mercury as antiseptics and germicides falls mainly in this century. Other

387

than silver acetate, no mention of organic silver or mercury compounds
was made in G. C. Wittstein's volume[1] (Practical Pharmaceutical Chemistry)
of 1853. Koch reported in 1880 [1] that mercury bichlorites inactivated most
microorganisms, and silver nitrate was introduced as a prophylactic
against gonorrheal ophthalmia neonatorum in 1901 [2]. Development and
use of the organic derivatives of these metals followed.

The organic compounds of silver have been used primarily for local
surface activity as antiseptics and germicides. Attempts to develop organic
silver derivatives that would be less caustic than the ionizable salts have
resulted in a number of useful preparations. These have been most often
products of unknown chemical composition formed from proteins by precipitation with soluble silver salts. They are classified as strong or mild
silver proteinates. Some of the inorganic and organic silver compounds
that have been used are listed in Table 10-1.

The strong silver proteinates consist of silver-protein complexes
which are believed to yield part or all of their silver content as ions when
in contact with serous fluids. They contain a lower percentage of silver
than do the mild silver proteinates but are designated as strong because
they liberate a higher concentration of silver ions. These silver preparations form colloidal solutions and do not possess the corrosive, astringent,
or irritant properties of the soluble salts. These have been used for the
treatment of infections of the mucous membranes, eyes, respiratory tract,
and urinary tract.

The activity of silver compounds most likely depends on the action of
silver ion, which can result in precipitation of bacterial proteins. Further
reactions are also considered possible. Silver ion also forms insoluble
complexes with deoxyribonucleic acid (DNA) and ribonucleic acid (RNA),
which may account for its antibacterial properties [3]. It also complexes
with riboflavin, a biochemical mediator in a number of important reactions
[4].

Mercury salts were used for the treatment of syphilis until the introduction of arsphenamine in 1905. Inorganic compounds such as red mercuric oxide, mercuric chloride, yellow mercurous iodide, or such preparations as mercury with chalk were administered orally for this and other
therapeutic uses. These preparations were abandoned because they caused
gastrointestinal distress, and a number of organic derivatives of mercury
were developed for use as antiseptics and disinfectants. Some of the organic compounds of mercury are listed in Table 10-2.

The organic mercurials in general ionize only very slowly, and most
of them do not show an immediate precipitate with hydrogen sulfide. Nevertheless, it is believed that those organic mercurials employed as antiseptics
act by combining with sulfhydryl groups of bacterial proteins. For instance,
cysteine and other thiols neutralize the antibacterial action of mercuric
chloride. Since the removal of the mercury ion from the bacterial protein

---

[1] Wittstein's Practical Pharmaceutical Chemistry (English trans. by S.
Darby), Churchill, London, 1853.

TABLE 10-1  Antibacterial Compounds of Silver

| Compound | Names | Uses |
| --- | --- | --- |
| Silver nitrate | Lunar caustic | Antiseptic, astringent, caustic; for infections of mucous membranes |
| Silver hexamethyl-enetetramine | Argentiform | As ointment for topical use, including the eye |
| Silver lactate | | Same as silver nitrate, but less caustic |
| Silver picrate | Picragol | For urethritis, vaginitis, and similar infections |
| Colloidal silver chloride | Lunosol | For infections of mucous membranes |
| Colloidal silver iodide | Neosilvol | For infections of mucous membranes |
| Mild silver protein | Argyrol, Lunargen, mild Protargin, Silvol | For rhinitis, proctitis, conjunctivitis, tonsillitis, other mucous membrane infections |
| Strong silver protein | Strong Protargin, Protargol | Same as mild silver protein, but more irritating |
| Colloidal silver | Colsargen | For infections of the eye, nose, throat, genitourinary tract |

results in reactivation of the inhibited enzymes and resumption of normal metabolism, the antimicrobial action of the mercurials is classified as bacteriostatic. Probably because of their poor ability to ionize, the organic mercurials are less irritating to the tissues than are the inorganic salts and can be applied to mucous membranes. The protein precipitant action of mercuric ion on human tissue, however, is quite irritating. This brings into question the assumption that mercury compounds are bacteriostatic because of their protein-precipitant effect. A number of them can also be injected for systemic effect, e.g., for diuretic action or for treating syphilis.

The alkylmercury halides are toxic compounds but can be used as skin-sterilizing agents for skin grafts. Solutions of 0.0036 M alkyl (methyl to heptyl) mercury chlorides in 70% ethyl alcohol gave sterile skin snips in general [5]. The inorganic mercurials are now seldom used as antiseptics but still have use as industrial preservatives. Ammoniated mercury ($NH_2HgCl$) is still used, however, for treatment of impetigo, ringworm, and psoriasis, and yellow mercuric oxide is used for the treatment of inflammations of the eyelid, conjunctiva, and external ear.

The organic mercurials that have seen the greatest amount of use as antiseptics and disinfectants are either aromatic mercurials or mercury complexes. These include Merbromin (Mercurochrome), Thimerosal (Merthiolate), and Nitromersol (Metaphen). These compounds are bacteriostatic rather than bactericidal. Thimerosal has low toxicity and uniform bacteriostatic action, and is used to disinfect tissue surfaces; it also has some use as a preservative. Nitromersol is especially effective against Gram-positive cocci and causes little irritation to mucous membranes or skin. It is used for treatment of ocular infections and for disinfecting skin.

B.  Copper and Zinc

Organic salts of both copper and zinc have mild antiseptic properties and have been used to a limited extent for this purpose. These salts also possess astringent and necrotizing properties which have given them greater usefulness. Copper citrate, for instance, has been employed in ointments for the treatment of granulations and trachoma; zinc acetate has also been used as an astringent and mild antiseptic for mucous membranes and skin. Zinc stearate, however, has been more generally employed as mild astringent and antiseptic in both ointments and dusting powders. The commercial product consists of variable proportions of zinc stearate and oleate and from 12 to 14.5% of zinc oxide. Its physical attributes of being water insoluble and unwettable and a good lubricant have accounted for its widespread use in pharmaceutical powders and ointments.

Organic salts and complexes of copper and zinc have been used more extensively for their action against fungi than against bacteria. Zinc propionate, zinc caprylate, and zinc undecylenate have all been employed in the treatment of fungus infections of the skin and mucous membranes, including dermatophytosis pedis. Copper soaps have also been used for antifungal effects. Organic copper and zinc derivatives used for antimicrobial effects are listed in Table 10-3.

Basic cupric calcium sulfate (Bordeaux mixture), introduced in 1885, has been used extensively for antifungal action in spray protection of crops. Iron and zinc complexes of dimethyldithiocarbamic acid have also been widely used for this purpose, although the copper complex has antifungal activity as well. The copper complex of oxine (8-hydroxyquinoline) has been employed as an antifungal agent, however, both in agriculture and for rot proofing of structural materials such as tents.

TABLE 10-2  Antibacterial Compounds of Mercury

| Compound | Composition | Uses |
|---|---|---|
| Mercuric oleate | Yellow mercuric oxide and oleic acid | For parasitic skin diseases (e.g., scabies); chronic arthritis |
| Mercuric salicylate | 3-Mercurisalicylic acid | Topical antiseptic; for syphilis |
| Mercuric succinimide | Mercuric imidosuccinate | For syphilis |
| Phenylmercuric chloride | Chlorophenyl mercury | Antiseptic, germicide, fungicide; also for veterinary use |
| Phenylmercuric nitrate | Phenylmercuric nitrate | Same as phenylmercuric chloride |

TABLE 10-2 (Cont.)

| Compound | Composition | Uses |
|---|---|---|
| Phenylmercuric acetate | Acetoxyphenyl mercury | As fungicide and herbicide; as algistat in swimming pools, for impregnating paper pulp |
| Phenylmercuric borate | (Dihydrogen borato)-phenylmercury | External antiseptic, for both skin and mucous membranes |
| Meralein Sodium | 2,7-Diiodo-4-hydroxymercuriresorcin sulfonphthalein sodium salt | Local antiseptic, sinus infections |

| | | |
|---|---|---|
| Mercocresols (Mercresin) | o-Hydroxyphenylmercuric chloride and sec-amyltricresol | For irrigation of deep wounds, wet packs, and mucous membranes |
| Acetomeroctol | 2-Acetoxymercuri-4-(1,1,3-tetramethyl-butyl) phenol | Topical antiseptic |
| Mercurophen | Sodium 4-(hydroxymercuri)-2-nitrophenolate | For sterilizing skin and surgical instruments |
| Merythrol | Sodium p-ethylmercurithiophenylsulfonate | Topical antiseptic |
| Mercurol | Mercurated nucleic acids from yeast | Topical antiseptic |
| Ethyl Mercury | Ethyl mercuric chloride | Fungicide for treating seeds (now largely outlawed) |

TABLE 10-2 (Cont.)

| Compound | Composition | Uses |
|---|---|---|
| o-Hydroxyphenylmercuric chloride | o-(Chloromercuri)phenol<br> | As antiseptic, in soaps and similar products; in eardrop preparation; as fungicide |
| Nitromersol | 3-(Hydroxymercuri)-4-nitro-o-cresol inner salt<br> | Disinfectant for skin and mucous membranes; also for veterinary use. |
| Merbromin (Mercurochrome) | Disodium 2,7-dibromo-4-hydroxymercuri-fluorescein<br> | Disinfectant for wounds, skin, and mucous surfaces |

Thimerosal (Merthiolate)

Sodium [(o-carboxyphenyl) thio] ethylmercury

For skin disinfection, wounds, denuded surfaces, urethral irrigation, nasal and ophthalmic use.

Thimerfonate sodium
(Sulfo-Merthiolate)

Sodium p-ethylmercurithiophenylsulfonate

Antiseptic surgical powder

TABLE 10-3   Antibacterial Compounds of Copper and Zinc

| Compound | Composition | Uses |
|---|---|---|
| Bordeaux mixture | $CuSO_4 \cdot 3Cu(OH)_2 \cdot$ $Ca(OH)_2 \cdot H_2O$ | Fungicide for plants and seeds |
| Cupric acetate | Cupric acetate | Fungicide |
| Cupric citrate | Cupric citrate | Escharotic, astringent, antiseptic; for treatment of granulations and trachoma |
| Cupric stearate | Cupric stearate | Fungicide; in antifouling paints |
| Cupric complex of oxine | Cupric 8-hydroxy-quinolinate | Fungistat |
| Zinc acetate | Zinc acetate dihydrate | Antiseptic and astringent; veterinary use; wood preservative |
| Zinc bacitracin | Bacitracin zinc salt | Topical antimicrobial; for veterinary use (e.g., enteric infections, mastitis, otitis, erysipelas, respiratory infections) |
| Zinc propionate | Zinc propionate | Fungicide, particularly on adhesive tape |
| Zinc caprylate | Zinc octanoate | Fungus infections of skin and mucous membranes |
| Zinc undecylenate | Zinc undecylenate | Fungus infections of skin and mucous membranes |
| Zinc phenolsulfonate | Zinc p-hydroxy-benzenesulfonate | Antiseptic and astringent |
| Zinc stearate | Zinc salts of stearic and palmitic acids | Mild astringent and antiseptic |
| Zineb | [Ethylenebis (dithiocarbamato)] zinc | Agricultural fungicide |
| Ziram | Bis(dimethyldithiocarbamato) zinc | Agricultural fungicide |

II.  METAL ION ACTIVATION BY CHELATION

A.  Oxine

Largely through the work of Albert and coworkers, it has been demonstrated that a number of organic antibacterial and antifungal agents exert
their effects on microorganisms through the formation of toxic chelates of
certain metal ions.  Although silver and mercury ions are not generally involved in these cooperative effects, copper ion frequently is.  Probably the
most thoroughly studied example of this kind is oxine, which was found in
1944 [6] to owe its antimicrobial effects to metal ion chelation.  The metal
ions which produced complexes toxic to microorganisms were cupric, ferrous and ferric ions.  This was shown by measuring the antibacterial action
of oxine in media which had been depleted of heavy metals; neither the oxine nor the cotoxic metals showed appreciable activity in such media, but
the metal complexes were bactericidal [7].  The six isomeric hydroxy-
quinolines, being incapable of metal ion chelation, were without antibacterial activity.  Also, blocking the ability of oxine to chelate metal ions by
methylation of either the O or N removed antibacterial activity.

   Further study showed that the 1:1 metal complexes were the toxic
species, or possibly the 2:1 iron complex, but not the saturated 2:1 copper
complex or 3:1 iron complex.  This became evident when an excess of
oxine, which would displace the normal equilibrium between all the possible
complexes toward a preponderance of 3:1 complex (see Fig. 10-1), was
present in the test media.  In the absence of the cotoxic metal ions, oxine
was found to enter the bacterial cell (Staphylococcus aureus) [8] and the
fungal cell (Aspergillus niger) [9] without harmful effects.  It was believed
that the unsaturated (1:1 with both copper and iron, and also 2:1 in the
case of iron) complexes were the toxic species, since they have the ability

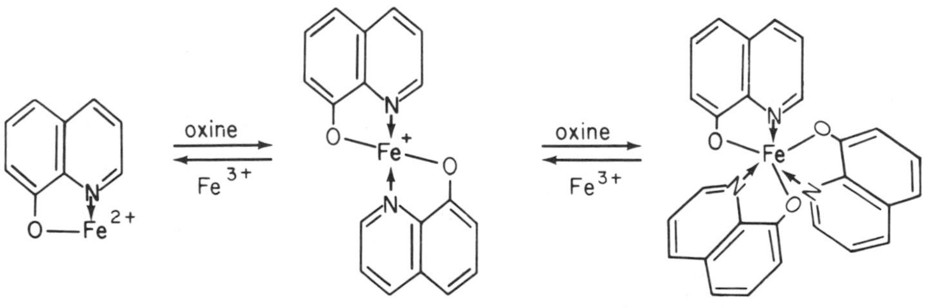

FIG. 10-1   Equilibria between the oxine-ferric ion complexes.

to combine with other vital molecules, but since they are charged species they cannot penetrate the cytoplasmic membrane. The saturated complexes (2:1 in the case of copper, and 3:1 in the case of iron) are considered the species that can penetrate the membrane but are nontoxic. Once inside the membrane, however, the normal equilibrium is reestablished, and the unsaturated, toxic complexes are formed.

Evidence for this belief was provided by the finding that oxine derivatives having low oil/water partition coefficients are not antibacterial [10]. A charged derivative, the 5-sulfonic acid of oxine, had essentially the same metal-binding ability as oxine itself, but it lacked the ability to penetrate a lipid phase and had no antibacterial activity. In addition, a series of uncharged aza derivatives of oxine showed a close correlation between oil/water partition coefficient and antibacterial activity [11]. Derivatives with high coefficients gave the greatest bacteriostatic effects (versus Streptococcus pyogenes); see Table 10-4. No correlation with metal-binding ability was evident.

The site of action of the toxic oxine-metal complexes is unknown. An indication of the type of destructive reaction in which the oxine-metal complexes may be involved is given by the fact that cobaltous ion prevents the bacteriostatic effect of oxine [12]. Whereas the toxic action of oxine-iron can be reversed by the presence of large amounts of other metals which give nontoxic complexes, such as cadmium, nickel, and zinc, by replacing the iron in the complex, traces of cobaltous ion are effective in reversing the toxic action of the oxine-iron. Cobaltous ion also protects yeasts [13] and trypanosomes [14] against oxine-copper but does not protect mycelial fungi. An explanation for this protective action of cobalt may lie in the fact that some vital cell constituents, particularly mercapto compounds such as thioctic acid, are easily oxidized by oxygen if traces of copper or iron are present. These oxidations produce hydrogen peroxide, which in turn oxidizes more of the substrate; a chain reaction is thus set up which can cause destruction of vital constituents in the cell. Traces of cobalt are known to inhibit these oxidations by acting as chain breaker [15].

A mixture of oxine and inorganic iron, however, was found to catalyze the oxidation by air of mercapto groups in nucleoproteins, whereas the iron alone was ineffective [16]. So it is quite possible that cobaltous ion may inhibit destructive oxidations of cellular thiols caused by oxine-iron or oxine-copper complexes in the same fashion. The enhanced catalytic ability of the oxine-iron complex, over inorganic iron, may be due to a rearrangement of the orbitals of the ferric cation caused by chelation [15].

The bactericidal action of 1:1 oxine-iron is quite rapid (about 3 min), but is slower if an excess of either iron or oxine is present. Generally, oxine exerts bacteriostatic or bactericidal effects on Gram-positive organisms; only weak effects are shown on most Gram-negative bacteria. Possibly, action against the latter does not involve metals. Oxine is

TABLE 10-4 Parallelism of Bacteriostatic Action of Oxine Derivatives with Partition Coefficient[a]

| Compound | Partition coefficient, oleyl alcohol/water | Lowest inhibitory dilution (1/M)[b] | Log first stability constant for $Ni^{2+}$ |
|---|---|---|---|
| Oxine | 67 | 200,000 | 9.8 |
| 5-Azaoxine | <0.02 | <800 | 5.8 |
| 7-Azaoxine | 0.1 | <800 | 6.7 |
| 6-Azaoxine | 1 | <800 | 5.9 |
| 3-Azaoxine | 5 | 13,000 | 7.6 |
| 2-Azaoxine | 6 | 13,000 | 7.8 |
| 4-Azaoxine | 8 | 6,400 | 7.6 |
| 4-Methyl-2-azaoxine | 16 | 25,000 | 8.1 |
| 4-Methyl-3-azaoxine | 17 | 50,000 | 7.9 |
| 4-Propyl-3-azaoxine | 135 | 100,000 | 7.9 |
| 7-Allyl-3-azaoxine | 310 | 100,000 | 7.9 |

[a] Reprinted from Ref. 11 by courtesy of Chapman and Hall, London.

[b] Carried out versus Streptococcus pyogenes in meat broth at pH 7.3 ($37^\circ$ C).

damaging to mycelial fungi only as the cupric complex, which is fungicidal. Oxine is used as a topical antibacterial, but cannot be injected into the bloodstream because it is inactivated by red blood cells; it is also used in dermatology for both local fungal and bacterial infections. The halogenated derivatives of oxine, such as haloquinol (a mixture of 5-chloro- and 5,7-dichlorooxine) and chlorquinaldol (5,7-dichloro-2-methyloxine), have similar uses. 5,7-Diiodooxine (Diodoquin) has been used for amoebiasis, and 5-chloro-7-iodooxine (Vioform) has been used for both amoebiasis and bacterial dysentry.

## B. Substances Acting Similarly to Oxine

Reference has already been made to the aza analogs of oxine (Table 10-4) prepared by Albert et al., where presence of the extra ring nitrogens

lowered both oil/water partition coefficients and bacteriostatic action.
Raising the partition coefficients by introduction of alkyl groups (e.g.,
structure 1), however, restored the bacteriostatic activity. Maximum activity from this type of compound appears to reside with oxine itself, since
an increase in the oil/water partition coefficient beyond that of oxine did
not raise the in vitro antibacterial action [17].

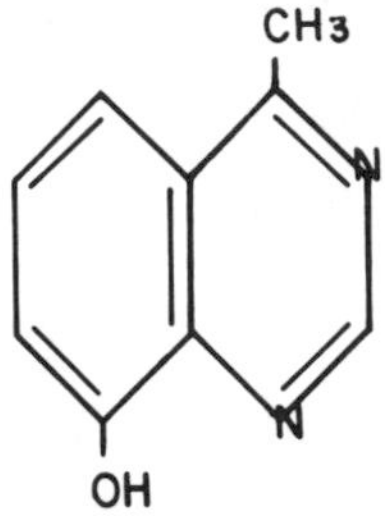

CH3

4 - Methyl – 3 – aza– oxine

(1)

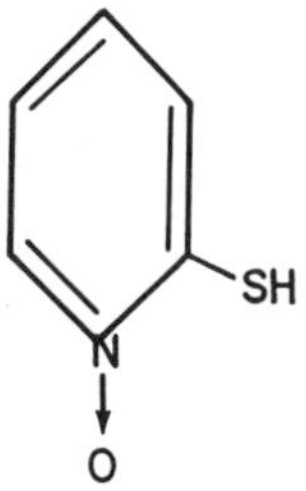

(2)

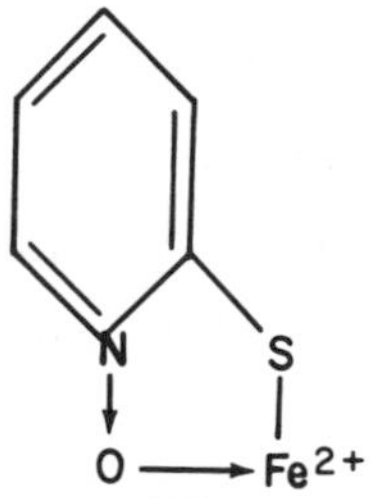

(3)

The N-oxides of pyridine, quinoline, and benzoquinoline that have an
ionizable, metal-coordinating group in the 2-position are also antibacterial.
Thus, 2-mercaptopyridine-N-oxide [Omadine (2)] has antibacterial activity
equivalent to that of oxine; it has a similar mode of action as oxine, since
it is effective as the iron complex (3) and its action is prevented by cobalt
as well as by an excess of the compound itself [17].


C. Other Antibacterials

Tetracycline (4) and its derivatives have metal-binding abilities of the
same order as that of the amino acids, except for the tervalent ions, $Fe^{3+}$
and $Al^{3+}$, which have higher stability constants for the tetracyclines. It
is doubtful that the tetracyclines exert their antibacterial effects as toxic
metal chelates, as do the oxine-like compounds. Metal chelation mechanisms of action for the tetracyclines have been proposed by several investigators, but convincing proof of the reality of these schemes has not been
provided. In the enzyme systems inhibited by tetracyclines, the inhibitions
are reversed by $Fe^{2+}$, $Mn^{2+}$, and $Ca^{2+}$ ions [18]; also, the tetracyclines

(4)

are active in iron-depleted media [19] versus S. aureus. It is more
likely, therefore, that the antibacterial action of tetracyclines involves the
binding of metals in metalloenzymes; it has been postulated that the inhibition of a nitro reductase takes place through the binding of the $Mn^{2+}$ ion
[20]. It has also been suggested that penetration of the bacterial plasma
membrane is due to the chelation of magnesium [21]; the tetracyclines accumulate in bacterial cells but not in mammalian cells. Their action is
believed to occur on the ribosomes, which contain magnesium.

The binding site in the tetracyclines for metal ions is considered to be
the tricarbonylmethane system in the 1, 2, 3 positions, probably between the
2-carboxamide and the 3-enol functions [22, 23]. Binding of the therapeutically active tetracyclines to conalbumin was increased greatly in the
presence of $Cu^{2+}$ ions [24], suggesting a metal bridge. No increase in
binding by $Cu^{2+}$ ion occurred with inactive tetracyclines.

Other antibiotics have metal-binding abilities, and some metal ion
activations and reversals have been noted; no clear evidence for a metal-
binding mechanism of action has been found for these compounds, however.
These cases include the following: penicillin [25-27], bacitracin [27],
polymyxin B [28], novobiocin [29, 30], neomycin [31], kojic acid [32],
cycloheximide [32], cycloserine [32], patulin (clavacin) [32], and asper-
gillic acid [33]. Cleavage of penicillins to penicilloic acids, which are
not antibacterial, is caused by cupric ion [34]; however, zinc ion also
reduces 100- to 200-fold the bactericidal action of penicillins for hemolytic
streptococci and pneumococci [35]. In this connection, penicillinase plasmids in S. aureus were found to carry determinants of resistance to a
series of inorganic ions, including zinc, as well as to the penicillins [36].

On the other hand, the antibacterial action of bacitracin against
S. aureus is diminished in the presence of ethylenediaminetetraacetate
(EDTA) but is restored by some bivalent cations, particularly zinc [37];
this behavior is indicative of a metal-binding mechanism of action. Polymyxin B is believed to affect synthesis of the cellular envelope by depletion
of magnesium [38]. The antibacterial effects of bacitracin are believed
not to be due primarily to an alteration of cell permeability, however [39],
since zinc ion potentiates the antibacterial action but not the effects on cell
permeability.

(5)                                              (6)

The chlorinated bisphenol antibacterials are metal-binding agents and
may exert their antibacterial effects through combination with metal ions,
although the evidence for this is not conclusive.  Both 2, 2'-thiobis(3, 5-
dichlorophenol) (5) and hexachlorophene (6) are metal-chelating agents,
particularly for $Fe^{2+}$, $Fe^{3+}$, and $Cu^{2+}$ ions [40].  The thiobisphenol is
bactericidal to S. aureus in distilled water and metal-depleted broth, and
it was therefore believed that it was not dependent on a metal ion for activity [41].  Both the copper and iron chelates showed the same antibacterial activity as the unchelated compound, however, and the activities of both
this compound and hexochlorophene were suppressed by $Fe^{2+}$ ion.  Since
this was the only metal ion found which suppressed their activity, the possibility remains that iron-containing enzymes may be inhibited by these
compounds.

1, 10-Phenanthroline (7) and some C-methyl derivatives have strong
metal-chelating abilities and are antibacterial as well.  Although the o-
phenanthrolines are toxic to bacteria in distilled water, it was believed
that they exerted their antibacterial effects as the cationic complexes with
metal ions.  The metal complexes, which included those with $Cu^{2+}$, $Zn^{2+}$,
$Fe^{2+}$, $Co^{2+}$, $Cd^{2+}$, $Mn^{2+}$, and $Rb^{2+}$, were more active than the hydrochlor-
ides or methiodides [42].  The $Cu^{2+}$ complexes were the most effective
against Erysipelothrix rhusiopathiae and Fusiformis nodosus [43].  Increasing the lipophilic character of the phenanthroline molecule by incorporation of methyl groups also enhanced activity [44].

(7)

(8)

Salicylic acid has been employed as an exfoliative agent in fungal infections, which may not be related to its strong metal-binding properties. Some halogenated salicylanilides, e.g., tribromsalan (8), are active against Gram-positive organisms and have been used in soaps, detergents, and cosmetics. Their metal-binding ability would be relatively weak, because of the low avidity of the amide linkage for metals, and their antibacterial action is most likely not dependent on metal ions.

Two compounds used as urinary antiseptics, 5-nitrofurantoin (9) and nalidixic acid (10), are both structurally capable of binding metal ions. No indications that their action is due to metal binding have appeared, however, although 5-nitrofurantoin is known to inhibit the metalloenzyme system responsible for sulfation [45], which requires magnesium ion.

A series of heterocyclic thiones, including 2-mercaptoimidazole (11), 2-mercaptobenzothiazole, and 2-mercaptouracil, which have bacteriostatic or bactericidal properties, showed a fairly good correlation between bacteriostatic activity and metal-binding ability for $Cu^{2+}$, $Al^{3+}$, and $Fe^{3+}$ ions [46]. 3-Acetyl-5-methyltetramic acid (12), reported to be bactericidal [47], also revealed relatively high binding constants for these metals. The copper, silver, mercury, and several other inorganic complexes of 2-mercaptobenzothiazole have bacteriostatic action in the presence of serum [48].

Some metal complexes of nonbacteriostatic agents have been found to exert greater growth-inhibitory effects against bacteria than do molar equivalents of the metal ions. This was demonstrated by the action of a cobalt complex of methionine and a copper complex of biotin against

(9)

(10)

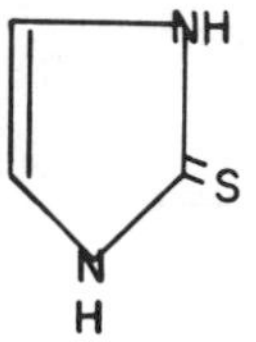

(11)

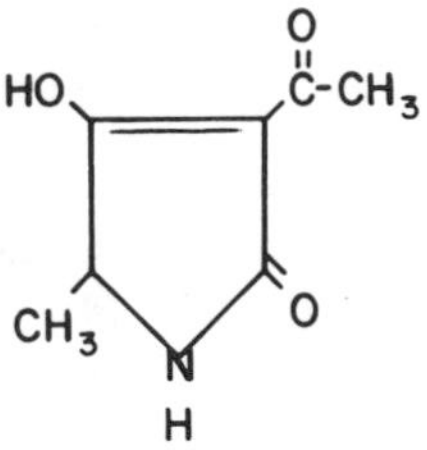

(12)

Escherichia coli and S. aureus [49]. A comparable experiment with a
copper complex of glycylalanine showed that the inhibitory effect of the
complex followed that exerted by the free metal ions alone over a period of
8 hr. The antibacterial action was believed to be due to the increased lipo-
solubility of the metal ion caused by complexation, which liberated the
metal ion or its complex into the cellular environment in larger quantities
than the cell can tolerate. The copper complex of methionine has since
been found to be a strong inhibitor of succinic dehydrogenase, which was
believed to cause its anthelmintic action against the lungworm Metastrongylus
elongatus [50].

It has been suggested that the mode of action of some antibacterial and
antifungal chelating agents may be linked to the effect they exert on the incorporation of copper into a porphyrin precursor of an essential heme pigment [51]. An increase in the rate of incorporation was observed for di-
ethyldithiocarbamate, 2-hydroxypyridine-N-oxide, oxine, kojic acid, and
salicyl aldehyde, whereas salicylic acid, glycine, terramycin, nitrilotri-
acetic acid, o-phenanthroline, bipyridine, histidine, and EDTA showed
inhibitory effects.

D.   Antitubercular Compounds

Since the postulation by Carl and Marquardt [52] of a direct relation between antitubercular activity and the ability of most antitubercular agents
to form copper complexes, nearly all of the compounds which have been
used against tuberculosis have been shown to be influenced by metal ions.
Convincing evidence of metal-binding mechanisms of action has been

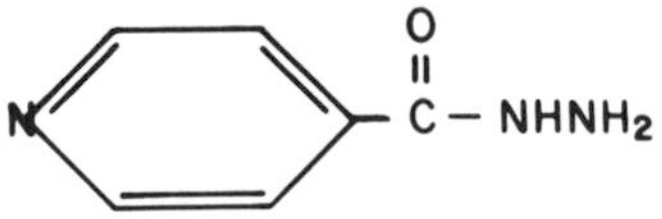

(13)

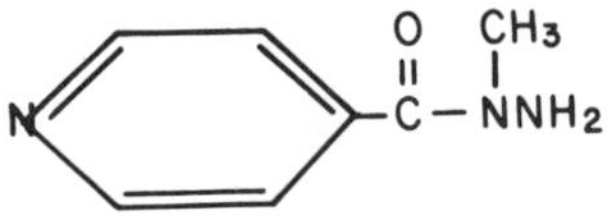

(14)

difficult to provide, however, but much indicative evidence of the involvement of heavy metal ions, particularly copper, in antituberculotic action
is known.

1.  Isonicotinic Acid Hydrazide

Isonicotinic acid hydrazide [isoniazid; INH (13)] would appear to act as a
metal-binding agent, since it has metal-binding constants of the same
order as those of the common amino acids [53], and INH derivatives incapable of metal ion complexation, such as 1-methyl-1-isonicotinoyl
hydrazine (14), are also devoid of antitubercular activity in vitro [54].
Growth inhibition of Mycobacterium tuberculosis by isoniazid, however, is
suppressed by the presence of $Fe^{3+}$, $Mn^{2+}$, $Cu^{2+}$, and $Co^{2+}$ ions in vitro
[55,56]. On the other hand, stimulatory effects by metal ions have been
noted in the ability of isoniazid to inhibit respiration of M. tuberculosis
[56] and to inhibit the action of mammalian or microbial catalase [57]. A
preformed copper complex of isoniazid has also been shown to be as effective in tuberculotic mice as is isoniazid itself [58].

Other studies have been carried out in vivo with preformed metal
chelates of isoniazid. Chelates of $Cu^{2+}$, $Co^{2+}$, $Fe^{2+}$, and $Zn^{2+}$ having
both 1:1 (15) and 2:1 (16) ligand-metal ratios all showed approximately the
same antitubercular activity when fed to mice as did isoniazid itself [59].
The toxicity patterns of the chelates in mice (see Table 10-5) in general
followed those found for the corresponding metal ions, although the toxic
effects of the chelates did not appear as rapidly as those from the metal
ions alone and, in fact, required a day or more to become evident. These
facts suggested that isoniazid-metal complexes liberate some, if not all,
of their metal ions in vivo, but apparently not rapidly enough to exclude
the chelate as an active entity in antitubercular action.

That the liberation took place within the cells is probable. This was
indicated by a parallel study in which a polar copper complex of p-aminosalicylic acid (17), incapable of penetrating the cells, had no antibacterial
activity, whereas the oil-soluble copper chelate [(18), the structure of
which could also involve both oxygen atoms of the carboxy group and not
the phenolic oxygen] was active [60]. Isoniazid is known to be taken up

(15)

(16)

TABLE 10-5  Toxicities of Isoniazid–Metal Chelates in Mice [a]

| Compound | Dose, i.p. (mg/kg) [b] | | | | | | |
|---|---|---|---|---|---|---|---|
| | 10 | 25 | 50 | 100 | 250 | 500 | 1,000 |
| INH-Cu⁺HSO₄⁻ · 2H₂O | | 3–4 days | 5 hr | 1 hr | | | |
| (INH)₂Cu · 3H₂O | | | 1–2 days | 3 hr | 23 min | | |
| CuSO₄ · 5H₂O | 2 days | 40 min to 1 day | 1–1/2 hr | 40 min | | | |
| INH-Fe⁺HSO₄⁻ · 2H₂O | | | | | 1 day | 3 hr | 2 hr |
| FeSO₄ · 7H₂O | | | | | 3 days | 2 days | 4 hr |
| (INH)₂Co · 4H₂O | | | | | | 1 day | |
| Co(NO₃)₂ · 6H₂O | | | | | 1 day | 9 min to 1 day | 4–6 min |
| INH-Zn⁺HSO₄⁻ · 2H₂O | | | | 1–2 days | | | |
| ZnSO₄ · 7H₂O | | | | 5 days | 17 min to 2 hr | 17–19 min | |

[a] From Ref. 59.

[b] Generally three mice were used for each determination. The time recorded indicates that required for death.

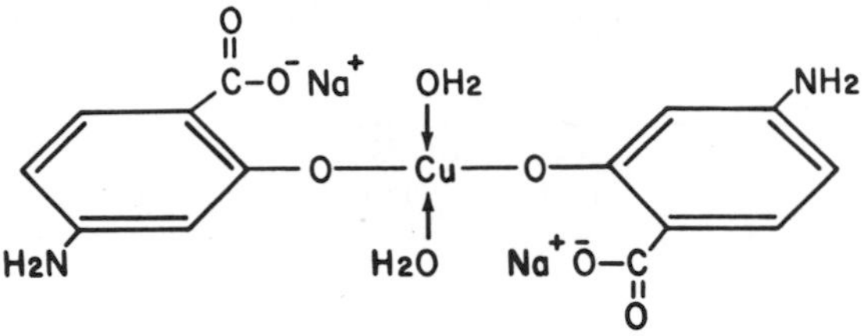

(17)

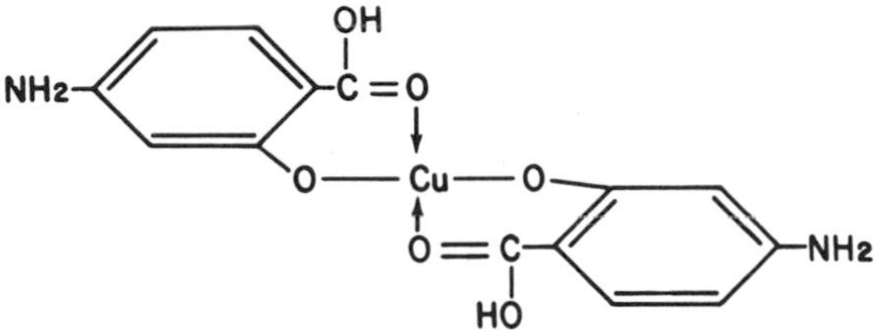

(18)

more rapidly by M. tuberculosis strain H₃₇ Rv in culture in the presence of cupric ions [61]. It is also known that isoniazid penetrates the cells of susceptible mycobacteria but does not enter cells of resistant strains [62]. However, salts of heavy metals alone inhibit growth of most atypical mycobacteria; the greatest inhibitory effects were caused by salts of cobalt, nickel, and bismuth, and only slightly less by copper and zinc salts [63].

The foregoing facts suggest that isoniazid is transported into bacterial cells as a nonpolar metal chelate or that it acts to liberate metal ions within the cells. Since the isomers of isoniazid have essentially equal metal-binding ability but no action against M. tuberculosis [64], it appears that a more structurally specific role is being exerted (see Table 10-6). One possibility is that isoniazid inhibits pyridoxal, which it resembles structurally. Transaminations and other reactions catalyzed by pyridoxal have been shown to occur by means of a metal chelation mechanism [65], and the antitubercular action of isoniazid is inhibited by pyridoxal, as well as by 2-ketoglutarate and pyruvate [66]. The cupric complex of isoniazid retained its activity against Mycobacterium smegmatis in the presence of equimolar amounts of pyridoxal, however [67]. Depression of the glutamic-oxalacetic transaminase levels of humans [68] and amino acid metabolism of rats [69] is also brought about by isoniazid. No correlation of antitubercular activity with transaminase activity was observed for a series of hydrazides [70], including the hydrazides of isonicotinic, nicotinic, picolinic, benzoic, and the N-methyl derivative of isonicotinic acid (14). This again indicated a structurally specific as well as metal-binding effect.

TABLE 10-6   Physical and Antitubercular Properties of Hydrazides Related to Isoniazid [a]

| Hydrazide | $pK_a$, hydrazide group | Log stability constant, 1:1 $Cu^{2+}$ | Log stability constant, 1:1 $Zn^{2+}$ | Relative activity vs. M. tuberculosis $H_{37}$ Rv In vitro | Relative activity vs. M. tuberculosis $H_{37}$ Rv In vivo |
|---|---|---|---|---|---|
| Isonicotinic | 10.77 | 8.0 | 5.4 | 1 | 1 |
| Nicotinic | 11.47 | 8.7 | | 0.001 | nil |
| Picolinic | 12.27 | 12.4 | 8.4 | 0.017 | nil |
| Benzoic | 12.45 | 9.0 | | 0.002 | nil |
| Cyanoacetic | 11.17 | 8.5 | | 0.008 | 0.2 |

[a] Reprinted from Ref. 64 by courtesy of Chapman and Hall, London

Other possibilities by which isoniazid may act against mycobacteria that involve metal binding may also exist.  It has been suggested that an isoniazid-metal chelate may compete with peroxide for sites on catalase [57]; this would result in accumulation of peroxide and death of the cells. It has also been postulated that isoniazid resistance or susceptibility in mycobacteria is more closely related to peroxidase than to catalase; peroxidase, for instance, is present in normal mycobacteria but absent in isoniazid-resistant mycobacteria [71].  Also, it has been proposed that isoniazid may interfere with cytochrome-catalyzed respiration through chelation of ferric ion [72].

Isoniazid is also known to inhibit the synthesis of the mycolic acids in the $H_{37}$ (human) strain of M. tuberculosis susceptible to isoniazid [73]. This inhibition would affect the waxy capsule that surrounds the mycobacterial cells and cause loss of essential metabolites.  In this connection, it has been demonstrated that isoniazid inhibits the growth of M. smegmatis, only after a lag of 6 hr, whereas the isoniazid-cupric complex almost immediately lysed the cells [67].  Cupric ion alone was also bacteriostatic. It has also been postulated that the isoniazid-copper system includes some Cu(I) complex, which may be the active species [74].  Regardless of the oxidation state of the copper, however, it does appear from the foregoing that an isoniazid-metal complex, in particular copper, is a toxic entity for mycobacteria.  In guinea pigs infected with M. tuberculosis, both the copper and cobalt complexes of isoniazid were more toxic to the tuberculous animals than was isoniazid itself; reduced toxicity was found in normal animals, however, with the copper complex reducing toxicity by 1/5 to 1/6 [75].

TABLE 10-7   Correlation of Antitubercular Action in Mice with Lipo-solubility of PASA-Metal Chelates and Complexes[a]

| Compound | Partition coefficient X $10^3$, oleyl alcohol/water | Mortality (%)[b] | Mean survival time (days)[c] |
|---|---|---|---|
| (PASA)$_2$Cu chelate (18) | 244 | 0 | 23* |
| (PASA)$_2$Fe chelate | 174 | 60 | 20.6* |
| PASA | | 0 | 21* |
| Controls | | 80 | 20.3* |
| (PASA)$_2$Cu complex (17) | 8 | 60 | 15.6 |
| (PASA)$_2$Fe complex | | 100 | 13.7 |
| Controls | | 90 | 14.3 |

[a] From Ref. 60.

[b] The compounds were fed to mice (groups of five) infected with M. tuberculosis strain H$_{37}$ Rv at dose levels of 0.1-0.5% in the diet.

[c] An asterisk indicates that the surviving animals were sacrificed at the indicated times for examination of lesions.


2.   Other Antitubercular Agents

Activations by both $Cu^{2+}$ [76] and $Co^{2+}$ [77] ions have been found for the tuberculostatic effect of p-aminosalicylic acid (PASA) in vitro, but most metal ions suppress its activity [78].   In vivo, the cupric chelate [18] was found to have antitubercular activity equal to or greater than that of PASA alone, whereas the ferrous chelate showed less activity [60].   Open-chain complexes of PASA with $Cu^{2+}$ and $Fe^{2+}$ (17) showed less activity, probably because of their much lower liposolubility (see Table 10-7).   It is possible, therefore, that PASA may involve copper ion complexation in exerting its tuberculostatic action, but in view of the finding that PASA replaces p-aminobenzoic acid in some species of bacteria [79], a metal-binding mechanism of action would more likely be involved in transport of the PASA or in a disruptive effect on the cell walls or waxy capsule of the mycobacteria.
   A number of thiosemicarbazones having antitubercular properties are also complexing agents for copper [80].   Thiacetazone [p-acetylamino-benzaldehyde thiosemicarbazone (19)], which is used as an adjunct to isoniazid for preventing the onset of resistant strains, was shown to be more active against mycobacteria in vitro as the copper complex [81].

(19)

That it acts in a different fashion from that of isoniazid is shown by the fact that strains of M. tuberculosis resistant to isoniazid are still sensitive to thiacetazone [82]. The anemia observed in humans after oral treatment with thiacetazone has been attributed to complexation of copper [83].

Other agents which have been used for treating tuberculosis, and which are also metal-binding agents, include ethambutol (20) and streptomycin. Ethambutol is highly active against M. tuberculosis and is used against strains resistant to isoniazid and streptomycin. It is nearly inactive against other species of bacteria and against fungi; this has been

(20)

(21)

(22)

(23)

**(24)**

attributed to the branched-chain structure, which allows 1:1 (21) and prevents 2:1 (22) chelate formation [84]. Ethambutol was presumed to form
a 1:1 chelate structure (23) in which the coordination with the hydroxyl
groups of the molecule prevented hydroxyl bridging from solvent hydroxyls.

Some evidence for metal ion inhibitions of the bacteriostatic effects of
streptomycin [24] has been reported [85,86]. Small enhancements of
activity by $Co^{2+}$ against Micrococcus pyogenes[1] [27] and Vibrio fetus [87]
have been observed, and like isoniazid, streptomycin suppresses catalase
activity, as well as dehydrogenase activity, in E. coli and Shigella species
[88]. This interference may be due to complexation of the metal (iron) involved in the activity of the enzyme. Streptomycin complexes with $Cu^{2+}$,
$Co^{2+}$, and $Ni^{2+}$ have been isolated and found to have a less potent but more
prolonged antitubercular effect in guinea pigs than that of streptomycin
[89]. This was most likely due to a slow liberation of streptomycin from
the complex rather than to a toxic effect of the complex per se. Streptomycin has another effect on metals, probably of greater importance, in
displacing magnesium from ribosomes, which does not appear to involve
metal binding [90].

Other compounds which have no essential antitubercular activity per
se do exert antitubercular effects in the presence of metal ions or as the
metal complexes. Erlenmeyer et al. [91] prepared a large number of
compounds which were capable of metal ion chelation and which resembled
oxine (8-hydroxyquinoline), and found their tuberculostatic activities in
vitro to be enhanced 400- to 800-fold in the presence of $Cu^{2+}$ or $Co^{2+}$ ions.
Cymerman-Craig et al. examined an extensive series of aromatic amines
showing tuberculostatic properties and found that the greatest activity re-
sided with those compounds capable of metal ion chelation [92]. It has

---

[1] Now known as Staphylococcus aureus.

also been observed by Garattini and Leonardi that most compounds capable of chelating heavy metal ions exert an inhibitory effect on M. tuberculosis and M. paratuberculosis [93]. Heavy metal chelates, including those of copper, of some o-hydroxyphenylazonaphthols and -phenanthrols, inactive in the unmetallized state, showed antitubercular effects in vivo [94], and copper chelates of phenylazocresols and phenylazoresorcinols are also tuberculostatic [95]. Another example where the copper complex showed a much greater antitubercular effect in mice than the unmetallized compound was p-aminophenyl-p-hydrazidomethylaminophenylsulfone [96].

Development of resistant strains of tuberculosis organisms may also be suppressed by cupric ion. Tests of eleven hydrazones of isonicotinic acid in vitro showed that M. tuberculosis strain $MK_2$ (isoniazid-resistant strain isolated from a human patient) was inhibited by the hydrazones in the presence of a nontoxic amount of $Cu^{2+}$. Since the inhibitions due to added cupric ions were independent of the concentrations of hydrazones, it was considered that cupric ion was responsible for uptake of the hydrazones by the bacterial cells [97]. It has already been mentioned that isoniazid is taken up by tuberculosis organisms (M. tuberculosis $H_{37}$ Rv) much more rapidly in the presence of cupric ions [61].

E.  Antifungal Agents

Several fungicides offer well-defined examples of the toxic action of metal chelates per se. The fungicidal action of oxine requires cupric ion [13], and the copper chelate of oxine is used as a fungicide [98,99]. Copper chelates of the halogenated oxines were found to have antifungal activity directly related to the size of the pores in the spore wall of the fungi tested [100].

A commonly used antifungal, which is effective as the 1:1 metal complex [101], is dimethyldithiocarbamate (DMDC). It has been used in agriculture as either the sodium salt (NaDDC) or the iron (Fermate or Ferbam) or zinc (Zerlate or Ziram) complex (25). The structure of the 1:1 complexes was determined by Chatt [102]. Studies of growth inhibition of Aspergillus niger by this compound showed that copper ion was essential for activity; several other molds showed the same metal requirement (103). As is true for oxine, equimolar proportions of metal and complexing agent are required for maximum activity. Cobalt ion also antagonizes the action of DMDC against A. niger. Like oxine, DMDC may exert its antimicrobial

(25)

TABLE 10-8   Solubilities and Inhibitory Concentrations for Molds of Various Copper Complexes[a]

| Complex | Solubility in water (ppm) | Minimum inhibitory concentration (ppm) | | |
|---|---|---|---|---|
| | | G. cingulata | A. niger | Fusarium oxysporum |
| (Oxine)$_2$Cu | 1 | 0.02-0.05 | 0.1 | 0.2 |
| (Pyridine-2-thiol-N-oxide)Cu | 0.2 | 0.1 | 0.2 | |
| (Dimethyldithio-carbamate)2Cu | 0.01 | 0.05 | No inhibition | No inhibition |
| Diethyldithio-carbamate-Cu | 0.002 | Partial inhibition | No inhibition | No inhibition |

[a]From Ref. 103.

action by catalyzing the oxidation of thioctic acid [103], which is involved in the oxidative decarboxylation of pyruvic acid.

Dimethyldithiocarbamate, as well as its disulfide, tetramethylthiuram disulfide, depend on the formation of 1:1 complexes with copper or zinc ions for their fungicidal action [101]. These complexes were found to bind with proteins, and succinic oxidase was inhibited by the zinc complexes but was unaffected by the copper complexes. It was concluded that the DMDC complexes acted as fungicides by combining with thiol groups in cell membranes and inhibiting local enzyme systems.

Stability constants and solubility products of the copper complexes of several dialkyl dithiocarbamic acids have been reported [104], and the growth inhibition patterns of these complexes were compared with those of the copper complexes of oxine and 2-pyridinethiol-N-oxide against A. niger and Glomerella cingulata [105]. The inhibition patterns were similar, indicating a similar mode of action to that of oxine (Table 10-8). Activity of the higher homologs of DMDC was limited because of the insolubility of the 2:1 complexes [103]. Some aromatic dithiocarbamates and their (1:1) copper complexes showed little antifungal activity [106], however.

Since the observation of Horsfall and Rich [107] that sulfur-containing fungicides can act by complexation with copper ion, and thus deprive the organism of $Cu^{2+}$, various explanations have been made regarding the manner in which metal ions are involved. Possible involvement with thioctic acid oxidation or with dehydrogenases has already been mentioned. In the case that destruction of thioctic acid is catalyzed by a metal complex, an accumulation of pyruvic acid should result. This has been observed in fungi treated with DMDC [108].

(26)

(27)

The effects of various metal chelates of o-phenanthroline [7] on the alkaline protease of fungi have been observed. The copper, zinc, cobalt, nickel, and cadmium chelates inhibited the caseinolytic activity, but only the copper chelate inhibited the esterolytic activity [109]. Both the o-phenanthroline-copper chelate and copper ion alone also inhibited alkaline protease as well as α-chymotrypsin [110]. An attempt to relate the antifungal activity of mixed copper chelates of oxine and arylhydroxycarboxylic acids with the stepwise dissociation constants led to the conclusion that the 1:1 oxine-copper chelate was the active toxicant [111], as Albert [6] had previously demonstrated for the antibacterial activity of oxine.

Some naturally occurring compounds toxic to fungi, presumably as toxic copper complexes, include a benzopyranone (26) isolated from Helminthosporium monoceras [112] and some derivatives of tropolone (27) found in the heartwood of some species of conifers [113]. Copper chelates of oxalic, pyruvic, and α-ketoglutaric acids are also toxic to fungi, as well as the copper chelates of malic, tartaric, and citric acid, which are less toxic [114].

Other chelating agents which have shown fungitoxic action include some di-o-hydroxy- and di-o-aminodiphenyl sulfides [107]. Copper complexes of variously-substituted salicyl aldehydes and their imines have fungicidal properties [115], as well as a number of metal chelates of 2-heptadecyl-imidazoline [116]. Salicylic acid has a high avidity for metal ions, including copper, but it is not known that the antifungal action is due to formation of a metal chelate. The salicylanilides (e.g., structure 8), which have a much weaker metal-binding ability, should have less tendency to act as toxic chelates.

F.   Antiprotozoal Agents

Biallylamicol (28) has metal-binding ability [117] and is used against amoebiasis; it is not known that the compound has a metal-binding mechanism of action. A number of compounds which are active against various types of protozoa have the ability to bind metals, but evidence for metal-binding mechanisms of action has not been revealed. The 4-oxo-1,4-dihydroquinoline-3-carboxylic acid derivatives (29), active as coccidiostats and antimalarials [118], constitute structures of this type.

CH₂=CH-CH₂

HO—

(C₂H₅) N CH₂

(28)

CH₂-CH=CH₂

—OH

CH₂N(C₂H₅)₂

R¹

R²

O
‖
COR³

N
H

(29)

REFERENCES

1.  R. Koch, Arb. Kaiserl. Gesundh., 1, 1 (1880).
2.  K. S. F. Crede, Ber. Klin. Wochenschr., 38, 941 (1901).
3.  T. Youmans and N. Davidson, Biochim. Biophys. Acta, 55, 609 (1962).
4.  A. T. Tu and J. A. Reinosa, Biochemistry, 5, 3375 (1966).
5.  M. T. Bush and A. D. Bass, J. Pharmacol., 74, 95 (1942).
6.  A. Albert, Med. J. Austral., 1, 245 (1944).
7.  A. Albert, M. I. Gibson, and S. D. Rubbo, Brit. J. Exp. Pathol., 34, 119 (1953).
8.  A. H. Beckett, A. A. Vahora, and A. E. Robinson, J. Pharm. Pharmacol., 10, 160T (1958).
9.  G. Greathouse, S. Block, E. Kovack, D. Barnes, C. Byron, G. Long, D. Gerber, and J. McLenny, Research on Chemical Compounds for Inhibition of Fungi, U. S. Corps of Engineers, Fort Belvoir, Virginia, 1954.
10. A. Albert, Biochem. J., 54, 646 (1953).
11. A. Albert, A. Hampton, F. Selbie, and R. Simon, Brit. J. Exp. Pathol., 35, 75 (1954).
12. S. D. Rubbo, A. Albert, and M. I.Gibson, Brit. J. Exp. Pathol., 31, 425 (1950).
13. B. Nordbring-Hertz, Physiol. Plant., 8, 691 (1965).
14. J. Williamson, Brit. J. Pharmacol. Chemother., 14, 443 (1959).
15. A. Albert, Selective Toxicity, 5th ed., Chapman and Hall, London, 1973, p. 375.
16. F. Bernheim and M. Bernheim, Cold Spring Harbor Symp. Quant. Biol., 7, 174 (1939).

17. A. Albert, C. Rees, and A. Tomlinson, Brit. J. Exp. Pathol., 37, 500 (1956).

18. E. D. Weinberg, Bacteriol. Rev., 21, 46 (1957).

19. A. Albert and C. Rees, Nature, 177, 433 (1956).

20. A. K. Saz and R. B. Slie, J. Biol. Chem., 210, 407 (1954).

21. T. Franklin, Biochem. J., 123, 267 (1971).

22. T. Sakaguchi, M. Toma, T. Yoshida, H. Omura, H. Takasu, Chem. Pharm. Bull. (Tokyo), 6, 1 (1958).

23. W. A. Baker, Jr. and P. M. Brown, J. Amer. Chem. Soc., 88, 1314 (1966).

24. J. T. Doluisio and A. N. Martin, J. Med. Chem., 6, 20 (1963).

25. J. W. Daniel, Jr., and M. J. Johnson, J. Bacteriol., 67, 321 (1954).

26. N. M. Ovchinnikov and K. S. Kutukova, Vestn. Venerol. Dermatol., 30, 34 (1956).

27. J. C. Trace and G. T. Edds, Amer. J. Vet. Research, 15, 639 (1955).

28. B. A. Newton, Bacteriol. Rev., 20, 14 (1956).

29. E. D. Weinberg, Antibiot. Ann., 1956-1957, 1056 (1957).

30. T. D. Brock, J. Bacteriol., 72, 320 (1956).

31. E. D. Weinberg, Antibiot. Ann., 1957-1958, 154 (1958).

32. E. D. Weinberg, E. A. Cook, and C. A. Wisner, Bacteriol. Proc., p. 42 (1956).

33. A. Goth, J. Lab. Clin. Med., 30, 899 (1945).

34. G. Günther, Pharmazie, 5, 577 (1950).

35. N. V. Vasil'eva, Teor. Prakt. Vop. Mikrobiol. Epidemiol. Khar'kov. Med. Inst., 53 (1965); Chem. Abstr., 66, 102622 (1967).

36. R. P. Novick and C. Roth, J. Bacteriol., 95, 1335 (1968).

37. E. D. Weinberg, Antibiot. Ann., 1958-1959, 924 (1959).

38. M. R. Brown and J. Melling, J. Gen. Microbiol., 59, 263 (1969).

39. P. R. Beining, C. L. Pinsley, and E. D. Weinberg, Antimicrob. Agents Chemother., p. 308 (1966).

40. J. B. Adams, J. Pharm. Pharmacol., 10, 507 (1958).

41. J. B. Adams and M. Hobbs, J. Pharm. Pharmacol., 10, 516 (1958).

42. F. Dwyer, I. K. Reid, A. Shulman, G. M. Laycock, and S. Dixson, Austral. J. Exp. Biol. Med. Sci., 47, 203 (1969).

43. G. Cade, M. Cohen, and A. Shulman, Austral. Vet. J., 46, 387 (1970).

44. H. Butler, A. Hurse, E. Thursky, and A. Shulman, Austral. J. Exp. Biol. Med. Sci., 47, 541 (1969).

45. W. O. Foye, M. C. M. Solis, J. W. Schermerhorn, and E. L. Prien, J. Pharm. Sci., 54, 1365 (1965).

46. W. O. Foye and J.-R. Lo, J. Pharm. Sci., 61, 1209 (1972).

47. C. O. Gitterman, J. Med. Chem., 8, 483 (1965).

48. S. G. Ong, Sci. Record (Peking), 3, 246 (1959); Chem. Abstr. 56, 10703 (1962).

49. F. T. Counter, Jr., R. N. Duvall, W. O. Foye, and R. W. Vander Wyk, J. Pharm. Sci., 49, 140 (1960).
50. C. Ishiyeki, Nippon Juigaku Zasshi, 24, 13 (1962); Chem. Abstr., 57, 11790 (1962).
51. M. B. Lowe and J. N. Phillips, Nature, 194, 1058 (1962).
52. E. Carl and P. Marquardt, Z. Naturforsch., 4b, 280 (1949).
53. A. Albert, Experientia, 9, 370 (1953).
54. J. Cymerman-Craig, S. D. Rubbo, D. Willis, and J. Edgar, Nature, 176, 34 (1955).
55. M. W. Fisher, Amer. Rev. Tuberc. Pulmonary Dis., 69, 469 (1954).
56. P. J. Pothman and G. Stuttgen, Z. Hyg. Infectionskrankh., 141, 359 (1955).
57. J. R. Maher, J.F. Speyer, and M. Levine, Amer. Rev. Tuberc. Pulmonary Dis., 75, 517 (1957).
58. S. D. Rubbo, J. Edgar, and G. N. Vaughn, Amer. Rev. Tuberc. Pulmonary Dis., 76, 331 (1957).
59. W. O. Foye and R. N. Duvall, J. Amer. Pharm. Assoc., Sci. Ed., 47, 285 (1958).
60. W. O. Foye and R. N. Duvall, J. Amer. Pharm. Assoc., Sci. Ed., 47, 282 (1958).
61. J. Youatt, Austral. J. Exp. Biol., 40, 201 (1962).
62. J. Youatt, Austral. J. Exp. Biol., 36, 223 (1958).
63. J. S. Chapman and M. Speight, Amer. Rev. Resp. Dis., 103, 372 (1971).
64. A. Albert, Nature, 177, 525 (1956).
65. D. E. Metzler, M. Ikawa, and E. E. Snell, J. Amer. Chem. Soc., 76, 648 (1954).
66. H. Pope, Amer. Rev. Tuberc. Pulmonary Dis., 68, 938 (1953).
67. M. Rieber and G. Bemshi, Arch. Biochem. Biophys., 131, 655 (1969).
68. M. Sass and G. T. Murphy, Amer. J. Clin. Nutrition, 6, 12 (1958).
69. T. Matsumoto, Kekkaku, 33, 283 (1958).
70. R. M. Hicks and J. Cymerman-Craig, Biochem. J., 67, 353 (1957).
71. M. O. Tirunarayanan and W. A. Vischer, Amer. Rev. Tuberc. Pulmonary Dis., 75, 62 (1957).
72. C. M. Coleman, Am. Rev. Tuberc. Pulmonary Dis., 69, 1062 (1954).
73. F. Winder and P. Collins, J. Gen. Microbiol., 63, 41 (1970).
74. A. F. Krivis and J. M. Rabb, Science, 164, 1064 (1969).
75. E. S. Sodikov, Med. Zh. Uzb., 1966, 45; Chem. Abstr., 66, 9824 (1967).
76. W. Roth, P. Zuber, E. Sorkin, and H. Erlenmeyer, Helv. Chim. Acta, 34, 430 (1951).
77. A. Tesi and D. Pisani, Sperimentale, 102, 298 (1952).
78. R. Bönicke and W. Reif, Beitr. Klin. Tuberk., 107, 379 (1952).
79. A. Wacker, H. Grisebach, A. Trebst, M. Ebert, and F. Weygand, Angew. Chem., 66, 712 (1954).

80. R. Behnisch, F. Mietsch, and H. Schmidt, Amer. Rev. Tuberc.,
    61, 1 (1950).
81. K. Liebermeister, Z. Naturforsch., 5b, 79 (1950).
82. V. Barry, J. Proc. Roy. Inst. Chem., 78, 313 (1954).
83. P. Marquardt, Arch. Klin. Chir. Langenbecks, 264, 431 (1950).
84. R. Shepherd and R. Wilkinson, J. Med. Pharm. Chem., 5, 823
    (1962).
85. G. Alva, Rev. Fac. Cienc. Quin., Univ. Med. La Plata, 30, 101
    (1957); Chem. Abstr., 54, 11142 (1960).
86. T. Kinoshita, Nagoya J. Med. Sci., 21, 323 (1958); Chem. Abstr.,
    54, 13262 (1960).
87. H. L. Gilman and D. E. Hughes, Rept. N.Y. State Vet. Coll.
    Cornell Univ., 1955-1956, 28 (1956).
88. E. L. Liebfried, Antibiotiki, 2, 21 (1957).
89. W. O. Foye, W. E. Lange, J. V. Swintosky, R. E. Chamberlain,
    and J. R. Guarini, J. Amer. Pharm. Assoc., Sci. Ed., 44, 261
    (1955).
90. Y. Choi and C. Carr, Nature, 217, 556 (1968).
91. H. Erlenmeyer, J. Baumler, and W. Roth, Helv. Chim. Acta, 36,
    941 (1953).
92. J. Cymerman-Craig, S. D. Rubbo, and B. J. Pierson, Brit. J. Exp.
    Pathol., 35, 478 (1954).
93. S. Garattini and A. Leonardi, Giorn. Ital. Chemioterap., 2, 18
    (1955).
94. W. O. Foye and J. G. Jeffrey, J. Amer. Pharm. Assoc., Sci. Ed.,
    44, 257 (1955).
95. N. Kaneniwa, Kanazawa Daigaku Kekkaku Kenkyusho Nempo, 7, 56
    (1957); Chem. Abstr., 52, 6233 (1958).
96. S. K. Gupta, I. S. Mathur, and M. C. Khorla, Arch. Int. Pharmaco-
    dyn, 114, 373 (1958).
97. V. A. E. Voyatzakis, G. S. Vasilikiotis, G. Karageorgiou, and
    I. Kassapoglou, J. Pharm. Sci., 57, 1255 (1968).
98. J. Fath and G. J. Leitner, U.S. Patent 2,745,832 (1956).
99. R. Feigin and M. P. Schwartz, U.S. Patent 2,755,280 (1956).
100. H. Gershon, J. Med. Chem., 17, 824 (1974).
101. J. Goksøyr, Physiol. Plant., 8, 719 (1955).
102. J. Chatt, L. A. Duncanson, and L. M. Venanzi, Nature, 177, 1042
     (1956).
103. A. K. Sijpesteijn and M. J. Janssen, Antonie van Leeuwenhoek
     J. Microbiol. Serol., 25, 422 (1959).
104. M. J. Janssen, J. Inorg. Nucl. Chem., 8, 340 (1958).
105. A. K. Sijpesteijn and M. J. Janssen, Nature, 182, 1313 (1958).
106. W. O. Foye, I. B. Van de Workeen, Jr., and J. D. Matthes,
     J. Amer. Pharm. Assoc., Sci. Ed., 47, 556 (1958).
107. J. G. Horsfall and S. Rich, Science, 119, 582 (1954).

108.  G. Grisebach, Angew. Chem., 68, 554 (1956).
109.  Y. Otani and Y. Ishikawa, Hakko Kogaku Zasshi, 47, 424 (1969); Chem. Abstr., 71, 98498 (1969).
110.  Y. Ishikawa, Hakko Kogaku Zasshi, 48, 723 (1970); Chem. Abstr., 74, 71950 (1971).
111.  H. Gershon, S. G. Schulman, and D. Olney, Contrib. Boyce Thompson Inst., 24, 167 (1969).
112.  D. C. Aldridge, D. Broadbent, H. G. Hemming, W. B. Turner, and K. J. Bent, German patent 1,953,205 (1970); Chem. Abstr., 73, 44273 (1970).
113.  J. Raa and J. Goksoyr, Physiol. Plant., 19, 840 (1966).
114.  Z. A. Avakyan and I. L. Rabotnova, Mikrobiologiya, 40, 305 (1971).
115.  H. G. Shirk, Nat. Acad. Sci.-Nat. Res. Counc. Publ., 514, 23 (1956).
116.  J. N. Hogsett, U.S. Patent 2,739,115 (1957).
117.  W. Dill, R. Fiskcn, T. Reutner, J. Weston, and A. Glazko, Antibiot. Chemother., 7, 99 (1957).
118.  J. F. Ryley and W. Peters, Ann. Trop. Med. Parasitol., 64, 209 (1970).

# AUTHOR INDEX

Numbers in brackets are reference numbers and indicate that an author's work is referred to although his name is not cited in the text. Underlined numbers give the page on which the complete reference is listed.

## A

Aasa, R., 256[11], 285
Aaronson, S., 309[135], 316
Abd-El-Malek, Y., 366[202], 367[202], 371[202], 383
Abe, M., 307[121], 309[121], 315
Abeyakoon, K. F., 353[83], 378
Abraham, E. P., 157[54], 202
Abrahamsson, K., 307[101], 309[101], 314
Abram, D., 156[36], 201
Abrams, A., 72[1], 93
Abrams, R., 163[138], 165[138], 205
Abramson, R., 157[62], 202
Acker, R. F., 296[52], 299[52], 313
Ackerson, L. C., 133[35], 142
Adams, F., 318, 337
Adams, J. B., 402[40,41], 416
Adams, M. H., 69[2], 75[2], 93
Adams, W. C., 159[89], 203
Adcock, K. J., 256[20], 278[20], 286
Adelstein, S. J., 156[48], 201
Adeshina, H., 256[20], 278[20], 286

Adler, J., 93
Adler, S. P., 116[1a], 117[1a], 140
Adye, J. C., 296[30], 298[30], 312
Agate, A. D., 329[125], 341
Agrawal, B. B. L., 139[1], 140
Ahkong, Q. F., 91[4], 93
Ahluwalia, G. S., 324[91], 340
Ahmed, K., 39[40], 43
Aikens, D. A., 10[45], 12[45], 43
Ajl, S. J., 307[104], 309[104], 315
Akene, J., 256[20], 278[20], 286
Albagli, L., 174[225], 208
Albert, A., 397[6,7], 398[10,11,12, 15], 400[17], 401[19], 405[53], 407[64], 415, 416, 417
Aldon, E. F., 351[54], 378
Aldous, E., 107[93], 146
Aldridge, D. C., 414[112], 419
Alexander, M., 326, 335[185], 340, 342
Alexander, M. B., 263[30], 286
Alexander, U., 362[158,159,160, 161,162], 364, 381, 382
Ali, S. H., 120-122[2], 140, 332[150], 341
Allen, P. J., 179[282], 180[282], 209
Allen, R. D., 67[144], 102

421

deWolf, A., 155[109], 204
Dhahir, H. I., 319[12], 337
Dhar, S. K., 2[20], 4
Dibella, F., 67[93], 99
Dickenson, C. J., 199[389], 213
Dickinson, C. H., 365[180,181],
 371[214,217,218,219,225,228],
 372[181,217,225,232], 375,
 382, 383, 384, 385
Dickinson, F. M., 199[389], 213
Dicks, J. W., 9[91], 46
Dieckert, J. W., 139[27], 142
Diekman, H., 219[19], 245
Diem, H. G., 371[224], 372[224,
 234], 373[245], 384
Dietrich, G. G., 307[109],
 309[109], 315
Diggelmann, H., 199[394], 214
Dill, W., 414[117], 419
Din, G. A., 329[123], 341
Dive, C., 263[34], 286
Dixson, S., 402[42], 416
Dixon, M., 71, 95
Döebereiner, J., 344[5], 376
Doft, F. S., 196[370], 213
Doherty, R. A., 319[12], 337
Doi, R. H., 118[44], 119[44],
 120[44], 143
Doi, S., 155[101], 160[101], 203
Doluisio, J. T., 401[24], 416
Donaldson, J., 174[229], 208
Dondero, N. C., 328[105], 340
Donohue, J., 323[56], 338
Doolin, L. E., 296[29], 297[29],
 304[29], 312
Dorn, F., 9[32], 12, 42
Doskocil, J., 303[80], 304[80],
 314
Douglas, W. W., 175[238], 208
Dove, W. F., 232[66], 247
Downer, D. N., 222[33], 223[33],
 224[45,46], 226[33], 234[45,76],
 245, 246, 247
Dransfield, P., 335[183], 342
Dreosti, I. E., 136[62], 144

Dring, G. J., 64[35,52], 65[34,52],
 95, 96
Drouet, F., 363[170], 364[170], 381
Druyan, R., 162[120], 204
Dubey, D. P., 185[340], 187[340],
 211
Duff, R. B., 163[130], 164[130],
 189[130], 204
Duffus, J. H., 9[22], 29[22], 31[21,
 22], 41, 91[36], 95
Dugan, P. R., 328[116], 329[122],
 340, 341
Duncan, D. W., 330[129], 341
Duncanson, L. A., 412[102], 418
Dunkley, W. L., 264[37], 286
Dürr, M., 32[13], 34[13,23], 41
Duvall, R. N., 404[49], 405[49,60],
 406[59], 417
Dvorak, H. F., 155[99], 160[99],
 203
Dwyer, F., 402[42], 416

E

Eady, R. R., 238[91], 247
Eagon, R. G., 8[24], 42, 175[232],
 176[253-256], 177[254], 208, 209
Eaton, J. W., 176, 208
Eaton, M. W., 63[41], 95
Ebert, M., 409[79], 417
Ebner, E., 130[61], 144
Eckert, R., 68[37], 71[37], 95
Edds, G. T., 401[27], 411[27], 416
Edgar, J., 405[54,58], 417
Edmisten, J. A., 354, 363, 369,
 379, 381
Edward, T., 319[18], 322[18], 337
Edwards, H. M., Jr., 71[24], 94,
 177[263,265], 186[263,265],
 195[361], 209, 212
Efthymiou, C. J., 106[28], 142
Eglite, A. K., 352[79], 378
Egorov, N. S., 303[77], 304[77],
 313

J

Jaakkola, T., 359[132], 380
Jackobs, J. A., 353[82], 378
Jacks, G. V., 359[131], 380
Jackson, N. E., 354[89], 379
Jackson, S., 243[142], 249,
  260[26], 277, 286, 288
Jackson, T. A., 359[130], 380
Jacob, H. S., 176[247], 208
Jacob, J. L., 32[72], 45
Jacob, M., 172[207], 207
Jacobs, A. A., 265[44], 286
Jacobs, E. E., 172[207], 207
Jacobson, K., 171[187], 206
Jaffe, L. F., 67[119], 100
Jagannathan, V., 156[32], 201
Jain, M. K., 352[74], 378
Jansen, E. F., 296[64], 299[64],
  313
Jansey, E. R., 198[387], 213
Janssen, M. J., 413[103,104,
  105], 418
Jasaitis, A. A., 172[208], 207
Jasper, P. L. P., 12[38], 19-
  22[38], 34[38], 42, 53[69],
  58[69], 97, 122[64], 133[64],
  144
Jefferson, B. L., 135[118], 149
Jeffrey, J. G., 412[94], 418
Jeffreys, E. G., 307[125],
  309[125], 315
Jennings, D., 30[76], 45
Jennings, W. G., 264[37], 286
Jensen, H. L., 69[104], 99
Jensen, S., 319[14], 337
Jernelov, A., 318[10], 319[10,
  14], 322[53], 323[53], 337, 338
Jindra, A., 303[89], 304[89], 314
Joffe, A. Z., 307[118,122],
  309[118,122], 315
Johnseine, P., 17[86], 18[86],
  27[86], 45, 107[109], 122[108,
  109], 123[108, 109], 124[109],
  147

Johnson, A. M., 353[88], 379
Johnson, C. E., 307[98], 309[98],
  314
Johnson, C. M., 325[98], 326[98], 340
Johnson, C. R., 351[61], 378
Johnson, G. T., 296[37], 298[37],
  312
Johnson, J. H., 39[39], 42
Johnson, M. J., 401[25], 416
Johnson, O., 174[218], 207
Johnson, W., 199[401], 214
Johnston, R., 182, 210
Joiris, C., 319[16], 337
Jones, H. E., 331[138b], 341
Jones, K., 368[208], 369[208], 383
Jones, P. J., 266[53], 287
Jooss, T., 174[225], 208
Jordan, D. C., 307[111], 309[111],
  315
Jordan, M. J., 153[13], 200
Joseph, S. W., 106[28], 142
Judah, J. D., 39[40], 43
Jude, A., 307[107], 309[107], 315

K

Kaback, H. R., 50[70,71,72], 53,
  97, 132[65,66], 144
Kadomtseva, V. M., 31[66], 32[66,
  67], 44
Kagawa, T., 139[56], 143
Kagi, J. H. R., 197[374,376,377],
  199[391], 213
Kahl, G. F., 167[156], 205
Kairis, M. V., 155[95], 160[95],
  203
Kaiser, W., 373[243], 384
Kakiuchi, K., 156[44], 161[44], 201
Kallio, P., 362[156,163], 364, 381
Kallio, S., 362, 364[163,178], 381,
  382
Kalyanasundaram, R., 296[36],
  298[36], 312

Y

Z

# SUBJECT INDEX

## A

A22765, ferrioxamine analog, 219

A23187, affinity for magnesium, 29
calcium ionophore, 89-91

Acanthamoeba encystment, effect
of phosphate, 304

Acetic-butyric acid fermentation,
in iron deficiency, 241

Acetomeroctol, as topical
antiseptic, 393

p-Acetylaminobenzaldehyde
thiosemicarbazone (see
Thiacetazone)

3-Acetyl-5-methyltetramic acid,
metal-binding activity of, 403

Achromobacter, oxidation of
arsenite, 334

Achyla, requirement for calcium,
65-67

Acid mine drainage, reduction of
ferric ions, 331

Acinetobacter, metals in attachment of surface protein in,
177

Aconitase, iron requirement for
activity in B. subtilis, 243

Acrodermatitis enteropathica,
malabsorption of zinc in, 196

Actinomycin D, lack of effect on
magnesium uptake in germination, 25

Actinomycin synthesis, effect of
metals, 298

Actinorubin synthesis, effect of
iron, 298

Active transport, essential cations,
5

Adenyl cyclase, inhibition by zinc,
174

Adipocytes, magnesium accumulation
in, 36

Aerobacter aerogenes, active transport of magnesium, 12, 21
enterochelin and aerobactin
synthesis, 229, 234
formation of 2,3-DHB, 221
growth response to magnesium,
10-11
loss of ribosomal proteins, 11

Aerobacter species, formation of
dihydroxamates, 219

($^3$H)Aerobactin($^{59}$Fe), transport in
A. aerogenes, 234

Aerobic cells, use of ferric iron,
216

Aeromonas, selenium granule
accumulation, 325

S